Bigger than Chaos

Bigger than Chaos

Understanding Complexity through Probability

MICHAEL STREVENS

Harvard University Press

Cambridge, Massachusetts

London, England

2003

Library of Congress Cataloging-in-Publication Data

Strevens, Michael.
Bigger than chaos : understanding complexity through probability / Michael Strevens.
p.   cm.
Includes bibliographical references and index.
ISBN 0-674-01042-6 (alk. paper)
1. Probabilities.   I. Title.

QC174.85.P76 S77 2003
003—dc21   2002192237

To Joy

Acknowledgments

It has been eleven years since Barry Loewer, in response to my very first graduate school paper on probabilistic explanation in biological and social systems, said, "Yes, but where do the probabilities come from?" Thanks to Barry for raising the question and for much subsequent aid and encouragement. Thanks also to the other people who have provided helpful comments in the course of this project, in particular David Albert, Joy Connolly, Persi Diaconis, Peter Godfrey-Smith, Alan Hájek, Tim Maudlin, Ken Reisman, and anonymous readers for Harvard University Press.

Contents

Figures

Note to the Reader

The technical nature of this study creates two problems for the author: a large number of new concepts with accompanying terminology are introduced, and a number of claims regarding some sort of formal justification— that is, proof—are made, in the course of the book. Almost all the concepts are presented in the main text of the study; to help the reader I have provided a glossary of the important terms coined and then used in more than one place. Terms included in the glossary are, when first defined, set in **boldface**. The proofs are included in appendices to chapters two and three. I have tried to confine necessary but unremarkable aspects of the definitions and the arguments—for example, requirements that various sets be measurable—to the notes and the appendices. References to the more formal aspects of many of the mathematical results invoked concerning probability are also secreted in the notes and appendices. Readers in search of these and other details should be sure not to confine their attention to the main text. Certain especially important notes—involving justifications of, qualifications of, and interesting generalizations of results stated in the main text—are indicated by underlining, like so.[1]

Some extended discussions of points raised in this book can be found on the *Bigger than Chaos* website, at **www.stanford.edu/~strevens/bigger**. References to the website are of this form: *see website section 3.6B*.

For the most part, the book is designed to be read from beginning to end. Several notions, however, are introduced some time before they are put to use. Examples include degrees of microconstancy (section 2.23) and effective and critical ic-values (section 2.7). I have structured the material in this way for ease of later reference. Where such passages occur, I suggest that the first-time reader skip ahead. For more advice on reading the book, see section 1.34.

. . . it's only soup staring up at the moon.

Laura Riding, "Forgotten Girlhood"

1

The Simple Behavior of Complex Systems

An ecosystem is a tangle of a thousand lives, each tracing an intricate path sometimes around, sometimes through the paths of others. A creature's every move is dependent on the behavior of those around it—those who would eat it, those who would eat its food, and those who would mate with it. This behavior in turn depends on other behavior of other creatures, and so on, with the general disorderliness of the weather and the rest of the world adding further convolutions. All together, these knotted histories and future histories make up a fantastically irregular filigree of life trajectories.

Now stand back. Individual paths blur each into the other until all that can be resolved are the gross patterns of existence, the ups and downs of population and little else. At this level of observation, something quite new emerges: simplicity. The sudden twists and turns of individual lives fall away, leaving only—in many cases—a pattern of stable or gently cyclic population flow.

Why? How can something as intensely complex as an ecosystem also be so simple? Is this a peculiar feature of living communities, to be explained by the flexibility of life, its diversity, or the fine-tuning of natural selection? Not at all. For there are many non-living complex systems that mingle chaos and order in the same way: a vastly complicated assemblage of many small, interdependent parts somehow gives rise to simple large scale behavior.

One example is a gas in a box, the movements of its individual molecules intractably complicated, their collective behavior captured by the stark and simple second law of thermodynamics, the ideal gas law, and other generalizations of kinetic theory.

Another example, well known from the literature on complex systems, is a fluid undergoing Bénard convection, in which hexagonal convection cells form spontaneously in a honeycomb pattern. Still another is a snowflake, in which a huge number of water molecules arrange themselves in patterns with sixfold symmetry. Although the patterns themselves are quite complex, the rule dictating their symmetry is very simple.

The phenomenon is quite general: systems of many parts, no matter what those parts are made of or how they interact, often behave in simple ways. It is almost as if there is something about low-level complexity and chaos itself that is responsible for high-level simplicity.

What could that something be? That is the subject of this book.

The key to understanding the simplicity of the behavior of many, perhaps all, complex systems, I will propose, is probability. More exactly, the key is to understand the foundations of a certain kind of probabilistic approach to theorizing about complex systems, an approach that I will call enion probability analysis, or EPA, and that is exemplified by, among other theories, the kinetic theory of gases and population genetics.

It is not enough simply to master EPA itself, as EPA makes probabilistic assumptions about the dynamics of complex systems that beg the most important questions about the ways in which low-level complexity gives rise to high-level simplicity. What is required is an understanding of why these assumptions are true. It is the pursuit of this understanding that occupies the greater part of my study: chapters two, three, and four.

1.1 Simplicity in Complex Systems

Simplicity in complex systems' behavior is everywhere. For this very reason, it is apt not to be noticed, or if noticed, to be taken for granted. There is much scientific work attempting to explain why complex systems' simple behavior takes some particular form or other, but very little about the reasons for the fact of the simplicity itself. I want to begin by creating, or re-creating, a sense of wonder at the phenomenon of simplicity emerging from complexity. Along the way, I pose, and try to answer, a number of questions: How widespread is simple behavior? What *is* simple behavior? What is a complex system? Why should probability play a role in understanding the behavior of complex systems? Most important of all, why should simple behavior in complex systems surprise us?

1.11 Some Examples of Simple Behavior

Gases   A gas in a box obeys the second law of thermodynamics: when the gas is in thermodynamic equilibrium, it stays in equilibrium; otherwise, it moves towards equilibrium. In either case, its behavior is simple in various ways. At equilibrium, its pressure and temperature conform to the ideal gas law. Moving towards equilibrium, gases observe, for example, the laws of diffusion.

Ecosystems Ecosystems exhibit a number of simple behaviors at various levels of generality. Three important examples:

Population levels: For larger animals, such as mammals, predator/prey population levels tend to remain stable. Occasionally they vary periodically, as in the case of the ten-year population cycle of the Canadian lynx and its prey, the hare. Such systems return quickly and smoothly to normal after being disturbed (Putman and Wratten 1984, 342).[1]

Trophic structure: If a small ecosystem is depopulated, it is repopulated with organisms of perhaps different species, but forming a food web with the same structure (Putman and Wratten 1984, 343).

Microevolutionary trends: Species of mammals isolated on islands tend to evolve into dwarf or giant forms. Mammals of less than 100 grams usually increase in size; those of greater than 100 grams usually decrease in size (Lomolino 1985).

Economies Not all simple generalizations made about economies turn out to be true, but when they do, it is in virtue of some kind of simple behavior. Perhaps the most striking example of such a behavior, and certainly the most keenly observed, is the phenomenon of the business cycle, that is, the cycle of recessions and recoveries. So regular was this alternation of sluggish and speedy growth between 1721 and 1878 that the economist W. Stanley Jevons wondered if it might not be related to what was then thought to be the 10.45-year cycle in sunspot activity (Jevons 1882).

Weather The weather, generated by interacting fronts, ocean currents, convection areas, jet streams, and so on, is an immensely complicated phenomenon. The best modern simulations of weather patterns use over a million variables, but even when they make accurate predictions, they are valid only for a few days. It might be thought, then, that there are no long-lived simple behaviors to be found in the weather.

This is not the case. One class of such behaviors are roughly cyclic events such as El Niños (which occur every three to ten years) and ice ages (which have recently occurred at intervals of 20,000 to 40,000 years). Another class of simple behaviors concerns the very changeability of the weather itself. Some parts of the world—for example, Great Britain—have predictably unpredictable weather (Musk 1988, 95–96). In other parts of the world, the meteorologist enjoys more frequent success, if not greater public esteem.

Chemical Reactions   The laws of chemical kinetics describe, in a reasonably simple way, the rate and direction of various chemical reactions. Even complicated cases such as the Belousov-Zhabotinski reaction, in which the proportions of the various reactants oscillate colorfully, can be modeled by very simple equations, in which only variables representing reactant proportions appear (Prigogine 1980).

Language Communities   A very general law may be framed concerning the relationship between the speed of language change and the proximity of speakers of other languages, namely, that most linguistic innovation occurs in regions that are insulated from the influence of foreign languages (Breton 1991, 59–60). As this rule is often phrased, peripheries conserve; centers innovate.

Societies   I will give just two examples of the regularities that have characterized various societies at various times. The first is the celebrated constancy of suicide rates in nineteenth-century Europe. Although different regions had different rates of suicide, effected differently (nineteenth-century Parisians favored charcoal and drowning, their counterparts in London hanging and shooting), in any given place at any given time the rate held more or less constant from year to year (Durkheim 1951; Hacking 1990).

The second example is the familiar positive correlation between a person's family's social status or wealth and that person's success in such areas as educational achievement. That such a correlation should exist may seem unremarkable, but note that in any individual case, it is far from inevitable, much to parents' consternation. If parents cannot exercise any kind of decisive control over the fates of their children, what invisible hand manufactures the familiar statistics year after year?

So that such a grand survey will not convey a false grandiosity, let me say what I will not do in this study. First, I will not establish that every kind of complex system mentioned in the examples above can be treated along the lines developed in what follows. I am cautiously optimistic in each case, but it is the systems of statistical physics and of population ecology and evolutionary biology on which I will focus explicitly.

Second, I do not intend to explain the details of each of the behaviors described above. Rather, I will try to explain one very abstract property that all the behaviors share: their simplicity. I will not explain, for example, why one ecosystem has populations that remain at a fixed level while another has populations that cycle. That is the province of the relevant individual science, in this case, population ecology. What interests me is the fact that, fixed or cycling, population laws are far simpler than the underlying goings-on in the systems of which they are true. Whereas science has, on the whole, done well in explaining the shape of simple behaviors, the question answered by such explanations is usually which, of many possible simple behaviors, a system will display, rather than why the system should behave simply at all. It is this latter question that I aim to resolve.

1.12 What Is Simple Behavior?

What does it mean to say that a system has a simple dynamics? The systems described above exhibit two kinds of dynamic behavior that may be regarded as canonically simple. First, there is *fixed-point equilibrium* behavior, where a system seeks out a particular state and stays there. Examples are thermodynamic equilibrium and stable predator/prey populations. Second, there is *periodic* behavior, where a system exhibits regular cycles. Examples are the lynx/hare population cycle and, at least during some periods of history, the business cycle. To these may be added two other somewhat simple behaviors: quasi-periodic behavior, in which there is an irregular cycle, as in the case of El Niño's three- to ten-year cycle; and general trends, such as insular pygmyism/gigantism in mammals, or the linguistic rule that peripheries conserve while centers innovate.[2] (For more on the relation between particular laws and general trends, see section 5.24.)

Rather than cataloguing various kinds of simple behavior, however, it will be illuminating to adopt a very general characterization of simple behavior. I will say that a system exhibits a **simple behavior** when it exhibits a dynamics that can be described by a mathematical expression with a small number of variables (often between one and three). In such cases I will say that the system has a simple *dynamic law* or *law of time evolution*. The canonical cases of simplicity tend to fit this characterization. Simple equations can be constructed to describe the behavior of almost all systems that exhibit fixed-point equilibrium, periodic or quasi-periodic behavior, and a family of such equations can describe systems that exhibit general trends.[3]

The goal of this study can now be stated a little more clearly. I aim to explain why so many laws governing complex systems have only a few variables. I leave it to the individual sciences to explain why those few variables are related in the way that they are; my question is one that the individual sciences seldom, if ever, pose: the question as to why there should be so few variables in the laws to begin with.

Two remarks. First, my characterization of simple behavior includes cases that do not intuitively strike us as simple. Some dynamic laws with few variables generate behavior whose irregularity has justifiably attracted the epithet *chaotic*. Thus the central insight of what is called *chaos theory*: a system may behave in an extremely complicated manner, yet it may obey a simple deterministic dynamic law. Such a system has a hidden simplicity. The appeal of chaos theory is rooted in the hope of chaoticians that there is much hidden simplicity to be found, that is, that much complex behavior is generated by simple, and thus relatively easily ascertained, dynamic laws. If this is so, then it will have turned out that there is even more simple behavior, in my proprietary sense, than was previously supposed. I will go on to provide reason to think that *all* simple behavior (again in my proprietary sense) is surprising, and so in need of an explanation, when it occurs in a complex system (section 1.15). It will follow that complex systems behaving chaotically present the same philosophical problem as complex systems behaving simply, if the chaotic behavior is generated, as chaoticians postulate, by simple dynamic laws.[4]

Second, the characterization of simple behavior offered here is not intended as a rigorous definition. It takes for granted that we humans use certain kinds of variables and certain kinds of mathematical techniques to represent complex systems; it is only relative to these tendencies of ours that the characterization has any content, for the dynamics of any system at all can be represented by mathematical expressions of a few variables if there is no constraint on the variables and techniques of representation that may be used. The reader might complain that *simplicity* then means only *simplicity-for-us*, and that a more objective—that is, observer-independent—criterion of simplicity is called for. Given my present purposes, however, there is no real reason to construct an objective definition of simple behavior. Such a definition might perhaps tell us much about the nature of simplicity, but it will tell us nothing about the way that complex systems work. Readers who are unhappy with this attitude, and who are uninterested in any question about "simplicity-for-us," ought nevertheless to find that this study has much of interest to say about the behavior of complex systems.

1.13 What Is a Complex System?

The complex systems described in section 1.11 consist of many somewhat independent parts, which I will call **enions**. The enions of a gas are its molecules, of an ecosystem its organisms, of an economy its economic actors.[5] The term *enion* should not be thought to impute any precise theoretical properties to the different parts of various complex systems that it names; it is introduced, at this stage, for convenience only. The deep similarities in the behavior of the enions of different systems will emerge as conclusions, rather than serving as premises, of this study.

It is the way a complex system's enions change state and interact with one another that gives the system its complexity. On the one hand, the enions tend to be fairly autonomous in their movement around the system. On the other hand, the enions interact with one another sufficiently strongly that a change in the behavior of one enion can, over time, bring about large changes in the behavior of many others. For the purposes of this book, I will regard as complex just those systems which fit the preceding description. A **complex system**, then, is a system of many somewhat autonomous, but strongly interacting, parts.

This proprietary sense of complexity excludes some systems that would normally be considered complex, namely, those in which the actions of the individual parts are carefully coordinated, as in a developing embryo.[6] There ought to be some standard terminology for distinguishing these two kinds of systems, but there is not. Rather than inventing a name for a distinction that I do not, from this point on, discuss, I simply reserve the term *complex* for the particular kinds of systems with which this study is concerned.

1.14 Understanding Complexity through Probability: Early Approaches

The notion inspiring this book, that laws governing complex systems might owe their simplicity to some probabilistic element of the systems' underlying dynamics, had its origins in the eighteenth century, and its heyday in the nineteenth. The impetus for the idea's development was supplied by, on the one hand, the compilation of more and more statistics showing that many different kinds of events—suicide, undeliverable letters, marriages, criminal acts—each tended to occur at the same rate year after year, and on the other hand, the development of mathematical results, in particular the law of large numbers, showing that probabilistically governed events would tend to exhibit not just a short-term disorder but also a long-term order. The mathematics was developed early on, but, although its principal creators, Jakob Bernoulli and Abraham de Moivre, grasped its significance as an explainer of regularity, they were for various reasons unable to commit themselves fully to such explanations. These reasons seem to have included a lack of data apart from records of births, marriages, and deaths; the ambiguous status of the classical notion of probability as an explainer of physical events; and a propensity to see social stability as a mark of divine providence as much as of mathematical necessity.[7]

By the middle of the nineteenth century, the idea that statistical law governed a vast array of social and other regularities had, thanks especially to Adolphe Quetelet and Henry Thomas Buckle, seized the European imagination. There were, however, a number of different ways of thinking about the workings of statistical laws, many of which views are at odds with the kind of explanation offered by my preferred approach of enion probability analysis. I consider three views here.

The first view holds that statistical stability is entirely explained by the law of large numbers, the large numbers being the many parts—people, animals, whatever—that constitute a typical complex system (my enions). Just as many tosses of individual coins exhibit a kind of collective stability, with the frequency of heads tending to a half, so, for example, the individual lives of large numbers of people will tend to exhibit stability in the statistics concerning birth, marriage, suicide, and so on. Siméon-Denis Poisson (1830s) argued perhaps more strenuously than anyone until James Clerk Maxwell and Ludwig Boltzmann that probability alone, in virtue of the law of large numbers, could found statistical regularity. Poisson's position is similar in spirit to my own; its main defect, in my view, is a failure to appreciate fully the explanatory importance of whatever physical properties justify the application of the law of large numbers—in particular, whatever properties vindicate the assumption of stochastic independence—and an ensuing overemphasis of the explanatory importance of the mathematics in itself.

The second view, far more popular, seems to have been that, roughly, of Quetelet (1830s–1840s) and Buckle (1850s–1860s). On this approach to explaining large-scale regularities, probability is relegated to a subsidiary role. The stability of statistics is put down to some non-probabilistic cause; the role of probability is only to describe fluctuations from the ordained rate of occurrence of a given event. Probability governs short-term disorder then, but does not—by contrast with Poisson's view—play a positive role in producing long-term order. The law of large numbers is invoked to show that fluctuations will tend to cancel one another out. Probability in this way annihilates itself, leaving only non-probabilistic order.

The third and final view belongs to opponents of the above views who called themselves frequentists, the best known of whom was John Venn (1860s). The frequentists approached statistical stability from a philosophically empiricist point of view. It is a brute fact, according to Venn and other frequentists, that the world contains regularities. Some of these regularities are more or less perfect, while others are only rough. Statistical laws are the proper representation of the second, rough kind of regularity. The frequentists disagree with Quetelet and Buckle because they deny that there are two kinds of processes at work creating social statistics, a deterministic process that creates long-term order and probabilistic processes causing fluctuations from that order. Rather, they believe, there is just one thing, an approximate regularity. The frequentists disagree with Poisson because they deny that the law of large numbers has any explanatory power. It is merely a logical consequence of the frequentist definition of probability. As in modern frequentism, probabilities do not explain regularities, because they simply are those regularities.

Of these three views, the first had probably the least influence in the mid-century. But by the end of the century, this was no longer true. Maxwell's (1860s) and Boltzmann's (1870s) work on the kinetic theory of gases, and the creation of the more general theory of statistical mechanics, persuaded many thinkers that certain very important large-scale statistical regularities—the various gas laws, and eventually, the second law of thermodynamics—were indeed to be explained as the combined effect of the probability distributions governing those systems' parts.

The idea that simple behavior is the cumulative consequence of the probabilistic behavior of a system's parts is the linchpin of EPA. By the end of the nineteenth century, then, questions about the applicability of and the foundations of what I call EPA were being asked in serious and sustained ways, especially in the writings of Maxwell, Boltzmann, and their interlocutors. One might well have expected a book such as mine to have appeared by 1900. But it did not happen. Why not?

There are a number of reasons. First, the dramatic revelations of social stabilities made in the first half of the nineteenth century had grown stale, and it was becoming clear that social regularity was not so easy to find as had then been supposed. There were no new social explananda, and so no new calls for explanation. Second, the mathematics of probability was not sufficiently sophisticated, even by 1900, to give the kind of explanation I present in chapter four. In particular, the extension of the law of large numbers on which I there rely, the ergodic theorem for Markov chains, was developed only in the twentieth century. Third, J. W. Gibbs's influential formalization of statistical mechanics obscured the explanatory relation between probability distributions over the behavior of particular particles and the second law of thermodynamics. Fourth, philosophers of science were first preoccupied with relativity, and then, when they returned to probability, had become empiricists in the style of Venn. When discussing probability in science, their usual choices of examples—medical probabilities and quantum mechanical probabilities— seem almost deliberately calculated to distract attention from the explanatory power of probability in the framework that I am calling EPA. Finally, biology, both ecological and evolutionary, has even now not reached a consensus on the explanatory role of probability. Thus one finds, for example, Sober (1984) offering a conceptual model of regularity in population genetics in the spirit of Quetelet and Buckle: genetic change is the combined effect of two processes, a deterministic process, natural selection, accompanied by probabilistic fluctuation, drift. Probability plays no positive explanatory role in this picture, or where it does, its contribution is a certain arbitrariness, as in, say, the founder effect.[8]

It appears, then, that interest in the foundations of the probabilistic approach to the behavior of complex systems waned over the course of the twentieth century. There is one exception to this trend, however: the rich literature on the foundations of statistical physics. I discuss aspects of this literature having some kinship to my own project in section 1.25.

1.15 Microcomplexity and Macrosimplicity

Why should simple behavior in complex systems surprise us? There are very good reasons to expect a complex system to have a very complicated dynamics, in the sense—minimal, but adequate for my purposes—that its dynamic laws will contain many variables.[9]

To see this, consider two examples. In an ecosystem consisting of a thousand organisms, the state of each of which is characterized by ten variables (for example, position, health, water and food levels, whether pregnant), the state of the entire system will be represented by ten thousand variables. In a small container of gas at normal atmospheric pressure, the state of the entire system will be represented by approximately 10^{23} variables. Thus the laws governing the changes in state of both of these systems will have vast numbers of variables. The same is true for complex systems generally.

If, however, one is prepared to represent the state of a system a little less precisely, fewer variables are required. To represent the exact state of a baseball, for example, one must specify the position of every particle in the ball, an undertaking that would require unthinkably large numbers of variables. But if one is interested only in the approximate state, one might represent only the position and velocity of the center of mass of the ball, using just six variables. More generally, by representing statistical properties of enions—for example, temperature, population, or GDP—rather than the exact state of each enion, one can represent the state of any complex system using just a few variables.

There are, then, two levels at which a complex system may be described. First, there is a lower level at which the state of each individual enion is represented. This I call the **microlevel**. Second, there is a higher level at which only statistics concerning enions are represented. This I call the **macrolevel**. In a microlevel description, the state of the system is characterized by giving the values of all the microlevel variables, or **microvariables**. In a macrolevel description, the state of the system is characterized in terms of **macrovariables** that represent enion statistics, such as temperature and population. Some more terminology: call a law governing the behavior of microvariables a *microlevel law of time evolution*, or a law of **microdynamics**. Call a law governing the behavior of macrovariables a *macrolevel law of time evolution*, or a law of **macrodynamics**.

Turning back to section 1.11, the reader will note that the simple behaviors of complex systems described there are all macrodynamic behaviors. The microdynamic laws of the same systems are, as I have noted, monstrously complex combinations of the laws governing the interactions of individual enions with each other and with the other parts of the system. Thus the systems conform at the same time to microdynamic laws with vast numbers of variables, and to macrodynamic laws with only a few variables. There is no contradiction here, but there is much scope for puzzlement. One might simply ask: where do all the variables go?

This is not, of course, quite the right question. The many microvariables are assimilated into the few macrovariables by way of statistical aggregation, that is, by throwing away much of the microlevel information and keeping only averages. It is always possible to take an average. What ought to surprise us is that, although vast quantities of dynamically relevant information are discarded in the process of taking the average, there is a determinate, often deterministic, lawful relationship between the macrovariables that remain.

How can there exist a law stated entirely in terms of macrovariables—that is, a law to the effect that only macrovariables affect other macrovariables—when small changes in the behavior of a single enion can, at least in principle, drastically alter the behavior of a complex system as a whole?

The problem may be stated as follows. On the one hand, the behavior of the macrovariables is entirely determined by the behavior of the microvariables, or to put it another way, macrolevel behavior is caused by a complex microdynamic law. On the other hand, the behavior of the macrovariables seems to depend very little on the microvariables, hence very little on the microdynamic law. To resolve this tension, one must investigate and come to understand the circumstances under which a complex microdynamic law entails the existence of a simple macrodynamic law.

In certain special cases, the understanding is easy to come by. The microdynamic laws governing the vibrations of the particles in a baseball are very complex. Yet it is no mystery that the dynamics of the ball as it flies around the diamond are simple. The reason is, of course, that the particles in the ball are tightly bound together, so that where one goes, the rest follow. The position of one particle, then, is a reliable guide to the position of all the particles. The exceptions are the particles that become detached from the ball in the course of the game, but they are few enough to be ignored.

This kind of reasoning cannot, in general, be applied to complex systems. A complex system's enions, unlike a ball's particles, are free to wander more or less as far from the other enions as they like. Nor is there any set of enions that can, like the particles that become detached from the ball, be safely ignored.

How, then, can it be that the complex microdynamic laws manifest themselves in such an orderly fashion at the macrolevel? How does microlevel complexity not just coexist with, but in effect *give rise to* macrolevel simplicity?

1.2 Enion Probability Analysis

To explain the simple behavior of complex systems, what we need is a way of representing the dynamics of a complex system that allows us to discard vast amounts of information about the microlevel, while retaining enough information to derive simple generalizations about the behavior of high-level statistics. The mathematics of probability is ideally suited to this task.

Macrovariables represent statistical information about the many individual enions of a complex system. It follows that information about the behavior of macrovariables—that is, the information conveyed by the macrodynamic laws—may also be seen as a statistic, a summary of the microlevel behavior of individual enions. To inquire into this behavior is to inquire into the properties of a statistic. The theory of probability is the branch of mathematics that deals with the properties of statistics. And the greatest successes of probability mathematics have been explanations of properties of statistics that make very few assumptions about the properties of the underlying events.

How, then, to think in a probabilistic way about the relation between the microlevel and the macrolevel of a complex system? As follows. First, assign a probability distribution over the behavior of each of the system's enions. I will call the probabilities that describe the dynamics of individual enions **enion probabilities**. Second, aggregate these enion probabilities to obtain a probability distribution over the behavior of enion statistics, that is, over the system's macrolevel behavior. Third, deduce a macrolevel law from the macrolevel probability distribution. These three steps make up the technique that I will call **enion probability analysis,** or EPA.

To understand why EPA can be successfully applied to a complex system so as to derive a simple macrolevel law, I will argue in this section, is to understand why that system behaves in a simple way. Most of this study, then, will constitute an attempt to understand the foundations—mathematical, physical, and philosophical—of EPA's successes. In what follows I give a very simple example to show how probabilistic thinking can explain macrolevel simplicity (section 1.21), I describe some uses of EPA in the sciences (section 1.22), I examine the three steps of EPA more closely (section 1.23), and I argue, in section 1.24, that in order to understand the success of EPA, it is necessary to understand certain special properties of enion probabilities, in particular, their satisfaction of a requirement that I will call the *probabilistic supercondition*. The section concludes with a few pages on the foundations of statistical physics (section 1.25), looking for parallels to my treatment of EPA.

1.21 Understanding Simplicity through Probability

A few very general probabilistic assumptions about a complex system suffice to entail that the system behaves in a simple way, that is, that it obeys a macrolevel law with a small number of variables. The assumptions do not imply any particular form for the simple behavior, but that is not my aim here. What I want to show is that, if they hold, then the system will obey some simple law or other.

Consider the following very straightforward example. We have an ecosystem of rabbits. The rabbits do not reproduce, but they are occasionally eaten by foxes. What law governs the change in the rabbit population over time?

In principle, this problem seems as difficult as any involving a complex system. To trace the change in population, one might think, it is necessary to trace the life history of each rabbit in the population, as it roams around the ecosystem looking for especially lush patches of grass, avoiding hungry foxes, and, for the sake of the example, scrupulously resisting the urge to procreate.

But suppose you have the following information: the probability of any rabbit's dying over the course of a month is 0.95, and the deaths are stochastically independent. It is easy to calculate from these facts a probability distribution over the possible values for the rabbit population a month from now. The calculation is especially straightforward if the population is very large, for then the law of large numbers implies that, with very high probability, the population in a month's time will be about 0.95 of the size that it is now, that is, that the population will very likely undergo a decrease of almost exactly 5%. The probabilistic information entails, then, that the rabbit population conforms to the following law relating the population n' in a month's time to the current population n:

$$n' = 0.95n.$$

The derivation of this law is achieved without following the rabbit population's microdynamics at all; although we know that 5% of the rabbits will die, we have no idea which 5% die or how they meet their ends. The information in the 0.95 probability, conjoined with the assumption of stochastic independence, somehow picks out just those properties of the ecosystem that determine the system's simple population flow, and no more. It is for this reason that I see EPA as a powerful framework for understanding the macrolevel simplicity of complex systems.

Although this example gives a specific value for the probability of rabbit death and derives a specific form for the population law, these specifics are not what interests me in this study. What interests me is that, given the kind of information supplied in the example, it is always possible, using a certain method, to derive a macrolevel law with very few variables, that is, a simple macrolevel law. The form of the law will depend on the details of the probabilistic information supplied, but the simplicity of the law depends on only a few very general properties of that information.

The probabilistic method I refer to is enion probability analysis. As remarked above, EPA has three steps:

1. Probabilities concerning the behavior of individual enions in the system, the enion probabilities, are discovered or postulated. In the example, the sole enion probability is the 0.95 probability that, over the course of a month, a given rabbit dies.
2. These probabilities are aggregated, yielding a probability distribution that describes the behavior of one or more macrovariables only in terms of other macrovariables. In the example, the assumption of stochastic independence and the law of large numbers are used to aggregate the probabilities, yielding a very high probability of a 5% decrease in population.
3. The macrolevel probability distribution is taken to induce a macrodynamic law. This simple step leads, in the example, from the claim about the high probability of a 5% decrease to the law $n' = 0.95n$.

Suppose I make the rabbit example more complex. For each rabbit, I specify not only a probability of death over the course of a month, but also a probability that the rabbit reproduces, and a probability distribution over the number of offspring. (Never mind, for now, that some of the rabbits are male.) Then the population law will be a different one; in particular, it may not imply a steady decrease in the number of rabbits. But the law will be simple: it will, as in the original example, involve only a single macrovariable, the rabbit population itself. Add a further complication, a dependence between the probability of rabbit death and the population level (limited resources mean that the chance of any rabbit's dying increases as the population increases). The law changes again, now taking on a form familiar to population ecologists, but still it involves only one macrovariable.

More generally, provided that there exists probabilistic information about the behavior of individual rabbits of a certain form, it will always be possible to derive a population law for a rabbit ecosystem that has as its sole macrovariable the rabbit population. My aim in this study is, first, to generalize this observation as far as possible, finding a very abstract specification of a certain kind of probabilistic fact, such that systems for which probabilistic facts of this sort obtain always obey simple macrolevel laws, and second, to examine the circumstances under which the probabilistic facts do obtain. The first goal is to find the form of EPA, the second is to investigate the foundations of EPA.

To this end, section 1.22 will give some examples of the scientific use of EPA, in which probabilistic information about a complex system is used to derive a simple macrolevel law for the system; section 1.23 will determine what the probabilistic information and the derivation have in common across the sciences and across the different kinds of simple macrolevel law, yielding an abstract description of the form of EPA; and section 1.24 will provide the framework for what will occupy by far the greater part of this study, the study of the foundations of EPA.

1.22 Enion Probability Analysis in the Sciences

Although the name is new, the technique is not: EPA has played an important role in several of the major sciences dealing with complex systems. In what follows, I survey the role of EPA in statistical physics, in the historically important case of actuarial science, and in evolutionary biology, identifying in each case the three steps of EPA described in section 1.21.

KINETIC THEORY

Perhaps the most famous probabilistic account of a complex system is the kinetic theory of gases. The use of EPA is seen most clearly in the early treatments of Maxwell and Boltzmann. Both scientists began with some almost a priori assumptions about the probability distributions over the behavior of individual gas molecules, in particular, about the symmetry and the independence properties of distributions over individual molecules' positions and velocities (step 1). Using the independence assumptions, they were able to derive probability distributions over the statistical properties—the macrovariables—of gases, such as temperature, pressure, and entropy (step 2), and then to infer laws governing these macrovariables (step 3). Enion probability analysis was first used here not to discover the macrolevel laws, which were already well known, but to explain them.[10]

ACTUARIAL SCIENCE

Actuarial science has a good claim to have made the very first explicit application of EPA to a complex system (Hacking 1975, chap. 13). The actuary's job is to calculate a profitable price for a life insurance policy, something that can be done only given knowledge of the behavior of the macrovariables—in effect, statistics about death in all its forms—which determine the cost to the insurer of a batch of insurance policies. In 1671 Johann de Witt calculated the appropriate macrolevel regularity, concerning expected numbers of deaths, for a Dutch annuity scheme by estimating probabilities for individual deaths (step 1) and aggregating these to give the expected number of deaths per year (steps 2 and 3). Actuaries have ever since pursued the same basic strategy; progress has come through offering more sophisticated insurance packages based on more accurate estimates of the enion probabilities.

Evolutionary Theory

According to Darwin, evolution proceeds chiefly by the process of natural selection, which operates when one variant of a species survives and proliferates more successfully than another variant. Eventually, the more successful traits tend to take over a population; this constitutes an episode of evolution by natural selection. It is said that success-promoting traits confer *fitness* on their owner. (The missing part of the story, not relevant here, explains how such traits arise in the first place.)

Now let me put this another way: natural selection occurs because ecosystems obey one of a family of macrolevel laws of population dynamics having the following consequence: the subpopulation with trait T increases more quickly (or decreases more slowly) than the subpopulation without T. If one can predict or explain these macrolevel laws, one can predict or explain episodes of natural selection. (For examples of successful predictions concerning the course of natural selection, see Grant (1986) and Endler (1986, chap. 5).)

When putting together a Darwinian prediction or explanation, then, the problem is in part to discover that or to explain why a certain macrolevel law of population dynamics is true of an ecosystem. Given the thousands of microvariables at work in an ecosystem, this might appear to be an impossible task. However the evolutionary biologist often performs this task quite easily, by reasoning as follows (compare section 1.21). Possession of a certain heritable trait T confers a relative increase in the probability of its owners' survival (step 1, since the probability of survival is an enion probability). Aggregate these enion probabilities, and you obtain a macrolevel generalization concerning organisms with T, namely, that their population level will increase relative to the population level of organisms without the trait (steps 2 and 3). Much successful Darwinian reasoning concerning real episodes of evolution follows this outline, and is thus a case of the application, albeit informal, of EPA. A more formal probabilistic theory following the same outline may be found in population genetics.

It follows, by the way, that one cannot give an entirely Darwinian explanation of the simple behavior of ecosystems: the explanatory use of the fact of natural selection assumes the prior existence of simple macrolevel laws of population ecology.

1.23 The Structure of Enion Probability Analysis

I now examine more closely, with special reference to the explanation of simple behavior, the three steps of EPA: assignment of enion probabilities, aggregation of enion probabilities, and deduction of macrolevel laws.

ASSIGNMENT OF PROBABILITIES

An enion probability is the probability that a particular enion ends up in a particular state at a particular time, given the macrostate of the system. It may be the probability that a rabbit will die in the course of a month, given the number of foxes and other rabbits in the ecosystem, or the probability that a gas molecule will have a certain velocity at the end of a one-second interval, given the temperature of the gas. The first step in EPA is to assign these probabilities to individual enions.

If EPA is to avoid becoming mired in the microlevel, the values of the probabilities assigned ought to depend, for reasons explained in the next subsection, only on macrolevel information about initial conditions. That is, the information about the system on which the values of enion probabilities depend must not be information about individual enions, but statistical information about the system as a whole. For the rabbit, the probability of death may depend on the number of foxes in the area, but not on the positions of particular foxes. For the gas molecule, the probability of having a particular velocity at the end of a given time interval may depend on the temperature of the gas at the beginning of the interval, but not on the velocities of particular gas molecules. The physical basis for this lack of low-level dependence in complex systems is the topic of section 4.4.

As I will later show, it is not always necessary to satisfy entirely the stringent low-level independence condition. A small amount of dependence on microlevel information about initial conditions may be allowed in the first stage of EPA, provided that it is eliminated in the second or third stages. A way in which microlevel dependence may persist in the rabbit/fox case is described in section 4.43; methods for the elimination of this dependence are discussed in section 4.6.

A terminological aside: There are two ways to articulate the requirement that enion probabilities not depend on low-level information. First, one can say, as I do, that they must be functions only of macrolevel information. Second, one can say that conditionalizing on low-level information must not affect the value of the probability, or that they are conditional only on macrolevel information. I treat these formulations as equivalent.

AGGREGATION OF PROBABILITIES

In the aggregation stage, probabilities concerning the behavior of enions are combined to form probabilities concerning the behavior of enion statistics, that is, probabilities concerning the dynamics of macrovariables. For example, the probabilities of individual rabbit births and deaths may be combined to produce the probabilities of various possible future population levels of a community of rabbits. In many cases, one particular future population level will, given the current values of the relevant macrovariables, such as the current rabbit and fox populations, turn out to be overwhelmingly probable, as illustrated by the example in section 1.21.

The macrolevel probabilities that result from aggregation are functions of whatever information determines the enion probabilities. The dependence is simply passed up from the microlevel to the macrolevel. For example, if the probability of a particular rabbit's dying is a function of the current number of rabbits and foxes, the macrolevel probabilities over future population levels will also be functions of the current number of rabbits and foxes.

If, by contrast, the probability of a particular rabbit's dying depends on microlevel information about the particular positions of particular foxes, the macrolevel probabilities obtained by aggregation will be functions of—will depend on—this microlevel information about fox position. In order to determine the probability of some future rabbit population level, then, it would be necessary to know all of the microlevel information concerning fox positions. Invocation of the law of large numbers cannot remove this dependence. Since I want to use EPA to show that macrolevel behavior depends only on the values of macrovariables, I required, in the last subsection, that enion probabilities depend only on macrolevel facts.

It is in the aggregation stage, if all goes well, that microlevel information entirely falls out of the picture, leaving behind only probabilistic relations between macrolevel quantities. In the first stage, during which enion probabilities are assigned, the presence of the microlevel is already much diminished, because the assigned probabilities are not functions of microlevel information; microlevel information thus becomes irrelevant as an *input*. But the microlevel is still there in the *output*, since enion probabilities are probabilities of microlevel events involving individual enions, such as the death of an individual rabbit. The aggregation stage removes the microlevel from the output as well, by moving from information about events involving individual enions to information about enion statistics, and thus to the macrolevel.

The disappearance of microlevel information, then, can be explained as follows. In the microdynamic description, microlevel information determines microlevel outcomes. The move to enion probabilities produces a description in which macrolevel information determines the probability of microlevel outcomes. The aggregation of probabilities then produces a description in which macrolevel information determines the probability of macrolevel outcomes. In this way, the probabilistic premises of EPA enable us to go from a complex microdynamic description to a simple macrodynamic description.

The story about the disappearance of microlevel information supposes that the aggregation of enion probabilities to yield probabilities of macrolevel events does not reintroduce microlevel dependencies into the description. The basis for this supposition is the assumption that the enion probabilities are stochastically independent.[11, 12] In the rabbit/fox scenario, for example, it is assumed that one rabbit's surviving for a month is stochastically independent of any other rabbit's survival over the same time period. That is, the probability (given the total number of foxes and rabbits in the system) of one rabbit's surviving for a month must be unaffected by conditionalizing on the fates of any other rabbits.[13] If stochastic independence holds, the probabilities of survival for all of the rabbits can be combined in a simple way to obtain a probability that any particular number of rabbits survives the month, given the current number of foxes and rabbits. Independence guarantees, then, that enion probabilities assigned independently of microlevel information can also be combined without referring to such information. The result is a probability distribution over macrolevel properties that depends only on macrolevel information. The physical basis for stochastic independence in complex systems is the topic of sections 4.4 and 4.5.

Derivation of Macrolevel Law

If a probability distribution over all future macrostates can be derived from enion probabilities that depend only on the current macrostate—for example, the current number of foxes and rabbits—one can put together a complete macrolevel dynamics in which microvariables do not appear. The number of macrovariables in the resulting macrolevel law will be no greater than the number of macrovariables that determine the enion probabilities. In my eco-logical example, I have assumed that there are two relevant macrovariables: the number of foxes and the number of rabbits. Enion probability analysis will thus generate a law containing just these two variables. More generally, provided that enion probabilities are functions of only a few macrovariables, the macrolevel laws generated by EPA will contain only those same few macro-variables, and so will be simple in the proprietary sense earlier defined.

I will have very little to say about this third stage of EPA, with one exception: I note that at the third stage, it is still possible to eliminate small dependencies on microlevel information that, because of less than perfect satisfaction of the independence requirements imposed in stages one and two, have not been earlier excised. This topic is treated in section 4.6.

1.24 The Probabilistic Foundations of Enion Probability Analysis

Using enion probability analysis, it is possible to derive the fact of simple be-havior in a complex system. From an understanding of the principles under-lying such derivations, I suggest, emerges an explanation of how microcom-plexity in effect gives rise to macrosimplicity. The simple behavior of complex systems can, in other words, be understood by inquiring into the foundations of EPA.

Let me begin the inquiry with the following question about EPA: under what circumstances, exactly, can EPA be successfully applied? Answer: it can be ap-plied only when there exist enion probabilities with the properties identified in section 1.23. This raises another question, whether there are enough probabil-ities of the right kind to provide a foundation for EPA in the complex systems whose simple behavior was described in section 1.11. The greater part of this book—chapters two, three, and four—is an attempt to show that there are, and more importantly, to explain why there are.

The five principal properties that enion probabilities must have in order to serve as a basis for EPA are:

1. Enion probabilities must have the mathematical properties assumed in the calculations that underlie EPA, which is to say that they must satisfy the axioms of the probability calculus.
2. The values of enion probabilities must be functions of only macrolevel information about the initial state of the system. In most cases, they must be functions of only those macrovariables that appear in the macrolevel law that describes the simple behavior to be explained.[14]

3. Enion probabilities must be mutually stochastically independent.[15]
4. There must be a strong link between the enion probability of an outcome and the frequency with which the outcome occurs. This ensures some kind of connection between a probability distribution over a macrovariable and the actual behavior of that macrovariable, in particular, between a simple probabilistic law and the corresponding simple behavior.
5. If EPA is to explain macrolevel behavior, enion probabilities must be explanatorily potent; that is, they must explain the outcomes they produce, and perhaps even more important, they must explain the way the outcomes are patterned. I will not assume any particular philosophical account of explanation in this study, but I will have plenty to say about the explanation of patterns of outcomes all the same.

Of these, conditions (1), (4), and (5) are normally supposed to be true of any kind of physical probability. Conditions (2) and (3) are not; they impose additional demands on the probabilities that are to provide the basis for EPA.

The rest of this section, and indeed, much of the rest of this book, is about (2) and (3) and the reasons that they hold. I begin by observing that (2) and (3) have the same mathematical form, in that they can both be expressed as requiring that the value of an enion probability be unaffected by conditioning on certain kinds of microlevel information. (In the case of condition (3), the microlevel information is that concerning the outcomes of events involving other enions.) Although it is, for practical purposes, easier to deal with the two conditions separately, they can be combined into a single condition, requiring roughly that:

The values of enion probabilities are unaffected by conditioning on microlevel information.[16]

This is the **probabilistic supercondition**. The supercondition states the one and only property required of enion probabilities, over and above the usual properties of probabilities, if they are to serve as a foundation for EPA. To understand the reasons why, despite appearances, the supercondition holds of a complex system, is to understand the simplicity of the system's macrolevel behavior, or so I now argue.

I begin by showing how the supercondition powers the EPA explanation of simple behavior. The problem of explaining the existence of simple macrolevel laws is essentially the problem of explaining why microlevel details have no effect on macrolevel dynamics. To put it another way, it is the problem of explaining why microvariables do not turn up in macrolevel laws, that is, why the dependencies expressed by the laws are not dependencies in part on microlevel information.

The supercondition encodes this absence of dependence mathematically. When it is true, then, microlevel information can be shown, as explained in the last section, to fall out of the picture at a certain level of abstraction, leaving behind a dependence relation between macrolevel quantities, a simple macrolevel law. But because the supercondition simply states the fact of the absence of microlevel dependence, it cannot be said to explain the existence of simple macrolevel laws. What explains simple behavior in complex systems is what explains why the supercondition is true, that is, what explains the lack of dependence that the supercondition merely asserts. It is the aim of this book to explain the supercondition.[17]

I have two important remarks to make about the supercondition. First, the supercondition seems, on initial inspection, almost certain *not* to be satisfied in a complex system. The fate of an enion in a complex system, it is widely and rightly believed, is harnessed so closely to initial conditions that the slightest change in these conditions can precipitate a complete reversal of fortune. The supercondition, however, requires that the probability distributions describing the behavior of the enions of a system be quite indifferent to such microlevel details. How can this be?

The answer, presented in chapter four, is that the chaotic dynamics that makes the fates of individual enions so sensitive to initial conditions also ensures that the statistical distribution of the behavior of enions will be almost completely independent of the distribution of initial conditions. Although the details of the initial conditions matter greatly to the individual, then, they make very little difference to the population.

To make use of this fact, an enion probability must be understood as a statistical property, that is, as a property describing, not the dynamics of an individual enion, but the dynamics of an entire class of enions. Then, insensitivity of population level distributions to the vagaries of initial conditions will translate into the independence of enion probabilities from microlevel information. In this way, the lack of microlevel dependence can be reconciled with the fact of microlevel chaos. Indeed, as I will later show, one can do better than that: one can *mathematically derive* microlevel independence from the fact of microlevel chaos.

My second remark is that the power of enion probabilities as tools for understanding the behavior of complex systems lies in a certain double aspect: enion probabilities are extrinsic properties that behave, in a sense, like intrinsic properties. They are extrinsic because their values are determined by aspects of a complex system that mostly lie quite outside the boundaries of the particular enion to which they are attached. This is due in part to their being, as just noted, statistical properties, in the first instance attached to types of enion rather than to individual enions. More important, however, is the fact that their values depend not only on the properties of the enion, or type of enion, to which they are attached, but also on the properties of all the other elements in a system—as the probability of rabbit survival, for example, depends on the number of foxes and other rabbits in the local population. It is because of their extrinsic nature that enion probabilities are able to comprehend the influence of the many parts of a complex system. Yet, because they are stochastically independent, enion probabilities behave like individuals. They can be plucked out of a system, hence removed from their context—in reality because they already contain what is important about their context—and put together with other enion probabilities according to the very simple rules of aggregation that apply to independent probabilities. Thus enion probabilities contain what is important about all the interconnections in a complex system, yet they can be combined without any thought to those interconnections. These comments are amplified in section 5.5.

As I have said, the main argument of this book can be understood as an explanation of the supercondition. I divide the supercondition into two parts, corresponding to conditions (2) and (3) above. Chapter two lays the theoretical groundwork for explaining why certain probabilities satisfy condition (2), while chapter three lays the groundwork for explaining why certain probabilities satisfy condition (3). Chapter four then pulls the pieces together, as explained further in section 1.34.

Two more general comments. First, although a large part of this book consists in a philosophical investigation of the properties of probabilities, I do not assume any particular metaphysics of probability. What I have to say is compatible with any of the main metaphysical views about probability, such as the frequentist, propensity, and subjectivist theories. My reasons for avoiding metaphysics are discussed in sections 1.32 and 1.33.

Second, let me say something about the epistemic, as opposed to the explanatory, use of EPA. Enion probability analysis can be and has been used both as a method for discovering simple macrolevel regularities and as a method for explaining the existence of those regularities already known. An example of the first, predictive use is actuarial science, in which the insurance industry predicts the macrolevel regularities that determine its costs. An example of the second, explanatory use is Boltzmann's explanation of the second law of thermodynamics, which had been known for fifty years. My chief concern, stated at the beginning of this chapter, is with the explanatory use, but much of what I say is relevant to the predictive use as well. The predictive use makes two additional demands of enion probabilities. First, the values of enion probabilities must be epistemically accessible. Second, the fact of their stochastic independence must be epistemically accessible. In both cases, the access must be by some method other than the observation of the statistics that are to be predicted. For a discussion of non-statistical inferences about probabilities, see Strevens (1998).

1.25 Enion Probability Analysis in Statistical Physics

Very little attention has been paid to the foundations of EPA in biology and the social sciences. But the same cannot be said for statistical physics: the question as to, first, the legitimacy, and later, the undeniable success, of the probabilistic treatment of questions concerning heat and gases, has been on the agenda, sporadically, in physics, mathematics, and the philosophy of science for over one hundred years. As a consequence, there is an enormous literature on the topic.

This literature is concerned exclusively with the foundations of EPA's application to physical systems, as opposed to biological and social systems. But one might reasonably wonder if the techniques used to understand the foundations of EPA as applied to systems of one kind might not solve, if not all, then at least some, of the problems encountered in understanding the foundations of the EPA approach in other systems.

In following this line of thought one immediately encounters two very daunting problems. First, there are many different approaches to the question of the foundations of statistical physics, far too many to survey here in any fruitful way. Second, there is no general agreement as to which approaches are best, or whether any is adequate at all. I refer the reader to Sklar (1993) for confirmation of both these claims, and for a useful and comprehensive philosophers' introduction to the area.

What I propose to do in this section is to examine briefly the approach to understanding the foundations of statistical physics that is closest to my approach to understanding EPA. This is the ergodic approach, which, like my approach, focuses on the physical properties of systems in virtue of which they behave in random-looking ways. Despite the similarity between the two approaches, I will contend that there are several deep differences that render the ergodic approach, unlike my own, unsuitable as an entry point to the study of the general application of EPA. The remainder of this section is intended chiefly for the benefit of readers who already have some knowledge of the ergodic approach; it is a justification of my tabula rasa approach to understanding the foundations of EPA. Other readers may move on to section 1.3. I will later return to the question of the similarity of the two approaches at a slightly more technical level (section 4.87).

The ergodic approach to the foundations of statistical physics seeks to prove that certain systems will tend to behave in random-looking ways no matter what—with a vanishingly small set of exceptions—their initial conditions. In more recent ergodic theory, the randomness of the patterns shown to occur is close to that of what I will later call the *probabilistic patterns*.

I will distinguish three important limitations of the ergodic approach. None, except perhaps the first, constitutes an objection to the use of the ergodic approach to justify the probabilistic apparatus of statistical physics; what is limited, rather, is the potential of the ergodic approach to explain the success of EPA outside of physics. I will need something more flexible.

The first limitation of the ergodic approach is that, in its earlier versions, it delivers the desired patterns of behavior only in the long run. Thus it does not seem well placed to explain the everyday, year-to-year or month-to-month behavior observed in biological or social systems, or even, for that matter, the minute-to-minute behavior of systems treated by statistical physics. That this is a serious shortcoming is especially evident when treating systems that themselves may exist for only a short time, as in many biological and social cases.

More recent results in the ergodic tradition rectify this shortcoming by showing that certain systems behave in a random-looking way at all times (see section 4.87 for further discussion). These results are, however, limited in a different way: they have been proved only for a handful of very simple systems. The method is rather complicated, and there is no real prospect that it can be extended to, say, biological systems.

The second limitation of the ergodic approach is related to the point just made. Even the most general theorems of ergodic theory—the theorems that apply to the greatest number of different systems—make various assumptions about the systems they treat that, although plausible for certain physical systems, are certainly not satisfied by biological or social systems, or even most other physical systems. One example is the assumption that systems conserve energy. Another is the assumption that systems' microlevel laws of time evolution are continuous. Neither assumption is satisfied by, say, the biological processes that are responsible for the behavior of ecosystems.

The third limitation of the ergodic approach is that it has nothing to say about enion probabilities or their aggregation. The results it proves concern patterns in the behavior of systems as a whole, rather than in the behavior of individual enions. These system-wide behaviors do have implications—of course—for enion statistics, and so for patterns of enion behavior, but the direction of explanation is from patterns of system-wide behavior to patterns of enion behavior. If I am right that the success of EPA is due to our ability to put together enion probabilities so as to understand or predict system-wide behavior, then the ergodic approach cannot explain EPA as such, in the sense that it cannot explain why we can make progress in science by first assigning probability distributions to enion behavior and then deducing the whole system's behavior from the probabilities pertaining to the behavior of the parts. In an important sense, then, although ergodic theory may be able to explain the simple behavior of some complex systems, it cannot—it does not even try to—explain the surprising efficacy of EPA in predicting and understanding that behavior.

In conclusion, for all its successes, ergodic theory falls short of explaining the foundations of EPA in at least two important ways. First, its techniques cannot be applied to many of the systems to which EPA can be applied. Second, it cannot make sense of the special role played by enion probabilities in EPA. And in this respect, ergodic theory is truly representative of theories of the foundations of statistical physics, for they all share these same drawbacks.[18]

1.3 Towards an Understanding of Enion Probabilities

I will conclude this chapter by providing an idea of how an investigation of the properties of enion probabilities might proceed. My two conclusions are, first, that what is required is not a metaphysics, but (in a sense to be explained) a *physics* of enion probability, and second, that a central element of such a physics will be the explanation of what I call *probabilistic patterns*.

1.31 Simple and Complex Probability

I begin by distinguishing what I will call simple, simplex, and complex probabilities. The distinction will be useful in a number of places in the book; in

what follows, for example, it is used to contrast the metaphysics and the physics of probability.

A **simple probability** is a probability that is either explicitly mentioned in
the fundamental laws of physics, or is a straightforward consequence of such
probabilities. Examples of simple probabilities are the probabilities that appear
in the laws of radioactive decay, and on at least one interpretation of quantum
mechanics, the probability of the death of Schrödinger's cat.[19]

A **simplex probability** is a probability that is in some sense entirely reducible
to certain simple probabilities, but that depends on those simple probabilities
in a complicated way. Examples are controversial, but if a certain kind of
collapse interpretation of quantum mechanics is correct, then the probabilities
of statistical physics may be simplex (Albert 2000, chap. 7). The question
of what it is for a probability to be *entirely reducible* to some set of other
probabilities will not be answered in this section, as it requires techniques
developed in chapter two. The reader will find a more precise characterization
of simplex probability in section 2.63.

Two remarks. First, if fundamental physics is deterministic, then there are
no simple or simplex probabilities. Second, the definitions of simple and simplex probability do not presuppose any particular philosophical account of
the nature of simple probabilities. It is left open whether simple probabilities
are, for example, frequencies, propensities, subjective probabilities, or indeed,
whether there is any such thing as simple probability at all.

A probability that is neither simple nor simplex is a **complex probability**.
Complex probabilities may depend in part on simple or simplex probabilities,
but the purest kind of complex probability is a probability associated with a
process governed by laws that are all deterministic or quasi-deterministic.[20]
An example is the one-half probability of a tossed coin's landing heads. The
outcome of a coin toss is determined by quasi-deterministic laws of mechanics,
some facts about coins, and the initial conditions of the toss. The probabilistic
aspect of fundamental physics has, practically speaking, no effect on how the
coin lands. Thus the probability of heads does not depend at all on simple
probabilities. As in the case of simple and simplex probability, no particular
metaphysics of complex probabilities, not even a commitment to their being
"real" probabilities at all, is presupposed by the definition.

I have defined three notions: simple, simplex, and complex probability. For
the purposes of metaphysics, the distinction of paramount importance is the
ontological divide between the simple and simplex probabilities, on the one
hand, and the complex probabilities, on the other.[21] Thus it is natural, and I

think quite correct, for metaphysics to subsume the class of simplex probabilities under that of simple probabilities, and to distinguish just two classes:
the simple probabilities (including the simplex probabilities) and the complex probabilities. The contrast is between those probabilities that inherit their
probabilistic nature from certain elements of the basic metaphysical furniture
of the universe that are themselves probabilistic, and those probabilities whose
probabilistic nature emerges from the way that the non-probabilistic furniture
is arranged.

As will become apparent in later sections, however, the distinction that will
be most important to this study is that between the simple probabilities, on the
one hand, and the simplex and complex probabilities, on the other. Thus I will
organize things in a rather unmetaphysical way: I will classify simplex probability as a kind of complex probability (except where I wish to draw an explicit contrast between simplex and complex probabilities, as in section 2.63).
Throughout the book, then, a complex probability is a probability that is complex in the "pure" sense (it depends not at all on simple probabilities), complex
in some less pure sense, or that is simplex. On this classification, the contrast between simple and complex probability is not a contrast based on the
kinds of fundamental ontological units of which probabilities are composed,
but is rather a contrast based on the *way* probabilities are composed—on what
might be called the physical structure of the probabilities. Simple probabilities
bear very simple relations to their metaphysical building blocks, or are themselves metaphysical building blocks, whereas complex probabilities bear quite
complex relations to their building blocks. It is this distinction that is important for my purposes, because, in order to understand the simple behavior of
complex systems, what is important is not so much the intrinsic properties of
a system's parts, but the way that the parts are put together. For this reason, I
will speak of my project as an exercise in physics, not metaphysics.

1.32 Metaphysics versus Physics

Enion probabilities are complex probabilities. An understanding of the properties of enion probabilities requires, then, a set of techniques for understanding the properties of complex probabilities, and in particular, a set of
techniques for understanding the complex relations that complex probabilities bear to their metaphysical building blocks, whatever those building blocks
may be.

The nature of probability has been debated by philosophers since early modern times, and especially intensely since the advent of statistical and quantum mechanics. It is natural, then, to go to the metaphysicians for the key to the understanding of complex probability. But the metaphysical literature contains very little discussion of this topic (for some exceptions, see section 2.A). The problem is not that the metaphysical theories are incomplete, but that metaphysical theories are designed to answer a different set of questions than the questions I ask here. For want of better terms, in what follows I distinguish between a *metaphysics* of probability and a *physics* of probability, with *physics* to be understood in the broadest possible sense as the science of nature. Philosophical theories of probability, I suggest, are concerned with metaphysical questions, not physical questions, but it is physical questions that must be answered if my inquiry into the foundations of EPA is to succeed. As to the metaphysics, I will remain uncommitted throughout this study. What, though, is the physics, as distinct from the metaphysics, of probability?

One important physical question concerns the sources of stochastic independence, a topic about which metaphysics has had little to say. But there is also a physical question about the nature of probability itself that is distinct from the metaphysical question about the nature of probability that philosophers have traditionally sought to answer. The physical question is, roughly, this: what are the physical underpinnings of probabilistic processes? Let me explain what I mean.

Begin with a very general question: what sort of phenomena invite probabilistic treatment? The answer, it is generally agreed, is that one posits a probabilistic process when one sees a sequence of outcomes that is disordered (hence unpredictable) in the short term, but which exhibits a certain sort of long-term statistical order, namely, a stable frequency or set of frequencies. Examples are the patterns of outcomes generated by a series of tosses of a fair coin, by radioactive decay, by measurements of errors in astronomical observation, and by certain random walks. Call these patterns of outcomes collectively the **probabilistic patterns.**[22]

The examples I have given each exhibit somewhat different kinds of pattern: the tossed coin generates outcomes in a pattern typical of what mathematicians call a Bernoulli process; radioactive patterns are typical of what is called a Poisson process; patterns in measurement error are typical of what is called a normal or Gaussian process; and the patterns generated by random walks are typical of what is called a Markov process. The most important subclass of the patterns are those associated with what are called independent and identically distributed (IID) trials, which include the Bernoulli and some Gaussian processes. I am mainly concerned with the IID patterns, since understanding these is sufficient, or so I will argue, for understanding the simple behavior of complex systems. This is true, in part, because other kinds of probabilistic processes, such as Markov processes, can be understood in terms of IID processes (section 4.4). Indeed, the main results of this study present some reason to think that all probabilistic processes in complex systems can be understood in terms of Bernoulli processes.

Given this rough characterization of the probabilistic patterns, the physical question may then be posed as follows: what sorts of physical processes are responsible for the probabilistic patterns? Or to put it another way, in virtue of what property or properties do probabilistic setups produce probabilistic patterns of outcomes? These properties are central aspects of the *physical nature* of probability; understand them and you understand why probabilities do what they physically do.

From an understanding of the physical nature of some set of probabilities, it may be hoped, will proceed an understanding of all of the properties of those probabilities. I have in mind in particular, of course, the properties required for the successful deployment of EPA: obedience to the axioms of the probability calculus, independence from microlevel information, independence from other enion probabilities, a connection to the frequencies, and explanatory potency. Once it is shown how enion probabilities explain the probabilistic patterns, the last two properties will require no additional insight: since frequencies are part of the probabilistic patterns, to show how an enion probability explains the patterns is both to establish a connection to the frequencies and to establish the probability's explanatory power. Understanding the other properties constitutes, then, together with an understanding of the probabilistic patterns, the whole of the physics of enion probability.

The metaphysics of probability aims to understand few or none of these things; at the very least, it does not aim to explain conformance to the supercondition. Rather, the aim of metaphysics, it might be said, is to analyze probability without remainder, in the sense that it aims to give a definition of probability that itself makes no reference to probabilities.[23] The point of such a definition is twofold. First, it provides a clear boundary between those things that are probabilities and those things that are not. Second, it provides rules for determining what kind of world can contain probabilities, and what kind of world cannot. This second achievement then provides a basis for understanding how there can be probabilities in our world.

I have emphasized what a physics of probability aims to provide that a metaphysics does not. But it is also important to see what a physics does not aim to provide. Most of all, a physics of probability does not attempt to say what probability is. As I have already mentioned, there is no account of the nature of probability in this study; my conclusions are compatible with any of the usual metaphysical theories of probability (see section 2.3).

An important benefit of renouncing the metaphysical question is that a physics of probability may assume, as a part of its account of the properties of some set of probabilities, the existence, without any further argument or analysis, of other probabilities. As with any explanation, the success of such an account will depend on how much the other probabilities are asked to do. If they do everything, there is not much of an explanation, as when, say, the solidity of a table is explained by the solidity of its parts. (For this reason, it seems unlikely that there could be an interesting physics of simple probability.) But if most of the explanatory burden is borne by other elements in the account, and by the way that these other elements work together with the probabilities, then there may be considerable enlightenment despite the explanatory remainder, as when, say, one explains Kepler's laws by deriving them from other, more basic laws. Some examples are given in sections 2.14 and 3.5.

A physics of probability, then, does not necessarily answer the metaphysical questions, and a metaphysics of probability certainly does not answer the physical questions. The goal of the metaphysics of probability is to say what probabilities are; the goal of the physics is to understand what probabilities do.

1.33 The Metaphysics of Complex Probability

I conclude this section with a very brief survey of three metaphysical accounts of probability. My aim is not to criticize these accounts, but to support the very general claims of the last section by showing, case by case, that metaphysical answers do not resolve physical questions.

THE ACTUAL FREQUENCY ACCOUNT

The frequentist holds that probabilities are frequencies. The most straightforward variety of frequentism is the actual frequency account, on which the probability that a trial on a system of type X produces an outcome of type e is identified with the frequency with which all trials on systems of type X

produce outcomes of type *e*; if there are an infinite number of trials, the probability is the limiting frequency.

A prominent early frequentist was, as noted earlier, John Venn. Later versions of frequentism, more sophisticated than the actual frequentism characterized in the last paragraph, were developed by von Mises (1957) and Reichenbach (1949). Perhaps von Mises's most influential innovation was his requirement that the set of trials determining a probability be randomly patterned, in a sense more precise, and more controversial, than need be spelled out here (for the details, see Fine 1973).

Consider, for simplicity's sake, actual frequentism. The actual frequency account postulates a probability more or less wherever there is a probabilistic pattern, and holds that there is no more, physically, to the probability, than the pattern. In so doing it provides clear, though controversial, answers to the metaphysical questions rehearsed above. In particular, it provides a simple criterion for the presence of probabilities in the world. What kind of world contains probabilities? Simple; a world that contains probabilistically patterned outcomes.

What the actual frequency account does not and could not provide is any understanding of *why* any world, including our world, contains probabilistic patterns. Nor does it provide any basis for understanding independence. According to the actual frequency account, type *e* events are independent of type *f* events just in case the frequency with which *e* and *f* events occur together is equal to the frequency of *e* events multiplied by the frequency of *f* events. Why the frequencies should be related in this way is a question that the actual frequency account was never intended to answer.

THE PROPENSITY ACCOUNT

According to the propensity account, probabilities are a special kind of irreducible disposition. Fetzer (1971) calls them *statistical* dispositions; see also Giere (1973). Because they are irreducible, there is very little more that can be said about these dispositions, except that they usually produce probabilistically patterned outcomes, or at least, that they do so with a high probability.

Despite the brevity of the account, it, like the actual frequency account, offers clear answers to the metaphysical questions. In particular, we are told what kind of world contains probabilities: a world that contains statistical dispositions.

Because statistical dispositions are irreducible, the laws that concern them must be fundamental, or so the propensity theorists argue. It follows that statistical dispositions are always simple probabilities; since all probabilities are statistical dispositions, there are no complex probabilities.[24] What, then, do propensity theorists say about the complex probabilities that appear in statistical physics, evolutionary biology, and so on? Giere claims that these probabilities can be given what he calls an "as if" interpretation, provided that we realize "that this is only a convenient way of talking and that the implied physical probabilities really do not exist" (Giere 1973, 481). Again, the answers to the metaphysical questions are clear. What counts as a complex probability? Nothing. What kind of world contains complex probabilities? No world.

The physical questions, however, go without answers. This is particularly evident in the case of complex probability. The propensity theory is metaphysically stern—there are no complex probabilities—but methodologically indulgent—one is often allowed to pretend that there are. Why is this pretense allowed? How can pretense serve as the basis for successful scientific theories such as evolutionary biology? What is the physical nature of the processes underlying "as if" probabilities, such that they give rise to outcomes that are probabilistically patterned, stochastically independent, and so on? The propensity account is silent. Its goal is metaphysical legislation, not physical understanding.

The Subjectivist Account

According to the subjectivist account, the probabilities that appear in scientific theories are nothing more than subjective probabilities that have a certain privileged position within the epistemic framework of a scientist or a community of scientists (de Finetti 1964; Skyrms 1980).[25]

All subjectivist accounts are broadly agreed that what distinguishes the subjective probabilities that are considered "scientific" and perhaps "real" from those that are reflections of mere opinion is a kind of invariance under the influence of further information, what Skyrms calls *resilience*. This is very suggestive, given the form of the supercondition required of enion probabilities for the success of EPA (section 1.24). The questions that are suggested are: Why should further information, particularly microlevel information, be discounted when setting one's subjective probabilities? When does ignoring such information lead to discovery? What is the nature of the physical processes that are the subject of these theories, such that they produce outcomes to which resilient probabilities may fruitfully be assigned? The answers to these physical questions are irrelevant to the subjectivists' metaphysical question about the nature of probability. The metaphysical question has a psychological answer: probability is a human projection onto the world. The physics of that world is entirely another concern.

Metaphysical theories of probability, then, simply do not attempt to answer the physical questions that I have posed concerning, first, the production of the probabilistic patterns, and second, the satisfaction of the probabilistic supercondition. This is not to be taken as an objection to the metaphysical accounts; it is no defect that they do not yield answers they were not intended to give. Rather, it is a reason for someone with my goals to put aside the metaphysical accounts. And—as noted above—so I do: throughout the book I remain agnostic on all metaphysical questions.

The primary purpose of my metaphysical noncommittal is to avoid needless philosophical entanglements, but noncommittal brings with it two positive benefits, as well. First, I can provide an explanation of the simple behavior of complex systems that is compatible with any metaphysics of probability. Second, I can show how small a part questions concerning the metaphysics of probability play in our understanding of the simple behavior of complex systems—indeed, in any aspect of our scientific understanding of the way the world works.

The physics and the metaphysics of probability may be distinct inquiries, but their methods sometimes overlap. Though writing in a predominantly metaphysical vein, both Reichenbach (1949, §69) and Savage (1973) invoke a mathematical framework that is central to much of my probabilistic "physics," the method of arbitrary functions (see sections 2.22 and 2.A). And in my physics, I borrow some ideas from the metaphysics of probability; let me conclude by listing one feature of each of the three metaphysical theories described above that contributes to my own characterization of complex probability in chapter two. From frequentism, I take the idea that probabilities are properties of setup types rather than of individual setups (section 2.11). From the propensity account, I take the idea that one cannot understand probability without understanding the intrinsic physical properties of the setup to which a probability is attached. From subjectivism, I take the idea that the impression we have of physical probabilities as intrinsic properties of individual setups is an illusion, brought about by an unwarranted inference from stochastic independence (or more generally, Skyrms's resilience) to causal independence (section 2.24).

1.34 The Structure of the Argument and a Guide for the Reader

For two long chapters I will now ignore complex systems altogether. Chapters two and three develop a physics of complex probability that is principally concerned to show how complex probabilistic setups produce probabilistic patterns, and when and how a complex probability satisfies the probabilistic supercondition, that is, when and how the value of the probability is unaffected by conditioning on low-level information. The examples in these chapters are centered around simple gambling devices rather than complex systems.

The first of the two chapters, chapter two, has three functions. It develops the notion of complex probability, settling a number of foundational issues, it proposes a partial explanation of the probabilistic patterns, and it discusses some circumstances under which the first part of the supercondition, the requirement that the value of a complex probability be a function of only high-level information about the initial state of the corresponding probabilistic setup, will hold.

Chapter three is devoted almost entirely to the discussion of circumstances under which the second part of the supercondition, the requirement that complex probabilities are stochastically independent of one another, will hold. It also completes the explanation of the probabilistic patterns.

In chapter four, complex systems are once more the principal topic. I show that the enion probabilities of many complex systems satisfy the probabilistic supercondition for much the same reasons as do the probabilities attached to the simple gambling devices that are the subject of chapters two and three. The simple behavior of complex systems is thus explained.

The final chapter, chapter five, discusses some implications of my conclusions for philosophical work on the higher level sciences, that is, all sciences other than fundamental physics.

I envisage two ways to read this book. The first is, of course, to read the chapters in the order that they are presented. The second is to move from the end of chapter one straight to chapter four, in order to see right away and in detail how the understanding of the supercondition developed in chapters two and three is put to use in explaining the potency of EPA. Sections 4.1, 4.2, and 4.3 require only a look at the opening pages of chapter two and perhaps the beginning of section 2.4. Section 4.4 requires knowledge of sections 2.1 and 2.2 of chapter two. Section 4.5 draws heavily on chapter three, but the overall aims of this section will be comprehensible even to the reader unfamiliar with the content of that chapter. The same is true of section 4.6, by the end of which the reader should have a fairly good understanding of the role of chapters two and three in the explanation of complex systems' simple behavior, without having to read more than the first part of chapter two.

A reader who is primarily interested in complex systems, and who wants a feel for the overall shape of the project, then, may begin by reading chapter one, sections 2.1, 2.2, and 2.4 of chapter two, and sections 4.1 to 4.6 of chapter four.

2

The Physics of
Complex Probability

Simple behaviors exist when the dynamics of a system's macrovariables are influenced only by other macrovariables, and not by microvariables. To understand simple behavior, we must understand this lack of dependence; to understand simple behavior probabilistically, then, we must understand the independence of enion probabilities from microlevel detail, or as I put it above, we must understand the foundations of the probabilistic supercondition.

Chapter two contributes to this goal in three stages. It provides a set of techniques for thinking about the physical nature of complex probability. It explains why a certain kind of complex probability, *microconstant probability*, produces outcomes that are probabilistically patterned. And—just what is needed to understand the first part of the supercondition—it shows that microconstant probability is independent of microlevel facts about initial conditions.

Three preliminary comments. First, although my real concern is with enion probabilities, I concentrate almost exclusively on the probabilities associated with coin tosses, die throws, and other classic probabilistic experiments. The distinctive features of these simple cases carry over to the enion probabilities examined in chapter four. Indeed, I conjecture that the understanding of complex probability provided in this chapter extends to *all* important complex probabilities, including those relevant to gambling, sampling, and the theory of measurement error.

Second, in the initial exposition (sections 2.1 to 2.4), I assume that the laws underlying the processes in question are deterministic or quasi-deterministic; probabilistic processes come later in the chapter.

Third, much of this chapter is concerned with providing a system for describing complex probabilistic setups and their properties. Were this description an attempt at a metaphysics of probability, two questions would naturally arise: (a) whether the system is the uniquely correct system for describing complex probabilistic setups, and (b) whether the system succeeds in fixing unique descriptions of complex probabilistic setups.[1] Neither of these questions is of great importance: as explained elsewhere, the system of description that I offer is not a metaphysics, but an attempt to find a useful representation for understanding the physical properties of complex probabilities and, ultimately, a useful representation for understanding the physical properties of complex systems. There are many ways, both inside and outside my system, to describe complex probabilistic processes. The claim I make for my system is not that it provides unique descriptions, but that it supplies fruitful descriptions. It is just one way to represent the processes that go on in complex systems, but it is a way that allows us to understand why these systems so often behave in simple ways.

2.1 Complex Probability Quantified

2.11 Probabilistic Setups

The probability that a tossed coin lands heads is one half. An ascription of a complex probability such as this mentions three things: a physical process (the tossing of a coin), an outcome (heads), and the value of a probability, namely, the probability that the process produces the outcome (one half). First, let me develop a way of representing the elements of the process; then I will state a formula that relates these elements to the value of the complex probability for any outcome. This formula is what I refer to as the quantification of complex probability.

I will begin by defining some terms and developing some notation. (Many of the more technical aspects of the following treatment, both mathematical and philosophical, have been confined to the notes.)

The general form of an ascription of a complex probability is as follows:

The complex probability that a physical process of type X produces an outcome of type e is p.

I call the event e the **designated outcome**,[2] and I write the complex probability of e as cprob(e).

Note that the canonical complex probability ascription associates the probability with a *type* of process, not with an individual token process. As a metaphysical starting point, this does not sit well with the claim, standard among propensity theorists, that probabilities are properties of individual token processes. For my explanatory purposes, by contrast, there is a very good reason to begin the inquiry with a definition of complex probability that attaches in the first instance to process types. The reason is that the satisfaction of the probabilistic supercondition—the stricture that enion probabilities must be unaffected by conditioning on microlevel information—requires enion probabilities to be understood as in some sense statistical properties.[3] Because these statistical properties will be used to explain a statistical phenomenon, namely, the dynamics of macrovariables, that is, the dynamics of enion statistics, there is no point in my investigation at which token probabilities will be required.

Let me begin with the nature of the physical process X. It is customary to call such a process a **trial** on a probabilistic setup. This suggests that a specification of a probabilistic setup is just a specification of a process type. The term *setup* is used in the literature somewhat ambiguously, however, sometimes referring, as here, to a type of physical process, but sometimes referring to the central device in such a process, for example, to a roulette wheel. I use the term in its former sense. A **probabilistic setup** type, then, is a type of physical process that generates outcomes with some well-defined probability. Eventually, I will develop the notion of a *probabilistic experiment* to supersede the notion of a setup. This term, happily, has more of the connotation of a process than of a device.

A probabilistic setup can be notionally divided into two parts, which I call the *mechanism* and the *initial conditions*. In a simple case such as a gambling setup, the mechanism consists of a physical device together with an operation on that device, for example, a coin together with the operation of tossing. When dealing with, say, the probability that a radium atom undergoes a radioactive decay over a year-long interval, the device is the atom and the operation is simply waiting for a year. An enion probability, such as the probability of a rabbit's dying over the course of a month, has as its "device" the entire complex system, rabbits, foxes, trees and all, and as its operation, waiting for a month (see section 4.2).

Anything that is not a part of the mechanism but that is relevant to the occurrence or otherwise of the designated outcome is an initial condition. The nature of the distinction will become clearer as I consider various examples in various fields, but it has a plain enough intuitive foundation in the gambling devices that are the paradigmatic probabilistic setups. In the case of the roulette wheel, for example, the wheel is part of the mechanism, while the local gravity and air pressure are some of the initial conditions. Other initial conditions, concerning the state of the ball as it is introduced into the wheel, are contributed by the croupier.[4] In the next two sections I propose simple mathematical representations of the mechanism and initial conditions.

2.12 The Initial Conditions

Some of the initial conditions of a setup type are constant, in the sense that they are the same for every instantiation of the setup type. The rest are variable: they change from instance to instance of the type. In the roulette setup, for example, the air pressure is, for all practical purposes, constant, while the quantities determined by the croupier—the point at which the ball is tossed into the wheel, the velocity of the ball at this point, its spin, and so on—are variable. Representing a constant initial condition is easy: one simply states the appropriate value (for example, "normal sea-level air pressure" or "101.3 kPa"). Representing a variable initial condition—what I will call an **ic-variable**—is a more complicated undertaking.[5]

The natural way to incorporate an ic-variable into a representation of a setup type is to state a probability distribution over the possible values of the variable, which I call the **ic-values**. Exactly what kind of information this distribution provides—for example, whether the probabilities are simple or complex, or even real probabilities at all—is an issue that I postpone to section 2.3.

For the most part, I will assume that the distribution of a real-valued ic-variable ζ can be represented by a density function $q(\zeta)$, which I will refer to as the **ic-density** of ζ. A density function represents probabilities as follows: the probability of the ic-variable ζ taking on a value between x and y is equal to the area under the density function $q(\zeta)$ between x and y (see figure 2.1). The best-known kind of density function is the normal curve.

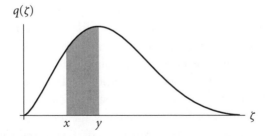

Figure 2.1 A density function over the ic-variable ζ. The function is defined so that the probability of ζ taking on a value between x and y is equal to the area under the function between x and y, shaded in the figure.

Formally, the area under a density function $q(\zeta)$ between x and y is written

$$\int_x^y q(\zeta)\, d\zeta.$$

I will make extensive use of this notation.

Not every probability distribution over a real-valued variable can be represented by a density function, hence the assumption that the distribution of an ic-variable can be so represented is a substantive one; the reader interested in the justification for the assumption should look to section 2.5.

When a variable takes on only discrete values (say, the integers from 1 to 6), the probability distribution over the variable cannot be represented by a density. Thus, strictly speaking, the treatment of initial conditions in what follows does not apply to ic-variables that take on discrete values. However, it is easy to generalize the treatment of complex probability presented here, replacing integration with summation, so that it applies to discrete ic-variables. (See section 2.72 for further details.)

In summary, the specification of the initial conditions of some probabilistic setup will consist of a specification of the constant initial conditions and a specification of an ic-density. If there is more than one ic-variable, the ic-density will have more than one argument, that is, it will be what mathematicians call a *joint density* over all the ic-variables.

2.13 The Mechanism

The other part of a probabilistic setup is the **mechanism**. A mechanism can be specified in a number of ways, of course, but in what follows, I want to isolate that aspect of the physics of the mechanism that is crucial to the occurrence or otherwise of the designated outcome.

In this initial exposition I will discuss for simplicity's sake a probabilistic setup with just one ic-variable ζ (for the multivariable case, see section 2.25). The question I pose is: how does the behavior of the mechanism depend on the value of ζ? I now invoke my assumption that the laws underlying the operation of the mechanism are deterministic. It follows that, given a set of constant initial conditions and a particular value for ζ, the laws completely determine the behavior of the mechanism. In particular, the initial conditions and laws determine whether or not the mechanism will produce an event of the designated outcome type e.

Call a value of ζ that causes the mechanism to produce an e outcome an e-value. Because I am interested only in e, all I need to know about the mechanism is which values of ζ are e-values and which are not. This information can be represented in a single function $h_e(\zeta)$ that maps onto the set $\{0, 1\}$ (that is, the value of $h_e(\zeta)$ is always either zero or one). The definition of this function is as follows: $h_e(\zeta)$ is equal to one if ζ is an e-value, zero otherwise. I call the function $h_e(\zeta)$ the setup's **evolution function** for e. I will drop the subscript when convenient.

2.14 *The Quantification of Complex Probability*

In this section I state a formula that relates the complex probability of an outcome to the relevant ic-density and evolution function. This is what I refer to as the quantification of complex probability. It is true of all the various kinds of complex probability surveyed in this study, and in this sense can be considered a kind of definition of complex probability.

It is natural to postulate the following relation between the complex probability of the occurrence of an outcome e and the initial conditions and mechanism of the setup to which the probability is attached: the complex probability of e is equal to the probability that the initial condition of any particular trial on the setup is an e-value. Thus, to obtain the probability of e, one should sum the probabilities of all of the e-values. This is equivalent to summing the areas under the ic-density that correspond to e-values; see figure 2.2.

In symbols, the summation operation can be written as follows:

$$\text{cprob}(e) = \int_V h_e(\zeta)q(\zeta)\,d\zeta$$

where V is the set of possible values for ζ. (See also definition 2.2.)

It is natural to posit this relation, but psychological compulsion is not enough. What is the objective status of the quantification? That depends on the true nature of complex probability, and thus on the metaphysics of probability. A doctrinaire frequentist, for example, will deny that the quantification is a *definition* of complex probability on the grounds that only information about frequencies can appear in such a definition.

To avoid metaphysics is to avoid the question of what kind of truth about complex probability, if any, is expressed by the quantification. I propose to take the quantification as a definition of a technical notion. I will refer to this

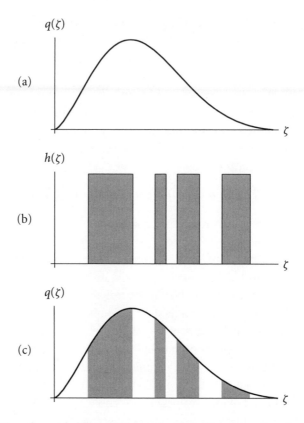

Figure 2.2 Complex probability. (a) An ic-density $q(\zeta)$ representing the distribution of the ic-variable ζ. (b) An evolution function $h_e(\zeta)$ representing the e-values, that is, the values of ζ that cause the mechanism to produce a type e outcome. (c) The complex probability cprob(e) is equal to the proportion of the ic-density spanned by the e-values (shaded area).

technical notion simply as **complex probability**. The metaphysical question of the relation between my technical notion of complex probability and *real* complex probability is left unanswered.

I have a number of remarks to make about the definition of complex probability. First, on the assumption that $q(\zeta)$ is a well-defined probability density and that $h_e(\zeta)$ is a suitably well-behaved function, it follows that the complex probability cprob(e) has a unique value and that complex probabilities will satisfy the axioms of the probability calculus.[6]

Second, the definition of complex probability assumes the division of experimental outcomes into two sorts: type e and type "not-e," which I will write $\bar{e}$. This raises the question of how to deal with probabilistic setups, such as a tossed die, that have more than two basic outcomes of interest. The simplest thing to do is to construct a separate evolution function for each basic outcome. In the case of the die, for example, the probabilities for the six basic outcomes will be derived from six different evolution functions. There are more elegant representations of the setup, but they make the development of the definition more complicated.

Third, I have not defined probabilities for compound events, such as the event of obtaining three type e outcomes in five trials. This topic is taken up in chapter three (section 3.2).

Fourth, for technical purposes, both physical and philosophical, define a probabilistic setup as a pair consisting of an ic-density and a mechanism. The purpose of this definition is to emphasize that the facts about the ic-density are part of the facts about the setup; as a result, a specification of a setup type uniquely determines a probability for any outcome of interest.

One might wonder if the definition could not be further formalized by substituting a set of evolution functions for the mechanism. It can, but this raises the question of which evolution functions to include in the definition. The obvious answer is: one for each basic outcome, for example, one for each number inscribed on a tossed die. But there is no entirely objective criterion for being a basic outcome.

The formalist's solution is to incorporate the set of basic outcomes into the definition. The result is a notion of a slightly different kind of thing, which I will call a **probabilistic experiment**. Call the set consisting of all the basic outcomes of interest the **designated set of outcomes**. (Normally, the outcomes in the designated set are mutually exclusive and exhaust the possibilities.) Then a probabilistic experiment is a triple $\langle q(\zeta), S, H \rangle$ consisting of an ic-density $q(\zeta)$, a designated set of outcomes S, and a set of evolution functions H, containing one evolution function for each outcome in S. No formal requirement is placed on the set of outcomes, except that it be non-empty. (See also definition 2.1.) The notion of a probabilistic experiment turns out to be very useful in what follows (see section 2.23), for which reason most of the rest of the discussion will concern experiments rather than setups. I often use the letter X to denote experiment types as well as setup types, but never, of course, in the same context.

Let me rehearse the steps by which a probabilistic experiment is built up out of its components: (a) adding an operation (for example, tossing) to a physical system (for example, a coin) yields a mechanism, (b) adding an ic-density (for example, the probability distribution over coin spin speed) to a mechanism yields a probabilistic setup, and (c) adding a set of designated outcomes (for example, heads and tails) to a probabilistic setup yields a probabilistic experiment.

Fifth, in a more philosophical vein, one might wonder exactly what sort of property a complex probability is. The short answer is that the complex probability of an event e is a quantification of the properties represented by the corresponding setup's ic-density and its evolution function for e, hence that the probability is itself a high-level property of the setup. That is how I will speak of complex probabilities: they are properties of their setups (not, note, of the outcomes to which they are attached). The short answer is not, however, metaphysically complete, for setups themselves are odd entities. One part of a setup is a physical mechanism—no problem there—but the other is an ic-density, an entity concerning the interpretation of which I remain agnostic throughout this study.

Finally, let me say something about the relationship between the technical notion of complex probability defined in this section and the probabilistic patterns. In section 1.32, I proposed that the physics of complex probability ought to begin with a search for the explanation of the probabilistic patterns. In this respect, it must be said that, although the definition of complex probability offered above is formally useful, it does not in itself convey any real understanding of the physical nature of complex probability, for the following reason.

Consider the significance of the claim that a process is a complex probabilistic process, that is, that it can be described as a trial on a probabilistic setup in the terms laid out above. What can be said about the outcomes of the process? Very little. If the initial conditions of the process are probabilistically patterned, then, in most cases, the outcomes will also be probabilistically patterned, but the process itself does not seem to contribute anything to the patterns: it simply passes them through the mechanism, from input to output.

The physics of complex probability aims to understand what kind of mechanism *produces* probabilistic patterns, not what kind of mechanism preserves them. In much of the rest of this chapter, I examine a special property of certain mechanisms in virtue of which they prove capable of taking a set of initial conditions that is not probabilistically patterned, and producing a set of outcomes that is probabilistically patterned.

This property of mechanisms I call *microconstancy*, and the probabilities attached to such mechanisms *microconstant probabilities*. All microconstant probabilities are complex probabilities, but not all complex probabilities are microconstant. It is microconstant probabilities, rather than complex probabilities as a whole, that will form the subject matter of much of the rest of the chapter and, indeed, of the book. At this stage, then, I turn my attention away from inclusive formal characterizations of experiments to specific physical characterizations.

2.2 Microconstant Probability

2.21 Insensitivity to Microlevel Information about Initial Conditions

The kind of probabilistic experiment that I call *microconstant* provides the kernel of the solution to two problems I have set myself: to explain the probabilistic patterns, and to explain why the values of enion probabilities do not depend on microlevel information about initial conditions.

I will use the second problem, the problem of insensitivity to microlevel information about initial conditions, to motivate the introduction of the notion of microconstancy. The rest of this section explains the problem in more detail, and draws a parallel between the microlevel insensitivity required of enion probabilities and a certain kind of insensitivity to initial conditions in gambling setups such as the roulette wheel. It is gambling setups that will serve as my examples throughout the discussion of microconstancy; for now, I will simply promise that the same conceptual apparatus used to treat gambling setups can also be used to treat enion probabilities.

Consider a deterministic probabilistic experiment, that is, an experiment in which the designated outcomes are generated by underlying laws that are deterministic. One wants to attach probabilities to the designated outcomes. To require that these probabilities are independent of microlevel information about initial conditions is to require that they remain unchanged by conditionalization on any amount of information about initial conditions. Thus it is required that the probability stay the same when conditionalizing on the values of an experiment's ic-variables in a particular trial.

As remarked in section 1.24, this demand looks at first to be impossible to satisfy. Conditionalizing on the ic-value of a particular trial ought, surely, to result in a probability of zero or one for any designated outcome. My solution to the problem is simply to attach complex probabilities to *types* of outcome, rather than to particular outcomes. In the quantification of complex probability in section 2.14, the ic-density is to be understood, not as a distribution over the ic-values for a particular trial, but as a distribution over the ic-values for all trials conducted on a given experiment type.

Understood in this way, the ic-density has the properties of a mean. Conditionalizing on a particular data point does not alter the value of a mean. For example, supposing that you know for sure that the mean weight of a male elephant seal is 3,530 kg, learning the weight of some particular elephant seal will not affect your belief about the mean, because the impact of all facts about elephant-seal weight relevant to the mean is—of course—already incorporated into the mean. In the same way, conditionalizing on the ic-value of a particular trial does not affect the probability distribution over a set of ic-values of which it is a member, because all the microlevel information relevant to the value of the probability is already contained in the distribution.

This definitional maneuver solves one aspect of the problem of microlevel insensitivity to initial conditions, but there is another, which will serve as the focus of the next few sections, and which the notion of microconstancy is introduced to solve.

The problem is as follows: given the quantification of complex probability proposed in section 2.14, it would seem that the value of a complex probability very much depends on the exact form of its ic-density. Information as to the exact form of the ic-density is going to count, for the purposes of EPA, as microlevel information, since the simple macrolevel laws surveyed in chapter one do not mention facts about distributions of initial conditions. There is a certain kind of microlevel information, then, conditionalization on which is going to make a difference to the values of enion probabilities, unless enion probabilities have some special property not possessed by just any complex probability.

That property is what I call microconstancy. Rather than showing right away that enion probabilities are microconstant, I will discover microconstancy first in probabilistic experiments from the gambling world that are, like enion probabilities, insensitive to changes in the shape of their ic-densities.

Consider a roulette wheel. The complex probability of the ball's ending up in a red section is determined, like all complex probabilities, by two things: the physics of the wheel, represented by an evolution function, and the distribution of the initial conditions, represented by an ic-density. The initial condition distribution will be determined by facts about the croupier who is spinning the wheel. Because the croupier changes from time to time, the relevant IC-density presumably changes from time to time as well. But, as everyone knows, the probability of obtaining red remains the same.[7] That is, the probability is independent of low-level information about the forms of the IC-densities induced by different croupiers.

How can this be? The answer to the question is that there are certain kinds of experiments for which almost *any* IC-density will determine approximately the same complex probability. Furthermore, given almost any distribution of actual initial conditions, such experiments will generate probabilistic patterns of outcomes. In the rest of this section, I describe these experiments—the microconstant experiments. Then, in chapter four, I will show that there are very general conditions under which enion probabilities are microconstant, and so have the insensitivity to microlevel information about initial conditions required by EPA.

2.22 The Wheel of Fortune

Let me begin with a worked example. I will explain why the probability of a certain outcome on a simple wheel of fortune remains approximately the same whoever spins it, and in so doing, I will give informal characterizations of some of the properties, including microconstancy, that play the leading roles in the more general treatment developed in section 2.23.

The results described in this section and generalized in section 2.23 are descended, with some modifications, from a mathematical technique known as the method of arbitrary functions, which was promulgated, if not invented, by Henri Poincaré, brought to a fully mature form by Eberhard Hopf, and further developed by Eduardo Engel and others. A description of the method, and the reasons for my modifications, along with bibliographical information, will be found in section 2.A.

The simple wheel of fortune consists of a rotating disk, painted with equally spaced red and black sections like a roulette wheel, and a stationary pointer. The disk is spun and allowed to come to rest. The outcome is determined by the color of the section on the disk that is indicated by the pointer. Assuming that the wheel is returned to the same position before each spin, the outcome of a trial on the wheel is determined by a single IC-variable ζ, the initial speed with which the wheel is spun.

Figure 2.3 A portion of the microconstant evolution function of a simple wheel of fortune, for the outcome *red*.

The contribution of the croupier to the probabilities of red and black, then, takes the form of a probability distribution over the initial spin speed ζ. Experience teaches us that, regardless of who spins the wheel, the probability of red is one half. That is, for any probability distribution over ζ—any ic-density $q(\zeta)$—produced by a normal human, the probability is the same. Why?

As Hopf noted, building on the earlier observations of Poincaré, the answer can be discerned in the evolution function for the wheel. A version of the evolution function is pictured in figure 2.3. Two properties of the evolution function should be noted: (a) the function oscillates rapidly between zero and one, corresponding to the function's regular switching between values of ζ that produce a black outcome and values that produce a red outcome, and (b) although they become thinner as ζ increases, adjacent "red" and "black" segments are of approximately equal size.

Consider two ic-densities for ζ, one heaped around lower values of ζ, the other around higher values, corresponding to, respectively, a mellow and an enthusiastic croupier. If the casino switches from one croupier to the other, hence from one ic-density to the other, what will be the effect on the probability of red? Very little, inspection of figure 2.4 suggests, provided that the ic-densities for both croupiers are fairly smooth, that is, provided that both ic-densities vary only slowly with ζ.

The reason is that a smooth density will be approximately flat over any neighboring pair of gray and white areas in the evolution function, for which reason the contribution made by that part of the ic-density to the probability of red will be approximately equal to the contribution made to the probability of black. The contributions made to the probabilities of red and black over the entire ic-density, then, will be approximately equal, and so the probability of

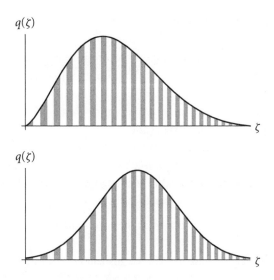

Figure 2.4 The wheel of fortune's evolution function shown with the ic-densities for two different croupiers. For either croupier, the proportion of the ic-density that is shaded gray is approximately 1/2.

red will approximately equal the probability of black: both probabilities will be about one half. Because this is true for any ic-density that is approximately flat over the very small regions spanned by neighboring gray and white stripes, and because most ic-densities produced by humans will tend to have this smoothness property—we do not expect croupiers, for example, to toss the ball frequently at around 0.7 ms^{-1} and 0.72 ms^{-1} but never at 0.71 ms^{-1}—when a new croupier comes on duty, though the ic-density will change, the probability of red will likely stay more or less constant. (The reasons for the prevalence of smoothness are discussed in section 2.53; see also Strevens (1998).)

The same reasoning applies to any of the classic probabilistic setups—the tossed coin, the roulette wheel, the thrown die, and so on—that tend to produce outcomes with the same probabilities regardless of the croupier. The evolution functions in each case have a similar aspect to that shown in figure 2.3: they alternate between gray and white very quickly, but with a constant rhythm, so that the ratio of neighboring gray and white stripes is always the same. This is the property that I call *microconstancy*. It can be shown—generalizing the reasoning in the last paragraph—that if the evolution function for an outcome e is microconstant, any smooth ic-density determines the same probability for *e*, equal to the ratio of gray to white. I call this ratio the *strike ratio* for *e*, and I call the necessary smoothness property of the ɪᴄ-density *macroperiodicity*. Putting this new terminology to work to state the mathematical principle behind the explanation of initial condition microlevel insensitivity: if the evolution function for an outcome is microconstant, then any macroperiodic ɪᴄ-density determines approximately the same probability for the outcome, equal to the strike ratio.

Assuming, then, that human croupiers tend to supply initial conditions that have a sufficiently smooth, that is, macroperiodic, distribution, it is microconstancy that explains the fact that a change in croupiers does not cause a change in probabilities or, more exactly, that it causes only a negligible change in probabilities.

2.23 Microconstancy

I now generalize the reasoning of the previous section. I provide semi-formal definitions of the strike ratio for an outcome, the microconstancy of an evolution function, and the macroperiodicity of an ɪᴄ-density, as well as some related terms, and I explain the theorem, stated at the end of the last section, that links the three notions: an outcome with a microconstant evolution function has a probability approximately equal to its strike ratio, provided that its experiment has a macroperiodic ɪᴄ-density.

Readers who require only an intuitive grasp of notions such as microconstancy and macroperiodicity may wish to skim this section on a first reading. Readers for whom semi-formal is too casual will find formal definitions and proofs in section 2.C1.

THE STRIKE RATIO DEFINED
I will begin by defining the notion of a strike ratio, which will then be used to define the notion of microconstancy. A **strike ratio** for an outcome *e* over some given set of ɪᴄ-values is the proportion of values in the set that produce *e*, that is, the proportion of the set that are *e*-values. In a graph of an evolution function for *e*, such as figure 2.3, the strike ratio for *e* is the proportion of a given region that is shaded gray. Sometimes I talk about the strike ratio over the entire range of the evolution function, sometimes over only a part of the evolution function, as in the last section when I was concerned with the strike ratio for red over a pair of neighboring gray and white regions. For a microconstant evolution function, such as the evolution function for the

wheel of fortune, the two strike ratios are the same: the strike ratio over any small (but not too small) region of the evolution function is the same as the strike ratio over the entire function.

Formally, the strike ratio for an outcome e over a given region U of the domain of the evolution function $h_e(\zeta)$ is the proportion of values in U for which $h_e(\zeta)$ is equal to one.[8] (See also definition 2.3.) If U is the interval $[x, y]$, that is, if U consists of all values of ζ between x and y, then the strike ratio for e over U is

$$\int_x^y h_e(\zeta) \, d\zeta \Big/ (y - x).$$

The numerator of this fraction is the total width of all gray stripes between x and y, and the denominator is the total width of all stripes, gray and white, between x and y.

MICROCONSTANCY DEFINED

A **microconstant** evolution function, such as the evolution function of the wheel of fortune, can be characterized as follows: the evolution function $h_e(\zeta)$ for an outcome e is microconstant just in case the domain of $h_e(\zeta)$ can be divided into many small, contiguous regions, each having the same strike ratio for e. The obvious way to divide up the domain of the wheel of fortune's evolution function, for example, is into intervals spanning pairs of neighboring gray and white bars in figure 2.3. Each of these intervals has a strike ratio of one half, equal to the strike ratio over the entire function.

Exactly how small the regions must be is a question discussed below; to indicate, however, that some kind of standard is required, instead of speaking of small regions I will speak of **micro-sized** regions. Use of this term will distinguish the smallness required by the definition of microconstancy from everyday smallness. Concerning the fact that properties such as microconstancy and macroperiodicity are relative to a choice of variable, or equivalently, a choice of measure on a given variable, see section 2.5.

Note that a microconstant evolution function does not have to swing from zero to one and back at the same tempo throughout the range of its ic-variable. All that is necessary is that it swing with the same rhythm, no matter how the tempo changes. The evolution function of the wheel of fortune pictured in figure 2.3 is an example of a function with quickening tempo but constant rhythm: moving from left to right on the figure, the oscillations become swifter and so the gray and white stripes become narrower, but because they become narrower at the same rate, they maintain their relative sizes, and thus maintain the strike ratio.

The Constant Ratio Partition

It is useful to give a name to the division of a microconstant evolution function's domain into regions with identical strike ratios; I call it a *constant ratio partition*. Because the strike ratio of each region is the same, talk of the strike ratio need not be relativized to a region; I write the strike ratio for an outcome e as srat(e).

Let me put this more formally, incidentally extending the notion of a constant ratio partition to non-microconstant evolution functions. A **constant ratio partition** for an outcome is a partition of the domain of e's evolution function $h_e(\zeta)$ into contiguous regions—that is, into intervals—each having the same strike ratio. (See also definition 2.4.) On this definition, the regions in a constant ratio partition do not have to be small. Consequently, any outcome has at least one constant ratio partition, namely, the "partition" consisting of a single interval spanning all possible values of the ic-variable.

Microconstancy, then, consists not in having a constant ratio partition, but rather in having a constant ratio partition with very small members. This observation makes for the following compact definition of microconstancy: the evolution function for e is microconstant just in case there is a constant ratio partition for e of which every member is micro-sized. In the evolution function for the wheel of fortune, for example, the intervals spanning pairs of neighboring gray and white bars are all micro-sized, so the function is microconstant.

Macroperiodicity Defined

I next define macroperiodicity, the smoothness property that most real ic-densities possess, and which in combination with microconstancy, determines a probability for an outcome approximately equal to the outcome's strike ratio.

When I say that an ic-density is smooth, I mean that there is only a slow variation (if any) in the density as one moves along the ic-variable axis. A smooth ic-density, then, has only a few, if any, sudden drops or rises.

What is mathematically important about the kind of smoothness I call macroperiodicity is that a macroperiodic density will be almost flat over any small region. I adopt this as a semi-formal definition of macroperiodicity: a function is **macroperiodic** just in case it is approximately constant over any contiguous, micro-sized region. (See also definition 2.5.)

The Theorem

The all-important theorem of this section states the relation between microconstancy, macroperiodicity, the strike ratio for an outcome, and its probability: if some designated outcome e of an experiment has a microconstant evolution function, and the experiment has a macroperiodic ic-density, then the probability of e is approximately equal to the strike ratio for e. That is, cprob(e) is approximately equal to srat(e). (This is the content, roughly, of theorem 2.2.)

Note that the relation between the probability and the strike ratio is only one of approximate equality. A change in ic-density, then, will likely bring about a change in probability, but a negligible change. This *almost* total lack of sensitivity to the details of the ic-density will be enough for the explanation of EPA.

One might wonder, however, whether, in a microconstant experiment, the probability ought not to be simply defined as equal to the strike ratio. For various reasons I think that this is not a good idea, in particular because there may be principled reasons why an experiment has, in some peculiar cases, a non-macroperiodic ic-density (see section 2.4). I want to preserve the intuition that, in such cases, the probability deviates from the strike ratio. Thus I stand by the quantification of complex probability in section 2.14; this has the additional benefit that the quantification is, as promised, true for all complex probabilities.

In the remainder of this subsection, I sketch the proof of the approximate equality of the probability and the strike ratio, and I comment on the proof. On a preliminary reading, the reader may wish to move on to the next subsection.

The proof of the theorem is as follows. Suppose that the evolution function for an outcome e is microconstant. Then there exists a partition of the domain of $h_e(\zeta)$ into small regions with identical strike ratios for e, that is, there exists a micro-sized constant ratio partition for e. Call this constant ratio partition $\mathfrak{U}$. For any set U in $\mathfrak{U}$, the conditional probability of e, given that the ic-value for the trial falls into U, is equal to the proportion of the part of the ic-density $q(\zeta)$ that spans U for which $h_e(\zeta)$ is equal to one. Because $q(\zeta)$ is approximately uniform over any such U, this proportion is approximately equal to the strike ratio for e, srat(e). Thus the conditional probability is approximately equal to srat(*e*). Now, the unconditional probability of *e*—that is, cprob(*e*)—is equal to a weighted sum of the conditional probabilities. But since the conditional probabilities are all approximately the same, the way that they are weighted makes almost no difference to cprob(*e*), which will then be approximately equal to the conditional probabilities, that is, approximately equal to the strike ratio srat(*e*).

I will make three remarks about this argument. First, the argument shows exactly what aspect of a macroperiodic ic-density it is that makes no difference to the value of a probability with a microconstant evolution function: it is the different probabilistic weights that attach to the different members of the constant ratio partition, that is, it is the way that the ic-density rises and falls, given that it does so only slowly. Provided that different ic-densities differ only in this respect, moving from one to another will not affect the values of probabilities with microconstant evolution functions.

Second, the argument shows that, for the sake of demonstrating the approximate equality of cprob(*e*) and srat(*e*), the *micro* in microconstancy and the *macro* in macroperiodicity are entirely relative. The absolute size of the regions in an outcome's constant ratio partition is not what matters; what matters is that the ic-density is approximately uniform over the regions, whatever their size. From this observation, I draw two lessons. The narrow mathematical lesson is that macroperiodicity is best defined as relative to a partition (see definition 2.5). The broader lesson is that the micro/macro distinction should be construed as a relative one. What is micro-sized in one context may be macro-sized in another.

Note, however, that if the exact properties of, say, an ic-density are not known, or are subject to change in ways that cannot be anticipated, then there is some virtue in showing that the outcome of interest has a constant ratio partition in which the regions are very small in absolute terms, since the smaller the regions, the more likely it is that the ic-density will turn out to be macroperiodic with respect to the partition. For this reason, I often use *micro* and *macro* as if there is some absolute standard for being micro-sized and macro-sized. In the face of uncertainty, striving to meet absolute standards is the surest feasible way of meeting the relative standard that is inherent in the micro/macro distinction.

Third, when dealing with experiments that have more than one ic-variable, it should be understood that a region's being micro-sized is not just a matter of its having very small volume, but also of its having a reasonably compact shape. A worm-shaped region that has a very small volume but that enters every part of IC-variable space is not, either intuitively or in my semi-formal usage in this study, *micro-sized*.

MICROCONSTANT EXPERIMENTS

Let me now give a reasonably formal definition of a microconstant experiment (with one IC-variable). A one IC-variable experiment $X = \langle q(\zeta), S, H \rangle$ is a *microconstant experiment* just in case, for every e in the designated set of outcomes S, the corresponding evolution function $h_e(\zeta)$ is microconstant. I say that the probability of X's producing an e outcome is a *microconstant probability* just in case $\langle q(\zeta), S_e, H \rangle$, where $S_e = \{e, \bar{e}\}$, is a microconstant experiment.

I have defined probabilistic experiments in such a way that they are individuated by, among other things, their IC-densities. If the IC-density for an experiment changes—if, for example, a new croupier comes on duty—then the experiment itself changes. The motivation for this feature of the definition is that a change in IC-density generally means a change in probability for the designated outcomes. But if an experiment is microconstant, this is not so. There is some reason, then, to extend the definition of an experiment in the microconstant case, so that a microconstant experiment may have not one IC-density but one of a set of IC-densities. Provided that all densities in the set are macroperiodic, the designated outcomes will each have, for practical purposes, a single, well-defined probability.[9] Although I will not formally adopt this suggestion, I will make use of it informally, in that I will often speak of a single microconstant experiment where there is a single mechanism but a changing, though always macroperiodic, IC-density.

RELAXING THE NOTION OF MICROCONSTANCY

The next two subsections concern, first, the need to relax the definition of microconstancy, and more generally, the issue of how to understand results that hold only approximately, and second, the quantification of microconstancy. Both subsections, especially the second, may be skipped on a preliminary reading; the reader may therefore proceed straight to section 2.24.

The definition of microconstancy I have presented above is very demanding. An evolution function is microconstant only if there exists a micro-sized partition in which the strike ratio over every region is *exactly* equal to the strike ratio over the function as a whole. For experiments other than idealized gambling devices such as the wheel of fortune, one may reasonably doubt that the condition of exact equality can be met. But this is obviously not a fatal objection to the approach taken in this section, as approximate equality will do almost as well, if the aim is to derive approximate equality of the strike ratio and the probability.

This sort of observation may be made about almost every result presented in this study. The conditions under which each result is obtained are often quite strict, in some cases obviously unrealistic, but if the conditions are relaxed in certain ways, the result will continue to hold, provided that an *approximately* or an *almost always* is inserted in the right place.

There are two ways of dealing with this feature of my material. I could present all results fully hedged, that is, with *approximately* and *almost* inserted as needed to derive results that have realistic conditions of application, or I could present results for idealized cases accompanied by discussion of the ways in which the results' unrealistic conditions of application might be relaxed without compromising the purpose of this study.

I have already taken the first route, in one respect, in the presentation above, by inserting an *approximately* into the definition of macroperiodicity. I could have defined macroperiodicity so as to require that a function be absolutely uniform over the regions in a relevant partition, but I have required only approximate uniformity. It is for this reason that I have derived only approximate equality of strike ratio and probability, and not exact equality.

In general, I will more often take the second route. I have two reasons for this. First, it is often much easier to understand in virtue of what mathematical principles a certain result holds by studying an especially idealized, and therefore relatively simple, version of the result. Second, there are many different ways that a condition of application may fail to be fully met. For example, a density may fail to be fully macroperiodic not only by varying slightly over every member of the relevant partition, but by varying wildly over a one or a very few members of the partition, while being quite uniform elsewhere. Tolstovianly, each kind of failure creates a different complication in the derivation of the result. Thus a result that takes all possible failures into account may well be extremely complex, if not quite as long as a Russian novel.

I propose to proceed as follows. In the appendices, I will prove idealized results. For example, in the appendix to this chapter (section 2.C) I define macroperiodicity in terms of absolute uniformity over the regions in a partition, and prove the exact equality of strike ratio and probability. These proofs are then followed by sections labeled *Approximation*, which discuss informally the effects of relaxing certain conditions of application. In the main text, I will tend to a more idealized presentation, but with some discussion where it is not obvious how the idealizations might be relaxed.

Degrees of Microconstancy

As noted above, when there is uncertainty as to the details of an ic-density, it may be desirable to show that an evolution function is as microconstant as possible. In this subsection I introduce some terminology that will be later used where it is necessary to consider the degree of microconstancy of an evolution function.

Roughly speaking, a function is more microconstant the smaller the regions of its smallest constant ratio partition. (A constant ratio partition of an evolution function is defined in the obvious way, as a constant ratio partition for its designated outcome.) What does it mean to talk about the size of a partition? Size could be measured in one of several different ways. A small partition might be defined as one in which the average size of the regions in the partition is small, or, more conservatively, as one in which the largest region in the partition is small.

I will opt for the more conservative definition. Consequently, I define the smallest constant ratio partition for an evolution function to be the partition that has the smallest largest member, and I define the degree of microconstancy of an evolution function to be the size of the largest member of the function's smallest partition.

Because the use I have made of the words *small* and *large* is not particularly graceful, I will introduce some technical terms to capture the relevant properties. Call the size of the largest member of a constant ratio partition the partition's **constant ratio index**, or CRI. For any evolution function, consider the constant ratio partition that has the smallest CRI. Call this the evolution function's **optimal constant ratio partition**. (There may be more than one; nothing depends on its being unique.) Then an evolution function's degree of microconstancy—which I will also refer to as the function's CRI—is the CRI of its optimal constant ratio partition (or partitions). If the evolution function is constant, as it will be for an inevitable or an impossible outcome, its CRI is defined to be zero. In such a case, there is no optimal constant ratio partition.

The CRI of an evolution function may be seen as a measure of how finely the function's domain can be partitioned in such a way that each member of the partition has the same strike ratio. If an evolution function $h_e(\zeta)$ has a CRI of r, then all intervals of $h_e(\zeta)$'s domain much longer than r will have the same strike ratio, since any long interval can be assembled from many shorter intervals belonging to the optimal constant ratio partition. Note that an evolution function is microconstant just in case it has a micro-sized CRI.

2.24 The Explanatory Power of Microconstancy

Several different kinds of phenomena are explained by the property of microconstancy. They are, in the order I will consider them, the independence of microconstant probabilities from low-level information about initial conditions, the probabilistic patterns, and the appearance some probabilities have of floating free of their physical context.

First, independence from low-level information. (What follows is a summary of the observations in section 2.22.) I have shown that a microconstant probability for an outcome is approximately equal to its strike ratio if the relevant experiment's ic-density is macroperiodic. If the experiment has more than one ic-density, as in the case of a roulette wheel operated by many croupiers, the approximate equality holds if all ic-densities are macroperiodic.

In most cases, then, the values of microconstant probabilities are determined by only a small part of the complete information that could be given about the workings of a probabilistic experiment, namely, the experiment's strike ratio for the designated outcomes and the fact of the macroperiodicity of the experiment's ic-density. The values of the probabilities are thus independent of all other physical properties of the experiment type, and in particular, of all other information about ic-variables. This independence from low-level information about initial conditions is more or less the quality desired of enion probabilities.[10]

As promised in section 2.21, microconstancy also casts considerable light on the probabilistic patterns. A microconstant experiment will produce the probabilistic patterns typical of Bernoulli processes, such as repeated tossings of a fair coin, given almost any sequence of ic-values, not just a sequence that is itself probabilistically patterned.

To see why this is so, conceive of microconstancy as having two aspects. First, a microconstant evolution function oscillates quickly between zero and one, so that each gray or white area is very narrow. In physical terms, this means that even small changes in the initial conditions of an experiment change the outcome from e to $\bar{e}$ and vice versa. The usual term for this property is *sensitivity to initial conditions*.

The swift oscillations of an evolution function that is sensitive to initial conditions may be quite irregular; what microconstancy adds to sensitivity is regularity. The second aspect of microconstancy, then, is the fact of the constant rhythm (or at least, the constant average rhythm) of the swift oscillations of the evolution function.

To explain a probabilistic pattern of outcomes is to explain both the short-term disorder in the outcomes and their long-term order. A microconstant evolution function tends to produce short-term disorder because it is sensitive to the exact value of its initial conditions. Even if there is a reasonably regular pattern in a microconstant experiment's ic-values, the mechanism's sensitivity to initial conditions will greatly exaggerate small variations in the pattern so as to produce irregular-looking outcomes. That is, sensitivity to initial conditions breaks up short-term patterns in the input, producing output with short-term disorder.[11]

A microconstant evolution function tends to produce long-term order, in the form of a stable frequency for a designated outcome, because it tends to produce outcomes, in the long term, with a frequency equal to the designated outcome's strike ratio.[12] This is a consequence of the regularity inherent in microconstancy, but not only the regularity: the result equating probability and strike ratio would not be possible without sensitivity to initial conditions. Thus, sensitivity—the generator of chaos—plays a role in creating, not just short-term disorder, but also long-term statistical order. Macroscopic order is explained in part by microscopic chaos. There will be much more to say about this idea.

Have I contradicted myself? I say that a microconstant process is *sensitive* to initial conditions, but I also call on microconstancy to explain the fact that microconstant probabilities are *insensitive* to microlevel information about initial conditions. There is no inconsistency in this: a particular microconstant process is sensitive to details about its particular initial conditions, but a class of microconstant processes is insensitive to details about the overall distribution of its initial conditions. By making complex probability a type notion, I make it insensitive; the insensitivity depends, though, in part on the sensitivity of the individuals that instantiate the type. This is a marvelous thing—it is the marvel that inspires this book—but it is not a paradox.

I will conclude by noting two related aspects of physical probability also explained, at least in part, by microconstancy. First, because a microconstant probability is so insensitive to changes in an experiment's ic-density, microconstancy creates the appearance that a probability is physically independent of the apparatus—for example, the croupier—that supplies the experiment's ic-values. It may even appear that the probability is an intrinsic property of the experiment's mechanism.

Second, observe that in a microconstant experiment, a designated outcome's strike ratio often depends only on a few, fairly high level facts about the physics of the mechanism. For example, the strike ratio for red on the wheel of fortune depends only on the wheel's paint scheme and on the rotational symmetry of the wheel's dynamics. The physical details underlying these facts are unimportant in themselves. In a wheel that comes slowly to a halt, for example, the precise facts about the frictional forces that slow the wheel do not matter. Only one fact about these forces matters, the rather abstract fact of the rotational symmetry of their combined effect. The same is true for a tossed coin: only the symmetrical distribution of mass in the coin matters. This aspect of microconstant experiments is responsible for several interesting facts:

1. It is possible to infer the value of a microconstant probability from a few facts about physical symmetries, even if one knows very little about physics (Strevens 1998).
2. The value of a microconstant probability may come out the same on many different, competing stories about fundamental physics. The probability of heads on a tossed coin, for example, is one half in Newtonian physics, quantum physics, and the physics of medieval impetus theory.
3. The irrelevance of physical detail to the values of microconstant probabilities, together with their apparent independence from initial conditions, creates the appearance, but only the appearance, of the truth of the central tenet of the propensity theory of probability: probabilities are intrinsic, metaphysically basic properties of the mechanisms to which they are attached.[13]

2.25 Multivariable Experiments

I now extend my treatment of microconstant experiments to experiments with more than one ic-variable. For the most part, the generalization is routine. One significant new concept appears, that of an *eliminable* ic-variable.

Because an entirely general treatment of multivariable experiments is a somewhat daunting exercise in the proliferation of subscripts, I will discuss the multivariable case mainly by way of a worked example. A more technical approach to the same example is taken in section 2.B, and the subscripts themselves may be found in section 2.C2, where formal definitions and proofs are given.

THE TOSSED COIN

Consider a toss of a fair coin. Assume that the tossing mechanism works as follows: the coin is given a certain angular velocity and allowed to spin at that velocity for a certain time. At the end of the allotted time, whatever face is uppermost, or closer to uppermost, determines the outcome of the toss. If the spin speed and the spin time are randomly determined, but the direction of spin and the orientation of the coin at the beginning of the toss are fixed, then such a mechanism has two IC-variables, the spin speed ω and the spin time τ. A coin-tossing experiment of this sort was considered within the framework of the method of arbitrary functions (see section 2.A) by Keller (1986); Keller treats the physics, and other important issues, in much greater detail than I do here.

The mechanics of the toss may be represented by an evolution function $h(\omega, \tau)$. (In what follows, I suppress the reference to the designated outcome; the evolution function is always the function for heads.) Given my assumptions concerning the tossing mechanism, $h(\omega, \tau)$ has the form shown in figure 2.5, where gray areas are those for which pairs of values of ω and τ yield heads, that is, areas for which the evolution function is equal to one.

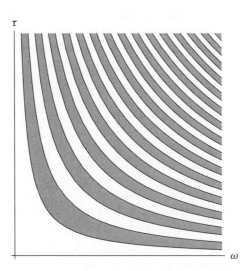

Figure 2.5 A graph of the evolution function $h(\omega, \tau)$ for a tossed coin. Shaded regions are those where $h(\omega, \tau) = 1$, meaning that the coin lands heads. After Keller (1986).

Let me briefly account for the main features of the evolution function. (Readers not particularly interested in the details of the coin toss per se may wish to move on to the next paragraph.) Figure 2.5 assumes that the coin starts out poised on its edge, in such a way that the slightest spin will move tails into the uppermost position. Each half revolution of the coin after that will reverse the outcome. The boundaries between heads and tails, then, are the values of ω and τ for which the coin completes a whole number of half revolutions. Since the number of revolutions completed by the coin is the product of ω and τ, assuming that ω is measured in revolutions per time unit, the boundaries between heads and tails are the curves $\omega\tau = n/2$ ($n = 1, 2, 3, \ldots$), which, as the figure shows, are hyperbolas.

The IC-density for the tossed coin is a joint IC-density over ω and τ, that is, a density having both ω and τ as arguments. As with any density, the probability that ω and τ take on one of a certain (measurable) set of values is given by the volume under the density corresponding to that set. Using the joint density, rather than individual densities for ω and τ, is necessary because ω and τ might not be independent.

A multivariable experiment may then be defined in a very similar way to a single variable experiment. Like a single variable experiment, it may be regarded as a triple consisting of a joint IC-density, a set of designated outcomes, and a set of evolution functions. The IC-density and the evolution functions have the same number of arguments as the experiment has IC-variables. The complex probability of a designated outcome e is defined in the obvious way, as the proportion of the IC-density that spans values of the IC-variables that give rise to e, that is, as the proportion of the IC-density for which the evolution function is equal to one. In the case of the coin,

$$\text{cprob}(e) = \iint\limits_{V} h(\omega, \tau)\, q(\omega, \tau)\, d\omega\, d\tau$$

where V is the set of all possible values for ω and τ. (See also definition 2.7.)

MICROCONSTANCY AND MACROPERIODICITY
It is evident from inspection of figure 2.5 that the evolution function for the tossed coin $h(\omega, \tau)$ has a property very much like microconstancy as defined for the single variable case. More precisely, the domain of $h(\omega, \tau)$ can be partitioned into micro-sized, contiguous sets, each with the same strike ratio, namely, one half (where the strike ratio for a multivariable evolution function

is defined in the obvious way).[14] If the IC-density is sufficiently smooth, so that it is approximately uniform over each member of such a partition, the probability of heads will be approximately equal to the strike ratio, that is, approximately equal to one half.

The concepts of microconstancy and macroperiodicity can be straightforwardly generalized, then, so as to obtain a result establishing the approximate equality of the probability and the strike ratio in the multivariable case. This result is exactly analogous to the result for one-variable experiments (theorem 2.2).

But an interestingly weaker result is possible, weaker in the sense that it requires less of the IC-density than complete macroperiodicity. To see this, consider a joint density for the tossed coin that is not macroperiodic. It will be easiest to deal with an extreme example. Pick one of the coin's two IC-variables, it does not matter which. I choose τ. Now suppose that, due to some idiosyncrasy in the tossing mechanism, there are only two possible IC-values for τ, that is, only two amounts of time for which the coin might possibly spin. Call these times t_1 and t_2, and call the probabilities that τ takes these values $Q(t_1)$ and $Q(t_2)$ respectively (with $Q(t_1) + Q(t_2) = 1$). If τ can take only two discrete values, the joint density $q(\omega, \tau)$ is not macroperiodic. (Indeed, it is not even strictly speaking a density, but I will ignore this nicety in what follows.) Nevertheless, on a certain plausible assumption about the distribution of spin speeds, the probability of heads is still equal to the strike ratio for heads, that is, one half, in virtue of a kind of microconstancy in the evolution function. Let me substantiate this claim. The argument is fairly informal; for a formal version, see section 2.B, and for a generalization of the formal argument, section 2.C2, theorem 2.3.

The probability of heads will be equal to the weighted sum of two conditional probabilities: the probability of heads given that τ equals t_1, and the probability of heads given that τ equals t_2. The weights are, respectively, $Q(t_1)$ and $Q(t_2)$, that is, the probabilities of t_1 and t_2. In what follows, I show that both conditional probabilities are approximately equal to 1/2. Because the weights must sum to one, the way that the conditional probabilities are weighted makes no difference to the probability of heads. The probability must be equal to the conditional probabilities, that is, approximately equal to 1/2.

More generally, provided that the probabilities conditional on each possible value of τ are all equal to 1/2, it does not matter how the values are weighted; thus the probabilities of the different possible values of τ do not matter. This means, roughly, that the topography of the ic-density in the direction of τ—the way that the ic-density changes as one travels along any line parallel to the τ axis—makes no difference to the probability of heads. Or rather, it makes no difference provided that the conditional probabilities are all 1/2. When will this be so?

The conditional probabilities are approximately equal to 1/2 in the example because of the kind of microconstancy referred to above. I present the argument for t_1; the argument for t_2 is identical. Suppose that $\tau = t_1$. Then the conditional probability is determined by the properties of the evolution function and ic-density for the coin with τ fixed at t_1. The properties for other values of τ do not matter. When τ is fixed at t_1, the coin's evolution function acts like a one-variable evolution function, the one variable being ω. The properties of this evolution function can be discerned by observing the properties of the full evolution function along the line $\tau = t_1$, that is, along a horizontal line on figure 2.5 that intersects the τ axis at t_1.

It will be seen that along such a line, the evolution function oscillates from zero to one and back in a way characteristic of a microconstant one-variable evolution function with strike ratio 1/2. The reason is that, for a constant value t of τ, the outcome changes from head to tails or back for whole-numbered values of $2t\omega$; the intervals between changes therefore take the constant value $1/2t$. Assuming that the experiment's ic-density is macroperiodic along the same line, the conditional probability for heads, given a spin time of t_1, will be approximately equal to the strike ratio for heads, that is, approximately equal to 1/2. The same is true for the probability of heads conditional on spin time t_2, so the unconditional probability for heads will be, for the reasons given above, approximately 1/2.

Observe that what matters in the derivation of this result is not the microconstancy of the entire evolution function for the coin, nor the macroperiodicity of the entire ic-density. Rather, what is necessary is a kind of point by point microconstancy and macroperiodicity. For each possible value of τ,

1. The *one ic-variable* evolution function obtained from $h(\omega, \tau)$ by holding τ constant must be microconstant, with the same strike ratio for heads, for all possible values of τ, and
2. The *one ic-variable* ic-density obtained from $q(\omega, \tau)$ by holding τ constant must be macroperiodic for all possible values of τ, or as I will sometimes say, the joint ic-density must be macroperiodic in the direction of ω.

All of the above holds if the roles of ω and τ are exchanged. That is, if the one IC-variable evolution functions obtained by holding ω constant are microconstant with strike ratio 1/2, and the joint IC-density is macroperiodic in the direction of τ, then, too, the probability for heads is approximately 1/2. Because of the symmetry of $h(\omega, \tau)$, the first of these conditions is of course satisfied. Thus the probability of heads is 1/2 if the joint IC-density is macroperiodic either in the direction of τ or in the direction of ω, or both.

In the appendix (section 2.C2), I generalize these conditions for the approximate equality of probability and strike ratio to cases in which the evolution function and IC-density are restricted by holding any number of IC-variables constant, up to one less than the total number of the experiment's IC-variables. Provided that an experiment meets the microconstancy condition holding at least one set of variables constant, for all possible values of those variables, I call it a microconstant experiment. A notion of what constitutes a constant ratio partition that complements this sense of microconstancy is introduced in section 2.B, and generalized in definition 2.12.

As shown by my example, (2) does not imply that $q(\omega, \tau)$ is macroperiodic. The question whether (1) implies that $h(\omega, \tau)$ is microconstant over all possible values of τ depends on some mathematical issues that I will not go into here. But the microconstancy of $h(\omega, \tau)$ and the macroperiodicity of $q(\omega, \tau)$ entail that suitably generalized versions of (1) and (2) hold.[15] Putting aside condition (1), then, the generalized version of (2) provides a weaker condition for the approximately equality of the probability and strike ratio than the macroperiodicity of the entire IC-density. In the next subsection, I investigate the advantages of this stronger result.

ELIMINABLE IC-VARIABLES AND STRIKE SETS

I have shown that the probability that a coin lands heads is 1/2 if the joint IC-density is macroperiodic either in the direction of the ω axis or in the direction of the τ axis; it need not be macroperiodic in both directions. Provided that the density is macroperiodic in one direction, then, its macroperiodicity in the other direction in a certain sense makes no difference to the probability of heads. In a case such as this, I say that one IC-variable can be eliminated in favor of the other. (This is a symmetrical relation in the case of the coin, but it is not always so; see below, especially figure 2.6.) If an IC-variable can be eliminated in favor of some other set of IC-variables, it is **eliminable.** An eliminated IC-variable is not literally removed, but it can be made unimportant to the determination of a probability, in that it may be, as it were, conceptually bracketed for the purpose of deciding the macroperiodicity of the ic-density.

A more general, though informal, definition of eliminability is as follows: an eliminable variable is one which, though it makes a difference to the outcome of any particular trial, does not by itself affect the *proportion* of trials in which the designated outcome is produced.

When the ic-variables for an experiment can be divided into two sets, the variables in one of which can be eliminated in favor of those in the other, I call the set of uneliminated variables a **strike set** for the experiment.[16] Because τ can be eliminated in favor of ω, for example, ω makes up a strike set for the tossed coin.

Just as τ is eliminable in favor of ω, so ω is eliminable in favor of τ. Thus τ also makes up a strike set for the tossed coin. But ω and τ cannot both be eliminated at the same time. In order to obtain the approximate equality of probability and strike ratio, then, the ic-density must be macroperiodic in one direction or the other. Equivalently, a strike set for the tossed coin must have at least one member. Hence the *in favor of:* when an ic-variable is eliminated, the ic-variable in favor of which it is eliminated retains its importance. The ic-density should be macroperiodic in the direction of that ic-variable.

A slightly more complicated example of eliminability is a roulette wheel with the following three ic-variables: the point on the wheel at which the ball is introduced, the velocity of the ball at the time of introduction, and the speed of the wheel at that same time. Any two of these can be eliminated in favor of the third. (Think about the consequences of keeping any two of the variables constant.) Thus all three ic-variables are eliminable. Furthermore, all three constitute, on their own, strike sets for the roulette wheel. For the probabilities of outcomes on the wheel to equal the corresponding strike ratios, then, the ic-density over the roulette wheel's three ic-variables need only be macroperiodic in the direction of one of the three. This is a much weaker condition on the joint density than the requirement that it be fully macroperiodic.

Eliminability is not always symmetrical. In the evolution function with two variables shown in figure 2.6, ξ is eliminable but η is not, since for any fixed value of ξ, the evolution function is microconstant in the direction of η with the same strike ratio, but not vice versa.

Let me give an overview of the significance of the notion of a strike set:

1. To derive the approximate equality of probability and strike ratio (theorem 2.3), the ic-density need be macroperiodic only in the direction of a strike set of ic-variables, and

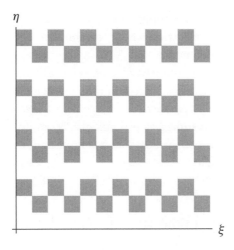

Figure 2.6 Asymmetric eliminability: ξ is eliminable but η is not.

2. As will be made clear in chapter three (section 3.45 and theorem 3.4), to prove certain independence results, the constraints put on the IC-densities by the relevant theorems need be satisfied only by densities over strike sets of IC-variables.

Eliminability, then, allows a significant weakening of various conditions required for the application of the central results in this study.

INDIVIDUATING THE COIN TOSS EXPERIMENT

The evolution function for the tossed coin is microconstant, in the relevant sense, for reasonably large values of its IC-variables ω and τ. However, as can be seen from figure 2.5, for small values of these IC-variables, the size of the smallest regions of constant strike ratio is larger than micro-sized. This is because, at low spin speeds, even quite generous amounts of extra time will not be enough for the coin to make an extra half-spin, and likewise, at short toss durations, even quite generous increases in spin speed will not suffice for an extra half-spin. The evolution function is microconstant almost everywhere, but not quite everywhere.

For practical purposes this does not matter, since we do not count as a "proper" coin toss any trial with low ω and τ. Such tosses strike us as too feeble to generate a truly random outcome. It seems, then, that our familiar probabilistic experiment type *proper coin toss* is defined so as not to include such tosses. In effect, our natural, pre-theoretic strategy is to redefine the experiment to include only trials whose ic-values fall into the microconstant region of the evolution function.

It is very suggestive that our everyday practice individuates experiments in the same way as the technical concept of microconstancy: both intuition and the criterion for microconstancy rule feeble coin tosses out of bounds. We humans are, it seems, in some way psychologically sensitive to the importance of microconstancy, an observation that has important consequences for the psychology of probability.

2.3 The Interpretation of IC-Variable Distributions

The notion of a microconstant probability is interesting only if many ic-densities are macroperiodic. But what kind of facts make it the case that a given ic-density is macroperiodic? More generally, what kind of information do ic-densities provide?

There need not be a single correct answer to this question. The role of an ic-density is to represent information that can play a role in explaining the probabilistic patterns and, later, the behavior of enions and the complex systems that contain them. Thus an ic-density must provide explanatorily relevant information about an ic-variable. I do not want to take a position as to what sort of information this must be, for two reasons.

First, there may be several kinds of explanatorily relevant information about a particular ic-variable, and any of these kinds of information may make for equally good explanations of the patterns. For example, statistical information about ic-values may be just as explanatorily powerful as information about how the ic-values were produced.

Second, what is explanatorily relevant will depend on the nature of scientific explanation, a subject of much disagreement. For example, if probabilistic explanation is an epistemic matter, as Hempel (1965) maintained, then a subjective probability interpretation of the ic-density may be acceptable (though this was not Hempel's view), but if not, not. I aim to avoid this controversy.

In this section I provide three reasonably uncontroversial ways of interpreting an ic-density. These are most definitely not exhaustive, but they are sufficient to provide interpretations of all the ic-densities that will appear when I apply the physics of complex probability to the study of complex systems in chapter four. The reader may regard them as suggestions only.

2.31 Simple Probability

If the fundamental laws of physics are probabilistic, it is possible that a probability distribution over an ic-variable is induced by a simple probabilistic process, as would be the case, for example, if the variable described the state of the atom whose decay causes the death of Schrödinger's cat. In such a case, it is natural to understand the corresponding ic-density as representing a simple probability distribution.

Such situations are not common, however. Even in an indeterministic world, the state of a typical ic-variable, for example, the spin speed of a tossed coin, is at most only partly determined by quantum effects within the tossing mechanism.

2.32 Complex Probability

In many cases an ic-density $q(\zeta)$ may be interpreted as representing a complex probability distribution. This option is available whenever the value of ζ is determined by one or more complex probabilistic experiments. I will call these experiments, when they exist, **feeder experiments**.

What does it mean to say that a feeder experiment determines an ic-density $q(\zeta)$? The presumption is that for any interval $[x, y]$ of ζ, there is a well-defined complex probability that the feeder experiment will produce a value for ζ that lies within that interval. The ic-density $q(\zeta)$ is just a compact representation of these complex probabilities. More formally, let $e_{x,y}$ be the event of the feeder experiment producing a value of ζ that lies between x and y. Then $q(\zeta)$ is to be defined so that the area under $q(\zeta)$ between x and y is equal to the complex probability of $e_{x,y}$. In symbols,

$$\int_x^y q(\zeta) \, d\zeta = \text{cprob}(e_{x,y}).$$

An example of an ic-density that can be interpreted as a complex probability distribution is the ic-density for the tossed coin (section 2.25). The initial spin speed of the coin (and perhaps the spin duration too) is generated by a human coin-tosser, whose action in imparting the spin speed can be regarded as a complex probabilistic experiment having the speed as its outcome. Some further examples of feeder experiments are given in section 2.4.

Note that not every complex probability distribution over the events $e_{x,y}$ can be represented by a density. In general, an experiment that takes one IC-variable ξ and transforms it into another variable ζ will produce a density over ζ if (a) there is a density for ξ, and (b) the transformation effected by the experiment—the function mapping ξ to ζ—has an inverse that is continuous, or piecewise continuous.

Let me conclude by saying what it is for a probability distribution over an IC-variable to be microconstant. A probability distribution over a real-valued variable is microconstant just in case the probability of every event $e_{x,y}$ is microconstant, that is, just in case the evolution function for every event $e_{x,y}$ is microconstant. Some toy examples of experiments that generate a microconstant distribution over a real-valued variable are discussed in section 3.7.

2.33 Statistical Information

An IC-density need not represent probabilities to play a part in the explanation of probabilistic patterns, provided that it represents some kind of explanatorily relevant information. The most obvious non-probabilistic interpretation of an IC-density is as representing statistical information about the distribution of actual IC-values.[17] On this interpretation, the IC-density for the initial conditions of a coin toss, for example, represents the frequency with which particular initial conditions—particular spin speeds and toss durations—appear in actual tosses. When there are an infinite number of instantiations of an IC-variable, the corresponding density might represent limiting frequencies. Whether it is actual or limiting frequencies that are represented, I will call the resulting density a **frequency-based IC-density**.

Some mathematical artifice is required to construct a density function from a presumably discontinuous distribution of actual frequencies. One method is to pick a small number r, and to define the density $q(\zeta)$ for an IC-variable ζ according to the following *smearing* operation:

$$q(z) = \frac{Q(z - \frac{r}{2}, z + \frac{r}{2})}{r}$$

where $Q(x, y)$ is the relative or limiting frequency of actual values of ζ that fall between x and y. Note that $q(\zeta)$ will conform to the probability calculus, as do all density functions, and as do the frequencies on which it is based.

In performing this operation one must be careful not to pick an r that will smear out oscillations in $q(\zeta)$ that ought to be counted as undermining ζ's macroperiodicity. This can be avoided by choosing an r that is smaller than the smallest oscillations of the evolution function, that is, smaller than the function's CRI. The rationale for smearing is revisited in note 28. As a consequence of the smearing operation, a frequency-based ic-density is not, strictly speaking, a frequency distribution. Rather, it is a device used to represent the salient facts about a frequency distribution in a canonical form, a density.

2.34 Hybrid IC-Variable Distributions

In a multivariable experiment, a single joint ic-density gives the distribution of all ic-variables. For this reason, some ic-densities may be hybrid, representing various combinations of simple probabilities, complex probabilities, and frequencies. A hybrid joint density may be constructed from conditional densities for each ic-variable. For example, if $q(\xi)$ represents a simple probability distribution and $r(\eta)$ represents a frequency-based distribution, then the hybrid joint density $s(\xi, \eta)$ for ξ and η may be defined as follows:

$$s(x, y) \;=\; q(x \mid \eta = y)\, r(y).$$

The conditional density represents the simple probability of ξ taking an ic-value x given the fact that the actual ic-value of η is y. In many cases, I take it, the fundamental laws of physics will entail that ξ is independent of η, so that $s(\xi, \eta)$ is equal to $q(\xi)r(\eta)$.

2.4 Probabilistic Networks

2.41 Probability Flow and Traffic Flow

That an ic-density can be interpreted as representing a complex probability distribution suggests the possibility of a network of probabilistic experiments feeding one another with ic-values. Looking to the real world, one finds many systems that can be interpreted as having such a structure.

Suppose, for example, that in investigating traffic flow, one wishes to understand the factors that determine the probability that a given vehicle is approaching a given intersection at a given time. This probability, it seems, will depend on various other probability distributions. These might include the probabilities of different kinds of weather, a probability distribution over the

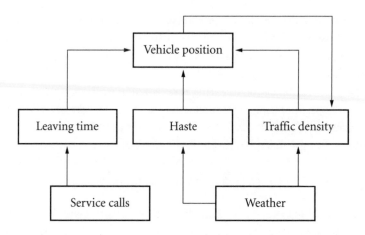

Figure 2.7 Probability flow determining the position of a vehicle at a time.

time that the driver of the vehicle leaves work, a probability distribution over the degree of haste with which the driver is heading home, and a probability distribution over the density of the traffic in the area. Some or all of these factors will depend on other probabilities in turn. The time that the driver leaves work may depend on the number of service calls that the driver's department has received that afternoon. The degree of haste may depend on the weather—when the weather is good, the driver may hurry home to enjoy the last of the sunlight; when bad, the driver may drive more cautiously. The probability distribution over the density of traffic at any point may also depend on the weather, and of course depends of the positions of all the vehicles, including that of the driver. If each of these factors has its own complex probability distribution, then the flow of probability has something like the form pictured in figure 2.7.

Underlying this flow of probability is a flow of ic-values traveling in the direction of the arrows in the figure, from one probabilistic experiment to another. I call such an arrangement of interconnected probabilistic experiments a **probabilistic network**. The basic unit in the network is a probabilistic experiment that takes one or more ic-variables as input, and produces an ic-variable for another experiment as output. The aim of this section is to investigate some of the properties of probabilistic networks, and eventually, to characterize a class of probabilistic pattern–producing experiments that is broader than the microconstant experiments, namely, the *true probabilistic experiments*.

2.42 A Generic Probabilistic Network

The network shown in figure 2.7 is of course incomplete, in that it leaves out many factors influencing the position of a vehicle at a time. Real probabilistic networks will tend to be extremely complicated, for which reason I will focus the discussion on the simple, uninterpreted network shown in figure 2.8. Figure 2.8 is more detailed than figure 2.7 in one respect: it represents each experiment as having two parts, its mechanism and its ic-density.

In the figure, the squares are the mechanisms of the experiments in question, represented mathematically by sets of evolution functions. The long rounded rectangles are the distributions of the associated ic-variables, represented mathematically by ic-densities. For simplicity's sake, all experiments have one ic-variable only. The arrows point from ic-variables to the mechanisms they feed, and from mechanisms to the ic-variables they generate. The arrows, then, represent a *formal* relationship between types of initial conditions and types of mechanisms, not a causal relationship between particular conditions and particular mechanisms (although causation in individual cases

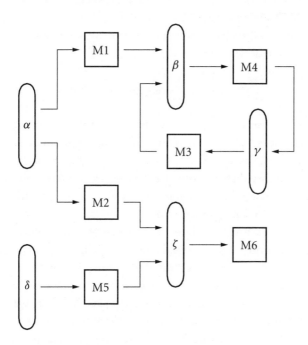

Figure 2.8 A probabilistic network.

will follow the direction of the arrows). There may be, for example, initial conditions at the "beginning" of the network, such as values of α, that are realized long after all of the mechanisms in the network have disappeared.

I have two remarks about the structure of probabilistic networks. In the network shown in figure 2.8, two ic-variable distributions, those for β and ζ, are generated by more than one feeder mechanism. Consider one of these, β. The two arrows leading to β are intended to represent a situation in which some values of β are generated by M1 and some by M3. It may be that M1 and M3 each generate β with slightly different distributions. An example of such a situation is the roulette wheel discussed earlier, different croupiers for which are characterized by somewhat different ic-densities. If the distributions are each macroperiodic, and the experiment they are feeding is microconstant, then the difference between the distributions may in practice be ignored. The network is drawn as if there is just one distribution.

Also shown in figure 2.8 is a situation in which one ic-variable, α, feeds more than one mechanism. The two arrows leading from α are intended to represent a situation in which some values of α feed M1 and some feed M2. Why, one might wonder, do I not distinguish between two ic-variables, one of which feeds M1 and one of which feeds M2? Simply because, provided that the distribution of ic-values fed to one mechanism is the same as that fed to the other, there is no reason to do so.

2.43 Basic Distributions

There are two ic-variables in the network, α and δ, that are not generated by feeder experiments. The ic-densities for these variables cannot, then, be interpreted as complex probability distributions. They must be interpreted as simple probability distributions, frequency-based distributions, or something else. I call these foundational ic-variables the **basic ic-variables** for the network. A basic ic-value is not necessarily uncaused, but it is not probabilistically caused. Distributions over basic ic-variables provide the raw material for all densities in a probabilistic network. Together with the facts about mechanisms, then, they determine the values of all of the probabilities in a network.

Note, however, that part of the network—the piece in the top right quarter of the diagram—is self-sufficient. Even if the mechanism M1 were removed, the two mechanisms M3 and M4 would continue to feed each other the ic-values they need. With M1 gone, the facts about $q(\gamma)$ depend only on $q(\beta)$ and the facts about $q(\beta)$ depend in turn only on $q(\gamma)$. But then what is it that these two IC-densities are telling us about their IC-variables? How can they represent explanatorily relevant information about the world, when they seem to float free of the world?

This is an important question, for loops such as this are invariably found in the probabilistic networks representing the workings of complex systems; see, for example, the loop involving vehicle position and traffic density in figure 2.7, or for the general case, figure 4.1. In what follows I consider the loop consisting of M3 and M4 alone, and show that wherever such loops are possible, there is a basic IC-variable of sorts that provides a real world foundation for the distributions in the loop.

If a loop is to get going in the first place, at least some IC-values must be supplied from outside. Once inside the loop these IC-values can be passed around as many times as you like—in the example, going from M3 to M4 and back again—but, assuming that the loop has some beginning in time, the IC-values must have come from somewhere. If the loop is connected to a network, then the network may supply the initial IC-values. In the network of figure 2.8, for example, initial values may come into the loop from M1. But what if the loop is isolated? Then, although many of the IC-values in the loop will have been generated by the mechanisms in the loop, there must have been some IC-values that were not. That is, there must have been some values of β or γ, or both, that started the loop off and that were therefore not themselves generated by a mechanism within the loop. These values are, in effect, basic IC-values for the loop. They may be the initial conditions of the system represented by the loop, but they need not causally precede all other IC-values in the loop: they may arrive from some non-probabilistic source throughout the loop's lifetime.

In the case of traffic, for example, the basic IC-values of the loop involving vehicle position and traffic density are the IC-values giving, first, the positions of those vehicles already inside the network at the time it is created (perhaps there are none, if the network is deemed to have been created when the roads are constructed), and second, the positions of any vehicles arriving, at any later time, from outside the network, that is, from outside the group of roads represented by the network. The first group of IC-values are initial conditions of the network in the temporal sense: they are present at the birth of the network. The second group continue to flow into the network throughout its life. Both groups are basic, because both groups are produced outside the network, or to put it another way, neither group has been influenced by the workings of the mechanisms within the network.[18]

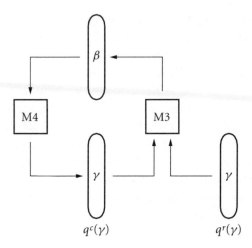

Figure 2.9 A loop founded by a raw IC-density.

I call the IC-values that arrive from outside to bootstrap an isolated loop the loop's *raw* IC-values. All other IC-values—that is, all IC-values whose causal history can be traced back to some earlier IC-value also in the loop—are *cooked* IC-values. Because cooked IC-values are processed by the mechanisms in a loop and raw IC-values are not, cooked and raw IC-values may be differently distributed. But even if they are not, it is useful to distinguish the probability distribution of the raw IC-values from the probability distribution of the cooked IC-values. Consider, for example, the isolated loop I have been discussing, consisting of just M3 and M4 from figure 2.8. Suppose, for simplicity's sake, that all of the loop's raw IC-values are values of γ (thus, all values of β can be traced back to values of γ). Let $q^r(\gamma)$ be the distribution over the raw IC-values and $q^c(\gamma)$ the distribution over the cooked IC-values. Then the loop can be represented as shown in figure 2.9, with the raw IC-density shown as a basic density and the cooked IC-density embedded in the loop. The raw density is a basic density, and should be interpreted as such—as, for example, a frequency-based distribution—while the cooked density represents a complex probability distribution.

Provided that $q^r(\gamma)$ and $q^c(\gamma)$ have the same distribution or, in the case where M3 is microconstant, are both macroperiodic, there is no harm in consolidating the two; still, it is important to see that they play rather different roles in determining the values of a loop's complex probabilities.

Even in a loop that is not isolated, some values of the ic-variables inside the loop may be raw. Suppose, for example, that the loop shown in figure 2.9 is attached to the network as shown in figure 2.8, so that some values of β are supplied by M1, a mechanism outside the loop. This does not rule out the possibility that some values of, say, γ are raw in just the sense described above: they are initial conditions of the loop, or arrive in the loop periodically from some external, non-probabilistic source (as do, for example, some of the vehicles in the traffic example). In such a case, any cooked values in the loop can be traced back either to M1 or to a raw value of γ. Raw ic-values may be found, then, in any loop, not just in an isolated loop.

I therefore define a **raw ic-value** as a value of an ic-variable in a loop that cannot be traced back either to another ic-value inside the loop or to another ic-value outside the loop.

2.44 True Probability

I will now advance a new notion, that of a *true probability*, in an attempt to group under one heading all those complex probabilities whose probabilistically patterned outcomes can be explained by way of microconstancy. The notion does not play an important role in the explanation of the simple behavior of complex systems in chapter four, so some readers may wish to move on to the next section. True probability is important, though, to my project of understanding the probabilistic patterns.

There are two ways that microconstancy might explain the probabilistic patterns of a complex probabilistic experiment X's outcomes. First, the mechanism of X itself might be microconstant. Second, the mechanisms of the experiments that feed X, either directly or indirectly, might be microconstant. In this second sort of case, the probabilistic patterns in the outcomes produced by X are simply reflections of the probabilistic patterns in X's ic-values, but the patterns in the ic-values are explained by the presence of some microconstant mechanism or mechanisms further upstream in the probabilistic network. Provided that every branch of the network leading to X includes a microconstant mechanism at some point, microconstancy will explain the probabilistic patterns just as well in the second sort of case as when the mechanism of X itself is microconstant.

A true probability is a probability attached to an experiment that satisfies either of these two descriptions. A true probability, then, is attached to an experiment that is either microconstant itself, or has an ic-density that is a

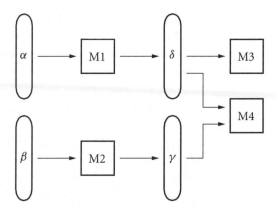

Figure 2.10 True probability. If M1 is the only microconstant mechanism in the network, then the distributions generated by M1 and M3 are true probability distributions, while those generated by M2 and M4 are not.

product, directly or indirectly, of microconstant experiments. To put it another way, a probability is true if, no matter what path you take from the probability's experiment X back up the probabilistic network, you eventually encounter a microconstant mechanism.

For an example of what does and does not count as true probability, consider figure 2.10. Suppose that M1 and only M1 is a microconstant mechanism. Then the distributions generated by M1 and M3 are true probability distributions, while those generated by M2 and M4 are not. The distribution generated by M1—that is, the distribution over δ—is true because M1 is microconstant. The distribution generated by M3 is true because M3's ic-variable δ is generated by a microconstant mechanism. The distribution generated by M2—that is, the distribution over γ—is not true, because M2 is not microconstant and M2's ic-variable β is basic and is therefore not generated by a microconstant mechanism. The distribution generated by M4 is not true because, although one of its ic-variables, namely δ, can trace its heritage to a microconstant mechanism, the other, namely γ, cannot.

Recalling that a complex probability distribution over a real-valued variable is microconstant if the mechanism that produces it is microconstant in the right sort of way (the details are spelled out in section 2.32), the definition of a **true probability** distribution can now be put succinctly in a recursive form: a probability distribution is true just in case either

1. It is a microconstant distribution, or
2. It is a complex probability distribution, and the ic-density of its probabilistic experiment is a true probability distribution.

A probability for a particular outcome is a true probability just in case

1. It is a microconstant probability, or
2. It is a complex probability, and the ic-density of its probabilistic experiment is a true probability distribution.

There may be explanations for probabilistically patterned outcomes other than microconstancy. For example, it may be that simple probabilities explain the probabilistic patterns that they generate. The definition of a true probability could then be broadened by adding a new condition to the recursive definition of a true probability distribution:

3. It is a simple probability distribution,

and a new condition to the definition of a true probability:

3. It is a simple probability.

A probability would satisfy this broadened definition if it was, or could be traced back to, either microconstant complex probabilities or simple probabilities. The definition could be further extended if other explanations of probabilistic patterns are discovered, so that a true probability, in a suitably extended sense, would be any probability that explains the probabilistic patterns of its outcomes. In the remainder of this study, however, I will be concerned with the narrower definition, on which all true probabilities are founded, directly or indirectly, in microconstancy.

2.5 Standard IC-Variables

In this section, I pause to consider what appears to be a foundational problem with the notion of microconstancy. Readers who are not so concerned with foundational questions may move on to the next section; this section does contain, however, some very important material about the explanation of the probabilistic patterns.

The foundational problem is that microconstancy, as I have defined it, is a property that an experiment has only relative to a particular way of quantifying the experiment's initial conditions. There are an unlimited number of ways to quantify initial conditions, and any experiment is always microconstant relative to some quantification schemes and non-microconstant relative to others. Since a quantification scheme, unlike the physics that is to be quantified, is chosen by the observer, this would seem to make microconstancy itself a matter of choice, and so to render the property of microconstancy explanatorily empty, not a happy result given the strategy of this book. My aim in what follows is to recover the explanatory power of microconstancy.

The discussion proceeds as follows. Section 2.51 presents the problem, which is caused by ic-variables that are quantified in non-obvious, rather devious, ways. I call these ic-variables *gerrymandered* ic-variables. Section 2.52 discusses some possible solutions to the problem. All aim in one way or another to rule out microconstancy relative to a gerrymandered ic-variable as a legitimate explanatory property. I call the non-gerrymandered ic-variables— the initial conditions quantified in normal ways—the *standard ic-variables*. The purpose of this section is to find a way in which microconstancy relative to standard ic-variables might somehow be explanatorily more significant than microconstancy relative to non-standard variables.

I settle on the following strategy: I will show that, given some minimal assumptions about the way the world works, standard ic-variables will tend to be macroperiodically distributed, while non-standard ic-variables will tend to be non-macroperiodically distributed. Because microconstancy relative to an ic-variable is an explanatorily potent property only if that ic-variable has a tendency to macroperiodicity, it follows that microconstancy relative to a standard ic-variable is an explanatorily potent kind of microconstancy, but that microconstancy relative to a non-standard ic-variable is not.

Section 2.53 implements this strategy. Microconstancy is then stipulated to mean, for the duration of this study, microconstancy relative to standard ic-variables.

2.51 A Problem with Microconstancy

For all the claims made in the preceding sections, microconstancy suffers from a problem apparently so severe as to render it useless as an explanatory property: *any* experiment can be described so as to have a microconstant evolution function. Worse, any experiment can be described so as to be microconstant with *any* strike ratio. This is because there are many ways of quantifying the physical properties that are the initial conditions of a probabilistic experiment,

$h(\tau)$

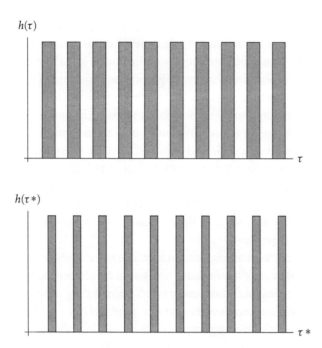

$h(\tau*)$

Figure 2.11 The evolution function for *red* on a simple wheel of fortune measuring time in the usual way *(top)*, and the evolution function for the same experiment with time differently measured *(bottom)*.

and the definition of the strike ratio is sensitive to the particular quantification used.

To see this, consider a wheel of fortune, divided as before into a number of equally sized red and black sections. Suppose that the wheel is spun at a constant speed for a variable time τ. (This wheel has a simpler physics than the wheel of section 2.22, which had a variable speed and was allowed to come to rest.) The wheel's evolution function $h(\tau)$ for red is shown in figure 2.11 (top). As one would expect, the function is microconstant with a strike ratio of one half.

Now measure the time the wheel is allowed to spin in a different way. Define a new IC-variable τ^* as follows:

$$\tau = \tau^* + \frac{\cos \pi \tau^* - 1}{\pi}.$$

(I regard τ and τ^* as different ic-variables, even though they represent the same physical quantity; see note 5.) This new variable is an odd and unintuitive way to represent time, but there is no formal obstacle to such a representation, and the definition of the complex probabilities for *red* and *black* goes through just as well using τ^* as it does using τ.

The evolution function $h(\tau^*)$ for red is pictured in figure 2.11 (bottom). As can be seen from the figure, the change of variable compresses the shaded regions and expands the unshaded regions.[19] The new function is also microconstant, but the strike ratio is now less than one half. Yet no physical fact has changed. The mechanism is the same and the physical quantity that is the only initial condition of any trial—the spin time τ—is the same. All that has changed is the way that the spin time is quantified, or as mathematicians say, the way that it is measured.

I conclude that facts about the microconstancy and the strike ratio of an evolution function are always relative to a way of measuring the physical quantities that are the initial conditions of the corresponding experiment. Microconstancy and the strike ratio, then, are not quite, as I have been treating them, intrinsic properties of an experiment.

2.52 Solving the Problem

Is microconstancy, then, meaningless? If so, why does it appear to explain the fact that the probabilities associated with the roulette wheel are independent of the idiosyncrasies of croupiers? And why does it appear to explain the probabilistic patterns? I will focus on the first question, though I could equally well resolve this difficulty by discussing the second.

An obvious answer to the question lies in the following observation: part of the explanation for the croupier-independence of the roulette probabilities is the assumption that the ic-density for each croupier is macroperiodic. This gives us reason to restrict our attention to those instances of microconstancy where an evolution function is microconstant relative to an ic-variable that is or has some tendency to be macroperiodic.

In what follows I formulate what looks like a straightforward solution, based on this observation, to the problem, and I show that the solution is not effective. I then canvass a number of other approaches, and decide on a particular strategy, to be implemented in section 2.53.

First, the attempt at a straightforward solution. Observe that the way that an experiment's initial conditions are quantified affects not only the microcon-

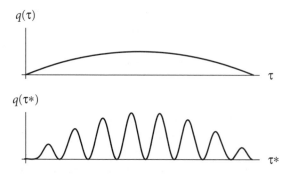

Figure 2.12 The probability distribution over spin time, first measured standardly *(top)* and then non-standardly *(bottom)*. The move from a standard to a non-standard quantification of time destroys the macroperiodicity of the IC-density.

stancy of the experiment's evolution functions but also the macroperiodicity of the experiment's IC-density. Suppose, for example, that the spin time τ of section 2.51's wheel of fortune has the density shown in figure 2.12 (top). Then the density over τ^*, the gerrymandered version of τ, will have the form shown in figure 2.12 (bottom). Gerrymandering the variable ruins the density's smoothness by introducing a series of micro-sized peaks and troughs.

As I said above, microconstancy relative to a variable is of explanatory interest only if the IC-density relative to the same variable is or has a tendency to be macroperiodic. But when an IC-density is macroperiodic relative to a standard variable, it is likely to be non-macroperiodic relative to variables that quantify the same underlying physical property in a non-standard way.

Indeed, a considerably stronger statement is possible. Suppose that, relative to a standard variable ζ, an experiment's evolution function is microconstant for some outcome e with strike ratio x, and that the experiment's IC-density is macroperiodic relative to ζ. Take some non-standard variable ζ^* constructed so that the evolution function for e relative to ζ^* is microconstant but with a strike ratio y not equal to x. Then it is not possible that the experiment's IC-density is macroperiodic relative to ζ^*. The reason is this: if the IC-density were macroperiodic relative to ζ^*, then the probability of e would have to equal the strike ratio y. But, by the same theorem, the probability of e must equal x. Since cprob(e) cannot vary with the description of the initial conditions (see note 6), this is not possible, and thus the IC-density for ζ^* must be non-macroperiodic.

If we ignore microconstancy relative to variables that are not macroperiodically distributed, then, we thereby ignore the worst kinds of gerrymandered IC-variables, namely, those variables, such as τ^*, relative to which the evolution function is microconstant with a strike ratio that differs from the strike ratio relative to the standard variable.

Ruling out this particular kind of gerrymandering does not, however, fully resolve the problem. There are other gerrymandered variables that continue to threaten the explanatory status of microconstancy. The difficulty is that, for absolutely any complex probabilistic experiment and designated outcome e, there exists some non-standard IC-variable or IC-variables relative to which (a) the evolution function for e is microconstant, and (b) the IC-density is macroperiodic. The microconstancy is numerically well behaved, in the sense that the gerrymandered evolution function's strike ratio will be equal to cprob(e),[20] but nevertheless, the fact that any experiment is microconstant relative to some IC-variable seems to debase utterly the explanatory value of microconstancy.

Let me spell out the procedure for constructing an IC-variable of this sort. Suppose that we are given an experiment X with one standard IC-variable ζ, and that X is non-microconstant relative to ζ for some outcome e. A non-standard variable ζ^* that is macroperiodic, and relative to which X is microconstant for e, is constructed by cutting up ζ into pieces and reassembling the pieces in a non-standard order, as follows.

1. Take the standard IC-variable ζ. Cut up the range of ζ into very small intervals, so small that many such intervals would fit into a single gray or white bar in the graph of the evolution function for e.
2. Sort the intervals into two piles: those intervals that yield e, and those that do not.
3. Arrange each pile in order of increasing probability. (By the probability of an interval, I mean the probability that an IC-value falls into the interval.)
4. Now build a new IC-variable from the old as follows. Take a few intervals from the top of the e pile, then a few from the top of the $\bar{e}$ pile, in such a way that the ratio of the total width of those taken from the e pile to the total width of those taken from the $\bar{e}$ pile is equal to cprob(e). Put the intervals together to make up the first small segment of ζ^*, that is, the segment corresponding to the very lowest values of ζ^*. Repeat this procedure until both piles are empty.

So constructed, ζ^* will be microconstant for e, and its density will gradually increase, and so will be macroperiodic. (An exercise for the reader: why can't the same operation be performed to produce a microconstant evolution function with a strike ratio other than cprob(e)?)

If microconstancy is to be something special, then, a criterion must be found for ignoring microconstancy relative to variables such as ζ^*. Such a criterion will identify a privileged set of ic-variables, including most or all of the standard ic-variables, but not the kinds of gerrymandered ic-variables discussed above, and it will declare that microconstancy counts only if it is microconstancy relative to a privileged ic-variable. I have shown that it is not enough just to require that the ic-variables in the privileged set have macroperiodic distributions. Some more stringent criterion must be identified as the mark of privileged ic-variables. Let me suggest three possible criteria.

The first is an epistemic criterion. On this proposal, microconstancy with respect to an ic-variable counts only if we have *some reason to believe* that the variable has a macroperiodic distribution. There are various arguments that can be advanced to show that we have no such reason in the case of nonstandard ic-variables. But the epistemic approach is, I think, too subjective, in the sense that it makes the explanatory power of microconstancy partly a power that probabilistic experiments have in virtue of our knowledge about them. Or putting my own views aside, the epistemic criterion would require the adoption of an epistemic view of explanation, contrary to my policy, laid down in section 2.3, of avoiding commitment to any particular view on the nature of explanation.

The second approach to the problem is to insist that it is a primitive fact about explanation that only microconstancy relative to standard ic-variables counts. That is, rather than holding that standard ic-variables, and only standard ic-variables, have some other property that makes them explanatorily special, standardness itself is taken to be the source of explanatory significance.

If the approach is to succeed, there must be some way of characterizing which are the standard ic-variables in a way that makes no reference to explanatory power. Because it is in any case useful to have such a definition, let me make a suggestion: the **standard ic-variables** are those variables that induce measures directly proportional to the units standardized for use by modern science, the si units. Thus, seconds are standard measures of time but *seconds** are not, where a period's duration t^* in *seconds** is related to its duration t in seconds as in section 2.51: $t = t^* + (\cos \pi t^* - 1)/\pi$.

One may then define **standard microconstancy** as microconstancy relative to standard ic-variables. The final step is to stipulate that only standard microconstancy, and not microconstancy relative to any gerrymandered ic-variable, is explanatory. This approach is not very satisfying. If at all possible, one wants to know why standard microconstancy has its special explanatory status, and, in defiance of the epistemic approach, one wants—or certainly, *I* want—the reason to concern only the physics of probability.

It is to this end that I introduce the third approach to the problem. At the beginning of this section I suggested that microconstancy counts only if it is relative to a variable whose ic-density is macroperiodic. I now amend that suggestion: what is required is that there be a *tendency* for the ic-variable to become macroperiodically distributed. I now attempt to show that there is a link between standardness and the tendency to macroperiodicity.

2.53 The Perturbation Argument

In a world such as ours, I will argue, standard ic-variables have a tendency to become macroperiodically distributed, while certain non-standard ic-variables have a tendency to become distributed in a non-macroperiodic way. This is true even for non-standard variables deliberately constructed to be macroperiodic, because, as I will explain below, they have a tendency to become *locally* non-macroperiodic.

Not every non-standard ic-variable will be considered in what follows; my focus is on the kinds of gerrymandered variable that threaten to undermine microconstancy's claim to explanatory power. Other, less unnatural, non-standard variables do not have a tendency to non-macroperiodicity, and some, those that are quite close to the standard variables, will be macroperiodic in many cases where the standard variables are macroperiodic. These mildly non-standard variables do not cause the problems discussed in sections 2.51 and 2.52, because evolution functions that are not microconstant with respect to standard variables are also not microconstant relative to the mildly non-standard variables. I therefore ignore them.

A preliminary note: the arguments that follow should be understood as primarily concerning basic ic-variables, that is, ic-variables that act as probabilistically uncaused causers in probabilistic networks (see section 2.43). The reason is that the tendency to macroperiodicity in any non-basic ic-variable will depend in part on the mechanisms upstream in the network, a complication that I wish to avoid. To take an extreme example, there might be a mechanism that takes one variable as its initial condition and produces another variable, in such a way that the density for the one is macroperiodic just in case the density for the other is not macroperiodic. Given such cases, it is impossible to make a generally valid argument that, say, all standard IC-variables have a tendency towards macroperiodicity. In the example, one of the IC-variables can have such a tendency only if the other does not. The simplest way to proceed is to present the argument for basic IC-variables only, so that the potentially confounding effect of mechanisms is entirely absent. Tendencies to macroperiodicity or non-macroperiodicity will then diffuse through a network in a manner determined by the network's mechanisms and structure. As the reader will see, however, the perturbation argument has a potentially broader application, on which I will remark below.

PERTURBATIONS

I will suppose that the instances of physical properties from which the basic IC-values are drawn are subject, over time, to various perturbations that affect the way they are distributed. (For reasons of economy, in what follows, I will talk about the IC-values themselves being perturbed, though it would be more accurate to say that *potential* IC-values are being perturbed.)

I assume that the distribution of actual values of a basic IC-variable in some way determines the IC-density for the variable. This is true if, for example, the relevant basic IC-densities are frequency-based densities. Given this assumption, perturbations of the IC-values have the power to change the corresponding density over time. The perturbation argument concerns the likely direction of these changes. The conclusion of the argument therefore concerns not the form of the global densities, the densities over all IC-values ever, but the form of various local densities. For standard IC-variables, there is a tendency for local densities to become macroperiodic, even if they start off non-macroperiodic. For gerrymandered IC-variables, there is a tendency for local densities to become non-macroperiodic.

Because I am interested in the effect of perturbations on IC-densities, I will think of perturbations as pushing IC-values around IC-variable space, that is, around the space over which the density is defined. In the case of a density over a single IC-variable ζ, this space corresponds to the ζ axis of the graph of the density. When a perturbation changes the magnitude of a particular IC-value from x to y, I think of the perturbation as moving the value along the axis from x to y. As a result, the density will sink a little in the vicinity of x and rise a little in the vicinity of y.

I assume that perturbations come in two sizes. *Small* perturbations change the magnitude of one or at most a few IC-values by micro-sized amounts, that is, they move a few IC-values a small distance in IC-variable space. *Large* perturbations change the magnitude of many IC-values by macro-sized amounts, that is, they move many IC-values a large distance in IC-variable space. This classification is not, of course, exhaustive, nor is the distinction between large and small a clear one, but it will do for the purposes of the qualitative argument I want to present.

I make the following further assumptions about perturbations:

1. Within a given small neighborhood—a neighborhood containing a number, but not a great number, of micro-sized regions—the chance of a given IC-value's being bumped out of a given microregion by a small perturbation bears a simple, well-behaved relation, for example, inverse proportionality, to the (standardly measured) size of the microregion. In particular, given two IC-values in neighboring microregions of similar size, small perturbations are roughly equally likely to push either value out of its region.

2. A given large perturbation may move many IC-values in the same general direction, but there will be a greater than micro-sized variation both in the regions acted upon and in the action, so that IC-values will be moved from all parts of a greater than micro-sized region to all parts of another greater than micro-sized region. Thus, if values are moved from a given microregion, values will also tend to be moved from the neighboring microregions, and if they are moved to a microregion, other values will also tend to be moved to neighboring microregions.

Standard IC-Variables

Perturbations, I contend, tend both to even out non-macroperiodic densities of standard IC-values, so that they become macroperiodic, and to work actively to preserve the distribution of those standard IC-variables that already have macroperiodic densities. I focus on the simplest case, that of a one-variable IC-density. Generalization is trivial: for the most part, simply read *region* for *interval* and *size* for *width*.

Suppose that a one-variable IC-density is highly non-macroperiodic, in particular, that it has a number of peaks of micro-sized width. Consider one of these peaks and a neighboring trough of equal width. By assumption (1) above, a given IC-value in the peak is about as likely to be bumped into the trough as a given value in the trough is likely to be bumped into the peak. But there are many more ic-values in the peak than in the trough, so the net flow will be from peak to trough. The two will end up the same height, at which point the net flow from one to the other will be zero. Equality of height over small distances is thus created and maintained by small perturbations. This will happen to almost every peak, creating and maintaining macroperiodicity. Large perturbations will not reverse this trend, because by assumption (2) they will be unlikely to create micro-sized peaks and troughs themselves; indeed, they are incapable of creating micro-sized detail at all.

Readers interested in statistical physics will note that the perturbation argument is analogous to Boltzmann's proof of his H-theorem. One could say that a non-macroperiodic distribution has low entropy, a macroperiodic distribution high entropy.

The perturbation argument is not, of course, intended to establish the prevalence of macroperiodicity a priori. It relies on probabilistic reasoning rather than creating a foundation for such reasoning. Specifically, something like a macroperiodic distribution over the perturbations themselves is tacitly assumed.[21] My aim is not to refute the skeptic, but to show that in a world, such as ours, where the assumptions made about perturbations hold, basic, standard ic-variables will tend to be macroperiodically distributed.

Because perturbations do not affect only the ic-values of basic ic-variables, the perturbation argument can be used also to argue for a tendency to macroperiodicity in some non-basic ic-variables. Suppose, for example, that a nervous croupier's hand quivers as he spins a wheel of fortune. If this quivering is thought of as a perturbation of the initial conditions of the spin, rather than as an ic-variable in its own right, and if the quivering satisfies the assumptions made above about the distribution of perturbations, then the perturbation argument can be used to establish that the quivering will even out any small departures from macroperiodicity in the ic-density for the wheel. I will not attempt to generalize this observation here; what I want the reader to see is that perturbations create a tendency to macroperiodicity anywhere that they are present.

Racked IC-Variables

Now to deal with the two kinds of gerrymandered ic-variable that earlier threatened to trivialize the notion of microconstancy. The first of these, introduced in section 2.51, was created by stretching and compressing a standard variable. I call this sort of non-standard variable a *racked* variable. The second, introduced in section 2.52, was created by cutting into pieces and reassembling a standard variable. I call this a *hacked* variable. For each kind of variable, I will show, first, that the argument that establishes a tendency to macroperiodicity in standard variables does not go through and, second, that a similar argument does go through that establishes a tendency to non-macroperiodicity.

The perturbation argument does not establish a tendency to macroperiodicity in racked ic-variables. Why not? One of the key premises in the argument is assumption (1) above, that there is a fixed relation between the size of a microregion and the chance of a given ic-value's being bumped out of the region. This assumption does not hold for racked ic-variables. Two equal micro-sized regions of a racked variable will often correspond to two unequal regions of a standard variable (one is compressed, one stretched). Thus assumption (1) cannot hold for both racked and standard variables; since it holds for standard variables, it does not hold for racked variables.

To see why there is a tendency towards non-macroperiodicity for racked ic-variables, consider in greater detail why, exactly, racked variables do not satisfy assumption (1). The reason that (1) holds for standard variables is that, for any small neighborhood (greater, but not too much greater, than micro-sized), the distance an ic-value is pushed bears a given relation to the size of the perturbation that pushes it. This is because the relevant local law of nature—the law that determines the relationship between the size of a perturbation and the change in the perturbed ic-value—stays roughly the same over a given small neighborhood. This last proposition is not true for racked variables. A law of nature phrased in terms of racked variables will change as one moves alternately through expanded and compressed microregions. If distance is racked, for example, *momentum** will not be conserved, but will increase in stretched regions (where things travel *faster** because *distance** means less) and will decrease in squashed regions. Perturbed ic-values will fly through stretched regions and struggle through squashed regions.

It is this behavior that tends to create non-macroperiodicity, as follows. Because, as I have tacitly assumed all along, the force with which small perturbations push ic-values around is macroperiodically distributed, perturbed values are equally likely to come to rest at any moment within a given small stretch of time. As shown above, perturbed ic-values travel faster through stretched than through squashed regions, so, because they are equally likely to stop in any small time interval, they will be more likely to settle down in a squashed region than in a neighboring stretched region of equal size. Thus they will tend to flow out of the stretched areas and gather in the squashed areas. This process will create peaks in the squashed areas and troughs in the stretched areas, as shown in figure 2.12 (bottom). The result is a non-macroperiodic distribution.[22]

HACKED IC-VARIABLES

The argument establishing a tendency to macroperiodicity in standard IC-variables has two parts: an argument that small perturbations will tend to create macroperiodicity, and an argument that large perturbations will not tend to destroy it. Neither part goes through for hacked variables, but in what follows, I focus on the second part, arguing that large perturbations destroy macroperiodicity in a hacked variable.

When a variable is hacked, microregions excised from one standard neighborhood end up alternating with microregions from another. By assumption (2), a large perturbation will generally increase the number of IC-values in a particular standard neighborhood. But such a perturbation thereby increases the number of IC-values in a hacked neighborhood in just those microregions corresponding to the swollen standard neighborhood, which will create a series of peaks and thus a non-macroperiodic distribution, as shown in figure 2.13.

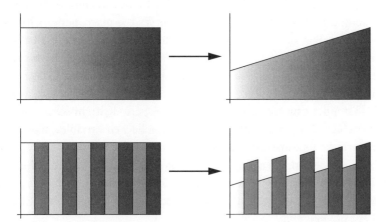

Figure 2.13 Large perturbations cause a non-macroperiodic distribution in a hacked variable. The upper half of the figure shows the effect of a large perturbation on a standard variable. The lower half shows the effect of the same perturbation on a hacked version of the variable. Levels of gray indicate the way in which the hacked variable is assembled from the standard variable.

This argument for a non-macroperiodic tendency in hacked variables turns on the fact that assumption (2) is true for standard neighborhoods but not for non-standard neighborhoods. This follows from the fact that it is with respect to the standard variables that the laws of nature are continuous, so that standardly contiguous groups of values are mapped by single perturbations onto other standardly contiguous groups. Continuity is not itself sufficient for this, of course—some other facts about perturbations, which I do not spell out, are also needed here. The important thing to see is that it is because the laws of nature are radically discontinuous with respect to hacked variables that assumption (2) does not hold for such variables.

There is one apparent difficulty with this argument: in section 2.52 I described variables hacked in just such a way as to have an overall distribution that is macroperiodic. Do variables hacked in this way have, in some sense, a tendency to macroperiodicity? The answer is, I think, no: these variables are macroperiodic, but they do not have a physical tendency to macroperiodicity.

But even if the answer were yes, it can be shown that such variables have no tendency to local macroperiodicity; on the contrary, distributions of ic-values in particular places and times, each subject to their own different large perturbations, will tend to be non-macroperiodic.

The reason is as follows. If the local densities of a standard variable are macroperiodic, then the local densities of a hacked version of the variable are macroperiodic only if the hacking preserves the macroperiodicity of any local density. For all practical purposes, this is possible only if all the local densities of the standard variable are similarly shaped. This is, because of the existence of large perturbations, very unlikely. Different local densities will be differently perturbed, and thus will be differently shaped, with the result that a hacking that preserves the macroperiodicity of some will fail to preserve the macroperiodicity of others. Some hacked variables, then, will be macroperiodic in some locales, and some in others, but no hacked variable will be macroperiodic in a majority of locales. The only reason that the overall distribution is macroperiodic is that the local distributions have been put together in such a way that their non-macroperiodic variations cancel one another out.

Note also that, even in a locale where a particular hacked variable is macroperiodic, the macroperiodicity is highly unstable: the first large perturbation to occur will cause the density to go non-macroperiodic. Macroperiodic locales are, then, of very short temporal duration.

In short, a hacked variable may be globally macroperiodic, but it will be, for the most part, locally non-macroperiodic. Crucially, local as well as global macroperiodicity is necessary to explain both the independence of probabilities from the microlevel and the probabilistic patterns, for the following reason.

Consider the robustness of the roulette wheel's probabilities. What explains the robustness of the probability of red is that, although each croupier brings a different distribution of initial conditions to the wheel, each such distribution is macroperiodic, and so induces the same probability for red. The explanation, then, depends on the macroperiodicity of each of the croupier's densities. These are local densities, thus the explanation depends on the fact that local densities tend to be macroperiodic. The local densities of hacked variables do not have this tendency.

Similar comments apply to the explanation of the probabilistic patterns. When occurrences of a designated outcome e are probabilistically patterned, the local frequencies with which e occurs tend to equal the long term frequency, that is, they tend to equal the probability of e. But because local distributions of hacked variables are not usually macroperiodic, experiments that are microconstant relative only to hacked variables will tend to produce local frequencies that *differ* from the probability. Thus, experiments that are microconstant relative to hacked IC-variables will tend to produce non-probabilistic patterns.[23]

CONCLUSION

To summarize the perturbation argument for the special status of standard IC-variables: In a world where perturbations are distributed in the ways I have assumed, standard IC-variables are likely to be macroperiodically distributed, both locally and globally, while racked and hacked IC-variables are not; on the contrary, they are likely to be non-macroperiodic, at least locally. Since microconstancy is explanatory only when local distributions are macroperiodic, microconstancy relative to a racked or hacked variable has no explanatory power. Thus, standard microconstancy explains probabilistic patterns but non-standard microconstancies do not. The underlying reason for this is certain fundamental physical properties of standard variables: only relative to standard variables are the laws of nature both symmetrical (in the sense required for the relevant conservation laws to hold over small locales) and continuous. This might even be taken as a new definition of what it is to be a standard variable, to replace the working definition framed in terms of the SI units. The explanatory primacy of the standard variables can be established, then, without any appeal to the idiosyncrasies of our explanatory practices or our epistemic situation.

The following objection might be made to the perturbation argument. In order to explain the macroperiodic distributions of the standard IC-variables, the perturbation argument must make certain assumptions about the distribution of the perturbations themselves. These assumptions are at least as strong as the assumption that standard IC-variables have macroperiodic distributions. Thus, in the same way that the mechanism of a non-microconstant probabilistic experiment does not really explain the probabilistic patterns generated by the experiment, so the mechanisms to which the perturbation argument appeals do not really explain macroperiodicity. The distribution of perturbations does all the explaining.

The analogy is a bad one. A non-microconstant probabilistic experiment simply transforms patterned inputs into patterned outputs. At no stage does something that is not probabilistically patterned become probabilistically patterned. To put it another way, no new probabilistic patterns are created; the number of probabilistically patterned events remains the same over time, though the events themselves come and go. But perturbations do create new macroperiodicity. The perturbation argument shows that macroperiodicity is *contagious*: the quantity of macroperiodicity in the world will, if the perturbations are macroperiodic, tend to increase—necessarily at the expense of non-macroperiodicity.

In the light of the perturbation argument, I take microconstancy from this point on to mean standard microconstancy.

2.6 Complex Probability and Probabilistic Laws

So far, I have discussed only the probabilities of outcomes generated by underlying deterministic laws. In this section I extend the treatment to cover probabilistic laws. The probabilities that occur in such laws—call them **nomic probabilities**—may be simple or complex. (For examples of complex nomic probabilities, see chapter four, especially sections 4.3, 4.8, and 4.9.)

There are two ways to represent the effect of probabilistic laws. The first, which gives the more elegant treatment, incorporates nomic probabilities into the evolution function, yielding what I will call an **explicitly probabilistic evolution function**. The second, which is more useful for making certain arguments, represents nomic probabilities by way of additional ic-densities, yield-ing what I will call an **implicitly probabilistic evolution function**. Both kinds of probabilistic evolution function are used in later chapters.

For simplicity's sake, I assume throughout the discussion that experiments have just one ic-variable. Generalization is straightforward.

2.61 The Explicitly Probabilistic Evolution Function

If a probabilistic experiment's behavior depends only on deterministic laws, then a given ic-value always causes the same type of outcome. If an experi-ment's behavior depends on probabilistic laws, then a given ic-value will cause different types of outcomes with differing probabilities. An explicitly proba-bilistic evolution function represents the values of these probabilities in the function itself. Rather than mapping to either zero or one, such a function maps to the probability of the designated outcome, given the particular ic-value in question. More formally, $h_e(z)$ is the probability of the designated outcome e, conditional on the ic-variable's taking the value z. As in the deter-ministic case, all information about the behavior of the mechanism relevant to the outcome of an experiment is contained in the evolution function.

I use the same formula as in the deterministic case to define complex prob-ability:

$$\mathrm{cprob}(e) = \int_V h_e(\zeta)q(\zeta)\, d\zeta.$$

Whereas in the deterministic version, the evolution function picks out or discards segments of the ic-density, in the probabilistic version it variably weights (perhaps infinitesimal) segments.

It is not immediately clear under what circumstances an explicitly proba-bilistic evolution function should be considered microconstant. Because such a function does not necessarily swing back and forth from zero to one, the in-formal characterization of microconstancy for deterministic experiments does not seem suitable.

Nevertheless, it turns out that the formal definition of microconstancy functions well in this new context. It can be carried over from the deter-ministic case, and—since the relevant theorem depends only on the formal definition—it will remain true that, when the evolution function for an event e

is microconstant and the corresponding IC-density is macroperiodic, the probability of e is approximately equal to the strike ratio for e.

Let me explain how the old definition of microconstancy is applied to the probabilistic case. First, the evolution function's strike ratio for an outcome e over an interval $[x, y]$ is defined, as in the deterministic case, to be

$$\int_x^y h_e(\zeta) \, d\zeta \Big/ (y - x).$$

This is just the mean value of the evolution function over $[x, y]$. Then the evolution function is defined to be microconstant just in case there is a partition of ζ into micro-sized intervals each with the same strike ratio, or more formally, the evolution function is microconstant just in case it has a micro-sized CRI. The apparatus for extending the definition to multivariable experiments in the deterministic case works just as well in the probabilistic case.

There are two obvious ways in which an explicitly probabilistic evolution function may turn out to be microconstant. (I will not discuss the ways that are less obvious, or at least more complicated.) The first is the analog of deterministic microconstancy; the second is peculiar to explicitly probabilistic evolution functions.

Oscillating microconstancy occurs when an explicitly probabilistic evolution function looks like a deterministic microconstant evolution function: it oscillates between values near zero and values near one with a constant rhythm, so that the proportion of time spent near values of one within any microregion is approximately the same.

Steady-state microconstancy occurs when an evolution function is everywhere approximately equal to a single value p. Thus, p is the approximate probability that any given IC-value will produce the designated outcome. The definition of the constant ratio index was arranged so that, in such cases, the evolution function has CRI zero.

Steady-state microconstancy does not in itself explain probabilistic patterns in an experiment's outcomes. Rather, the explanatory burden is handed off to the nomic probabilities. If the nomic probabilities can explain probabilistic patterns—if, for example, they are microconstant—then the patterns produced by an explicitly probabilistic evolution function with steady-state microconstancy are explained. Otherwise, they are not explained. I follow up on this comment in section 4.55.

An evolution function's steady-state microconstancy does, of course, explain insensitivity to details about initial conditions.

2.62 The Implicitly Probabilistic Evolution Function

The second kind of probabilistic evolution function represents nomic probabilities as ic-densities. The "initial conditions" whose values are the events over which these densities are defined represent whatever fluctuations or other effects are brought about by the action of the nomic probabilities. For example, if the effect of a setup's nomic probabilities is to perturb the outcome produced by the setup by a certain amount, then the implicitly probabilistic evolution function would take as one of its ic-variables a variable representing the magnitude of this perturbation in any particular trial. The density over the ic-variable, then, would be the nomic probability distribution over the perturbation magnitude. This ic-density will be simple if the nomic probabilities are simple, complex if they are complex.

Implicitly probabilistic evolution functions map only to zero and one, and so may be treated exactly like deterministic evolution functions, to which they are formally identical. In some cases, representing nomic probabilities in this way makes it easier to assess the properties of an evolution function. Note that if all other ic-variables can be eliminated in favor of the variables representing the effect of the nomic probabilities, the variety of microconstancy obtained will be steady state, that is, the sort in which the explicitly probabilistic evolution function is everywhere equal to some p. As remarked in the previous section, microconstancy of this sort does not in itself explain the probabilistic patterns in an experiment's outcomes.

2.63 Simplex Probability

The distinction between simplex and complex probability is chiefly of metaphysical interest; it is discussed here, however, to show that the physics of simplex probabilities is amenable to the same treatment as the physics of complex probabilities. Consequently, it is possible to discuss the physics of enion probabilities, in chapter four, without first settling the question whether they are simplex or complex.

A simplex probability, recall from section 1.31, is a probability that is in some sense entirely reducible to simple probabilities, but that depends on those probabilities in a complex way. Some examples of simplex probabilities, all of which depend on the assumption that the probabilities of quantum mechanics are simple, are:

1. The probabilities attached to a wheel of fortune for which the initial velocity is determined by quantum probabilities,
2. The probabilities of statistical physics, if quantum probabilities are able to affect the trajectories of individual particles (Albert 2000), and
3. The probabilities with which genetic mutations caused by exposure to radiation occur.

I offer the following definition of **simplex probability**. The probability of an experiment X producing an outcome e is simplex just in case either

1. All of X's ic-variables have simple or simplex distributions, or
2. Some of X's ic-variables have distributions that are neither simple nor simplex, but regardless of the values that these ic-variables assume on a particular occasion, the probability of e is the same. That is, the probability of e conditional on any particular values for the non-simple, non-simplex ic-variables is equal to the unconditional probability of e.

In either case, the only facts about ic-variable distributions that make a difference to the probability of e are facts about simple or other simplex distributions.

This definition is equivalent to the following more technical formulation: given the implicit representation of probabilistic laws described in the last section, simplex probabilities are those attached to an experiment for which there exists a strike set of ic-variables all of which have simple or simplex probability distributions, or in other words, to an experiment in which all ic-variables that do not have simple or simplex distributions can be eliminated. This definition has the advantage of offering a very clear division between simplex and complex probabilities; it says very little, however, about the difference between simple and simplex probabilities, which is in any case perhaps only a matter of degree.

As I have already observed in section 1.31, it may reasonably be claimed that, for metaphysical purposes, simplex probabilities ought to be placed in a category with simple probabilities rather than with other complex probabilities. But for the purpose of understanding complex systems, simplex probabilities are deployed in the same way, using the same technical apparatus—including microconstancy and so on, whenever possible—as any variety of complex probability, hence my broad use of the term *complex probability* to include the simplex probabilities.

2.7 Effective and Critical IC-Values

Certain aspects of an experiment's IC-values, I now show, are more important than others in determining the patterns of outcomes that the experiment produces; they are *effective* or even *critical*. On this observation are built the independence results in chapters three and four. I call the explanatorily salient parts of an IC-value the *effective IC-value* and the *critical IC-value*.[24]

Because the notions of an effective and a critical IC-value will first be deployed only partway through chapter three, in section 3.43, the reader may wish to skim this section on a first pass. Some idea of the explanatory role of effective and critical IC-values can be ascertained from the last paragraph in each section.

For the most part, the discussion assumes that probabilistic experiments have deterministic evolution functions and that they have a single IC-variable. What I say may be generalized to probabilistic evolution functions by considering the implicit version of these functions, which has the same form as a deterministic evolution function (section 2.62). The generalization to experiments with more than one IC-variable requires additional comment, for which see section 2.74.

2.71 Effective IC-Values

Consider the following probabilistic experiment. Standing against one wall of a rectangular room, I roll a ball towards the opposite wall. The ball bounces back and forth between the walls, eventually coming to rest. The floor is painted with red and black stripes perpendicular to the direction of the ball's travel. The outcome of a trial on the experiment is the color of the stripe within which the ball comes to rest, red or black. The experiment is, clearly, much like the wheel of fortune; the differences are for expository reasons.

Suppose further that the ball is rolled with a speed between 10 and 90 ms^{-1}, that the stripes are a meter wide, and that the friction of the floor is such as to slow the ball by 1 ms^{-1} for every meter that it travels, so that the distance traveled by the ball is equal to its initial speed. If the room is ten meters wide, and the ball starts on a red stripe, then the evolution function for the experiment is as follows. Round down the speed to the nearest integer. This integer will have two digits, representing the "tens" and the "ones" of the speed. For example, if the speed is 34.56 ms^{-1}, the "tens" digit is 3, representing the fact that the speed is between 30 and 40 ms^{-1}, and the "ones" digit is 4. The outcome of a trial is red if both the "tens" and the "ones" digit are odd, or both the "tens" and the "ones" digit are even. Otherwise it is black. For example, given a speed of 34.56 ms^{-1}, the outcome is black, since 3 is odd but 4 is even.

Now suppose that I wish to predict, given the speed of a particular roll, the outcome of the corresponding trial. How much information about the speed do I need? Clearly, I need only know the first two digits of the speed. This is enough to fully determine in which stripe the ball will come to rest, and so is enough to fully determine the outcome. Later digits will tell me where in a particular stripe the ball will stop, but if my only aim is to predict the color of the stripe, this information is of no interest.

Knowing only one of the first two digits is, by contrast, not enough, since the two digits must be compared in order to determine whether the ball comes to rest in a red or a black stripe. If information about the initial speed of a roll is dispensed in units of digits, then, I need to know no more and no less than the first two digits to determine the outcome of a trial.

This suggests that the decimal expansion of an ic-value for the rolling ball experiment be divided into two parts, the first part consisting of the first two digits of the speed, and the second part consisting of the remainder of the digits. The first part contains just enough information about the ball's speed to determine the outcome of the trial; the second part contains no information relevant to determining the outcome. What I mean by an effective ic-value of a trial is some mathematical entity that conveys just the information of the first sort.

Since an effective ic-value for a trial conveys a part of the information conveyed by the full ic-value for the trial, it is useful to talk, if only metaphorically, of the effective ic-value as being itself a part of the full ic-value. Using the metaphor, here is the basic definition of an effective ic-value: an **effective ic-value** of a trial is that part of the complete ic-value that fixes the values of the initial conditions with just enough precision to determine the outcome of the trial, and no more. The information conveyed by an effective ic-value I call the *effective information*. An effective ic-value, then, conveys just the effective information about a full ic-value.

The interest of effective information is as follows. Because the effective information about an ic-value is sufficient to determine the outcome of a trial, the distribution over the effective information about an experiment's

IC-values—that is, over the experiment's effective IC-values—is sufficient to determine the distribution of the experiment's outcomes. Properties of the distribution of outcomes, then, such as their being probabilistically patterned, can be explained by properties of the distribution of effective information.

I call the definition offered above the *basic definition* because it states succinctly the guiding idea behind the notion that I wish to develop. But it is also quite vague in a number of ways, for example, in its use of the part/whole metaphor. For this reason, I will replace the basic definition with what I call the *formal definition* of an effective IC-value. The formal definition clarifies what does and does not count as effective information, and it fixes on a particular scheme for conveying the effective information.

I have so far assumed that information about an IC-value is dispensed in the form of digits. In the case of the rolling ball experiment, this worked rather well: the first two digits of the ball's speed communicate the effective information. But this is only because of my careful construction of the experiment's parameters. In general, there is no number of digits that fixes a full IC-value with just the right precision to determine a trial's outcome, and no more. Suppose, for example, that the rolling ball slowed by only 0.5 ms^{-1} per meter rolled. Then, to predict the outcome of the trial, one would have to know the speed rounded down not to the nearest integer but to the nearest multiple of 0.5. It follows that the first two digits of the ball's speed would convey less than the effective information, because, for example, a ball with speed 12.1 ms^{-1} produces a red outcome but a ball with speed 12.6 ms^{-1} produces a black outcome. The first three digits of the ball's speed, by contrast, convey more than the effective information. They not only determine the speed to the next lowest multiple of 0.5, but they determine approximately how far from this multiple the speed lies. For example, the information that the first three digits of the ball's speed are 12.7 can be broken down into two parts:

1. The speed lies in the interval [12.5, 13), that is, it lies between 12.5 and 13, and
2. The speed lies in the middle fifth of that interval.

The first part of the information is alone sufficient to determine the outcome of the trial. The second part, then, is unnecessary and so is not part of the effective information. Two digits convey less than the effective information; three digits more. One can solve the problem in this particular case by representing the speed in binary notation, and giving all digits before and one digit after the radix point (the generalized equivalent of the decimal point). This will indeed communicate just the effective information, and so this particular set of digits satisfies the definition of an effective ic-value. But altering the base of the representation will not, in most cases, solve the problem.

The discussion above suggests a far more general solution. The effect of specifying the first however many digits of an ic-value is to specify an interval into which the value falls. For example, to be told that the first two digits are 3 and 4 is to be told that the speed lies in the interval [34, 35), that is, that the speed is somewhere between 34 and 35 ms^{-1}. The problems of the last paragraph arose because a set of digits in a given numbering system can specify only a limited set of intervals. The obvious solution to the problem is to identify an effective ic-value, not with a set of digits, but with an interval. This is the approach taken by my formal definition of an effective ic-value.

Which intervals, then, correspond to the effective ic-values for an experiment, relative to a designated outcome? There are several ways to answer this question, each corresponding to a different interpretation of the basic definition. The question of which to choose depends not so much on what are the "real" effective ic-values as on which interpretation works best for a particular purpose. I will identify two versions of the formal definition, that is, two schemes for deciding which intervals correspond to an experiment's effective ic-values for an outcome. The first scheme is more useful for non-microconstant experiments, the second for microconstant experiments.

On the first scheme, the intervals are those corresponding to the gray and white stripes in a graph of the outcome's evolution function, or more formally, to the maximally contiguous sets of ic-values over which the evolution function is the same (either zero or one). To know that the full ic-value for a trial is somewhere in such a set U is to know the outcome of the trial. If one knew the ic-value with any less precision—if one knew only that the ic-value falls into a slightly wider interval of which U is a subset—one could not predict the outcome. If one knew the ic-value with any more precision—if one knew that the ic-value falls into a narrower interval contained in U—one's predictions would in no way improve. To know that the full ic-value for a trial falls into U, then, is to know the ic-value with a precision both necessary and sufficient to predict the outcome of the trial.

The effective ic-value of a trial, relative to a designated outcome e, may therefore formally be defined as follows. Partition the domain of the evolution function for e into sets of ic-values that are the maximal, contiguous sets over which the evolution function takes the same value. Call this the **outcome partition** for e. The outcome partition for the rolling ball, for example, is the partition into the sets $[i, i + 1)$, for integers i between 10 and 89 that are not multiples of 10, and $[i - 1, i + 1)$, when i is a multiple of 10 between 20 and 80. (The reason for the complication is as follows. For a speed that is a multiple of 10, the ball reaches one of the two walls. If the speed were slightly greater, the ball would hit the wall and bounce back through the colored section that it just traversed. Speeds on either side of a multiple of 10, then, produce outcomes of the same color.)

Using the notion of an outcome partition, the effective IC-values for e may be defined as either:

1. The members of the outcome partition itself, or
2. The members of an index set that pick out (by way of some agreed bijective mapping) the members of the outcome partition. In the case of the rolling ball, for example, the indices might just be the integers between 10 and 89, with the integer i picking out the interval $[i, i + 1)$, or when i is a multiple of 10, $[i - 1, i + 1)$.[25]

The first option is of course the simpler, but as the reader will see in section 2.73, the second provides more continuity with the formal definition of a critical IC-value.

Now consider two other schemes for assigning effective IC-values. On the first alternative to the formal definition above, effective IC-values do not have to correspond to contiguous sets. After all, in order to predict the outcome of a trial on the rolling ball experiment, it is enough to know that the trial's IC-value falls into some one of a number of different intervals, all values in each of which bring about a red outcome. Why not define as an effective IC-value the union of all these intervals? In effect, there would be just two effective IC-values for any trial, 0 and 1. A trial with effective IC-value 1 would produce the designated outcome; a trial with effective IC-value 0 would not.

Although this definition captures, in a sense, the absolute minimum amount of information about a trial's initial conditions necessary to determine the outcome of the trial, it is not a very interesting or useful sense. The claims made at later points in this study concerning effective IC-values (for example, theorem 3.2) are all true on this definition, but they seem to have no practical significance. I put the definition aside.

The other scheme that I will describe for assigning effective IC-values is not quite true to the basic definition, but will prove to be useful all the same. It is this scheme that I will use whenever I deal with microconstant experiments, for reasons that will become clearer in section 2.73.

On the new scheme, effective ic-values are maximal, contiguous sets of outcomes that all lead to the same outcome and (here is the difference) fall entirely inside some member of a given optimal constant ratio partition. The definition, then, is relative to an optimal constant ratio partition, and so applies only to microconstant experiments. This new scheme for individuating effective ic-values slices effective ic-value space slightly more finely than the first scheme presented above: if a gray or white area straddles the boundary of a member of the specified constant ratio partition, it is cut into two parts along the boundary. The case of the rolling ball supplies an example: although the entire interval [19, 21) produces red outcomes, because the optimal constant ratio partition has as local members [18, 20) and [20, 22), the interval [19, 21) is divided into two effective ic-values corresponding to the intervals [19, 20) and [20, 21).

To make it easier to move back and forth between the two schemes I have offered for assigning effective ic-values, define an outcome partition for a microconstant experiment slightly differently than for a non-microconstant experiment. For a non-microconstant experiment, as before, an outcome partition is a partition into maximal, contiguous sets in which all ic-values map to the same outcome. For a microconstant experiment, require in addition that outcome partition members fall entirely within a member of some given optimal constant ratio partition. Now define an effective ic-value for an experiment as a member of the experiment's outcome partition, or as an index of the outcome partition. This results in a slightly different scheme for assigning effective ic-values in the microconstant case than in the non-microconstant case, as desired.

Let me repeat two remarks I have made along the way about the scheme for assigning effective ic-values in the microconstant case. First, the scheme is relative to a choice of optimal constant ratio partition. Any choice will do. Second, the scheme is not entirely faithful to the basic definition of effective information. On the microconstant scheme, effective ic-values will communicate a little more information than in the non-microconstant case, but this information does not play any role in fixing the outcome of a trial. The justification of the scheme, imperfections and all, is, as I have said, in its application.

Whether effective ic-values are taken to be intervals or indices of intervals, a very useful metaphor for thinking about effective ic-values and other parts of full ic-values suggests itself. In the case of the rolling ball, the effective ic-value gives **high-level information**, and leaves out **low-level information**, about the full IC-value. This is the same metaphor that is at work in talk of the macrolevel and microlevel, and so on. The guiding idea is that:

1. High-level information about a quantity x gives approximate information about the value of x, and
2. Low-level information in itself gives no information about the value of x, but in conjunction with all higher-level information about x, low-level information specifies the value of x in greater detail.

For example, given a real number x between zero and one, the first digit after the decimal point gives high-level information about x, the second digit after the point slightly lower level information about x, and the fifth digit after the point quite low level information about x. In the discussion in the main text, the notion of levels of information is used only qualitatively. But some of the proofs in chapter three use a formal definition, for which see section 3.B, especially definition 3.4, definition 3.13, and theorem 3.12.

The notion of levels of information should not be taken as relative to any particular system of notation (and thus, not to any particular scheme for locating a number in successively smaller regions, such as the decimal system). The level of a given kind of information, such as the effective information, corresponds to the size of the region within which that information, together with all higher-level information, locates the IC-value. If the region is relatively large, the information is high level; if it is small, low level.

Effective IC-values are put to use as follows. In chapter three I show that, even if two probabilistic trials have IC-values that are not stochastically independent, the outcomes of those trials may be independent if the effective IC-values are independent. The notion of an effective IC-value therefore allows a weakening of the conditions for stochastic independence (theorem 3.2), a weakening that is absolutely essential for the understanding of the success of EPA (sections 3.7 and 4.5).

2.72 Effective IC-Variables

In order to discuss the uses of probability distributions over effective IC-values, I introduce the notion of an **effective IC-variable**.

Given a continuous IC-variable ζ and an event e, define the corresponding effective IC-variable ζ^e as the discrete variable that takes on the effective IC-values for e. For example, in the rolling ball experiment, if the effective IC-values of the IC-variable are taken to be the integers between 10 and 89, as suggested in the last section, then the effective IC-variable is a discrete variable that takes as its values the integers between 10 and 89.

Only the probability distribution over effective information—the information conveyed by effective IC-variables—matters in deriving the equality of strike ratio and probability for a designated outcome on a microconstant experiment. The distribution over lower-level information makes no difference to the probability. Let me make good on this claim.

I begin with the observation that a probabilistic experiment's continuous IC-variables can be replaced with the corresponding effective IC-variables. In this way, every probabilistic experiment with continuous IC-variables can be redescribed as an experiment with discrete IC-variables. I will take advantage of this possibility in several later arguments, in particular, in section 3.7.

Complex probabilities attached to an experiment with an effective IC-variable are defined in the obvious way:

$$\text{cprob}(e) = \sum_{\zeta^e \in V} h(\zeta^e) Q(\zeta^e)$$

where V is the set of all possible values of ζ^e, that is, the set of all possible effective IC-values for e, and $Q(\zeta^e)$ is the probability distribution (not density) of the effective IC-variable ζ^e. Microconstancy is stipulated to be inherited from the continuous version of the experiment, in the sense that the discrete version of the evolution function is microconstant just in case the continuous version is microconstant. For most purposes, macroperiodicity may be stipulated to be inherited in the same way. But it is illuminating to give the following alternative definition of the macroperiodicity of an effective IC-variable distribution.

Define the *discrete density* of the effective IC-variable ζ^e as

$$\hat{q}(z) = \frac{Q(z)}{M(z)}$$

where z is any value of ζ^e and $M(z)$ is the measure of the interval corresponding to z.[26] Then $Q(\zeta^e)$ is macroperiodic just in case, for any two neighboring values z_1 and z_2 of ζ^e, $\hat{q}(z_1)$ is approximately equal to $\hat{q}(z_2)$.

For a microconstant experiment, if the effective distribution $Q(\zeta^e)$ is macroperiodic, then the probabilities for the designated outcomes are approximately equal to their strike ratios. I leave the proof as an exercise for the reader.[27] Because the macroperiodicity of $Q(\zeta^e)$ is a weaker condition than the macroperiodicity of $q(\zeta)$, this is a stronger result than the earlier strike ratio/probability result.

As promised, then, for the purposes of deriving the equality of strike ratio and probability, only the distribution of the effective information counts. Consequently, features of a system that depend only on the probability of an outcome or outcomes depend only on that aspect of the relevant initial conditions that is captured by the probability distribution over the corresponding effective IC-variables. Thus, in studying phenomena such as independence, one need attend only to the properties of the effective IC-variable distributions. As noted at the end of the last section, this provides the foundation for a series of independence results in chapter three that are crucial to understanding the underpinnings of EPA.[28]

2.73 Critical IC-Values

I introduced the notion of an effective IC-value to exploit the fact that the distribution of low level information about IC-values is not relevant to determining distributions of outcomes. The notion of a critical IC-value is introduced to exploit a parallel fact: in a microconstant experiment, the distribution of *high-level* information about IC-values is irrelevant to determining the distribution of outcomes.

To see this, consider some very high level information about a full IC-value for the rolling ball experiment described in section 2.71, say, the information conveyed by the ball's initial speed rounded to the next lowest multiple of 10. The distribution of this information is the distribution of the approximate value of the speeds. It tells you where, approximately, the full IC-values fall in speed space: so many between 10 and 20 ms^{-1}, so many between 20 and 30 ms^{-1}, and so on. Since the rolling ball experiment is microconstant, the distribution of approximate location in speed space makes very little difference to the probability of, say, red, because the evolution function for red looks relevantly the same in any locale. What makes a difference is the distribution within each approximate location: each such local density must be roughly uniform, which is just to say that the density as a whole must be macroperiodic. This suggests the division of the effective information about an IC-value for a microconstant experiment into two parts: a higher-level part, the distribution of which makes little or no difference to the probability of the designated outcome, and a lower-level part, the distribution of which determines whether or not the IC-density is macroperiodic,[29] and thus which makes a great difference to the probability of the designated outcome.

Let me put this more precisely. I propose to divide an effective IC-value for an outcome on a microconstant experiment into the following two parts: the part that determines into which member U of an optimal constant ratio partition a full IC-value falls, and the part that determines approximately where inside U the value falls. The second part is the **critical IC-value** for the outcome. I will go on to show that certain properties of the distributions of outcomes of microconstant experiments, and thus certain properties of microconstant probabilities, depend only on properties of the distributions of critical IC-values.

It is with the details of the division into critical and non-critical parts that the rest of this section is concerned. Begin with the case of the rolling ball. The effective IC-values for this experiment, recall, correspond to the intervals $[i, i + 1)$, where i is an integer between 10 and 89.[30] Consider a particular interval, $[26, 27)$. The fact that a full IC-value for a particular trial falls into this interval is equivalent to the following pair of facts:

1. The IC-value falls into the interval $[26, 28)$, and
2. The integer part of the IC-value is even.

The first of these facts determines into which member of an optimal constant ratio partition for the experiment the full IC-value falls. (The constant ratio partition in question consists of all intervals $[i, i + 2)$, for i between 10 and 88.) More exactly, it conveys the information that the IC-value falls somewhere within a particular pair of gray and white stripes in the evolution function for the experiment. The second fact conveys the information as to which member of this pair contains the full IC-value. This is the critical information. A critical IC-value, then, is a device for conveying the critical information.

The way I have set things up, neither the first nor the second fact is sufficient on its own to determine whether the outcome is red or black. Knowing only that the full IC-value for a trial falls into a certain member of an optimal constant ratio partition is no help, because every member contains an equal proportion of IC-values producing and not producing the designated outcome. But equally, knowing only that the full IC-value falls into the even-numbered interval in the constant ratio partition member is no help, because some even-numbered intervals produce red outcomes and others produce black outcomes. This is normal when dealing with critical IC-values, but, as I will show near the end of this section, it is not always the case. What is always the case is that neither knowing the optimal constant ratio partition member nor knowing the critical-level information is alone sufficient to determine the *effective ic-value* of a trial, that is, to determine into which member of the outcome partition the value falls.

I now provide both a basic and a formal definition of a critical ic-value. The basic definition has already been given, very briefly, above, but I will repeat it now at greater length.

Given a microconstant probabilistic experiment and a designated outcome e, take an optimal constant ratio partition for e. (Critical ic-values are defined only if such a partition exists, hence, only for microconstant experiments. If there is more than one optimal partition, there is more than one way to divide effective ic-values into critical and non-critical parts, but nothing in this study assumes the uniqueness of the scheme of division.) Each member U of the chosen optimal constant ratio partition can be further partitioned into maximal contiguous sets all members of which produce the same outcome. For any U, these sets are, of course, just those members of the outcome partition that belong to U. Call these sets the outcome partition of U.

Now, given a full ic-value, determine the corresponding effective ic-value. Divide the information in the effective ic-value into two parts, so that one part is sufficient to determine into which member U of the optimal constant ratio partition the full ic-value falls, and the other part is just sufficient to determine either

1. Into which member of the outcome partition for U the full ic-value falls, given that it falls into U, or
2. At least, whether it falls into a "gray" or a "white" member of the outcome partition—that is, into a member all of whose ic-values map to 1 or a member all of whose ic-values map to 0—given that it falls into U.

The reason for the either/or is discussed below. This division accomplished, the first part of the effective information is the non-critical information; the second part is the critical information. The critical ic-value corresponding to a full ic-value is any mathematical entity conveying just the critical information about that value. Observe that, by definition, neither the non-critical information nor the critical information is sufficient by itself to determine the effective ic-value of a trial.

As I did for effective ic-values, I now attempt to give a formal definition of a critical ic-value that makes a particular choice as to the scheme for conveying the critical information. There are, it turns out, complications, but they will be overcome.

Effective ic-value

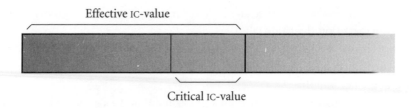

Critical ic-value

Figure 2.14 The parts of the decimal (or other) expansion of a full ic-value corresponding to the effective and critical ic-values.

The choice I made concerning effective ic-values—that they correspond to intervals—will not in general work for critical ic-values.[31] To see this, consider the case of the rolling ball. The critical information here is the information as to whether the integer part of the full ic-value is even or odd. But no interval can convey just this information and no more.

A better model for a critical ic-value is a digit in the decimal (or other) expansion of the full ic-value. In the rolling ball case, the critical information can be captured exactly by taking the last digit of the integer part of the binary expansion of the full ic-value. This last digit is 0 if the integer part is even; 1 if it is odd. On the assumption that the non-critical information is at a higher level than the critical information, one ends up with the sort of picture shown in figure 2.14.

As with effective ic-values, however, there is no guarantee that there will be some system for expanding a full ic-value so that a particular digit or digits conveys the critical information. The simplest solution to the problem is simply to number the members of the outcome partition for each member of the constant ratio partition, starting from zero each time. These numbers will convey the critical information.

Let me discuss this suggestion with the help of a simple example. Consider the microconstant evolution function pictured in figure 2.15. In the figure, intervals that are members of an optimal constant ratio partition are delimited by dashed lines. Each of these intervals is assigned a different number (in the figure, from 0 to 3). This number conveys the non-critical information in the effective ic-value, that is, it determines into which member of the optimal constant ratio partition a full ic-value falls.

Each member of the constant ratio partition has an outcome partition that has either two or three members, corresponding to its gray and white regions. My proposal, as applied to a simple one-dimensional function such as this,

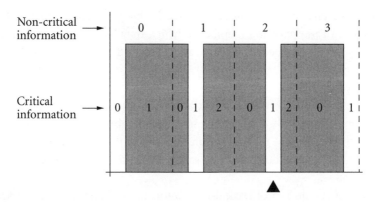

Figure 2.15 The ordinal scheme for assigning critical ic-values. The members of the constant ratio partition, delimited by dashed lines, are numbered from 0 to 3. These numbers convey the non-critical information. The members of the outcome partition for each member of the constant ratio partition—the gray and white regions within each constant ratio partition member—are numbered from 0 up within each partition member. These numbers convey the critical information, and so play the role of critical ic-values.

is to number the leftmost member of the outcome partition in each constant ratio partition member '0', the next leftmost '1', and so on. I call this the *ordinal scheme* for assigning critical ic-values.

The number of the outcome partition member into which the full ic-value falls—in figure 2.15, either 0, 1, or 2—conveys the critical information about the ic-value, as desired. For example, if the full ic-value falls into the white section indicated by the black triangle in the figure, the non-critical information concerning that value is conveyed by the number 2, the critical information by the number 1. Thus the critical ic-value for such a full ic-value is 1. The ordered pair (2, 1) containing both pieces of information picks out a particular member of the outcome partition, and therefore picks out the effective ic-value corresponding to a full ic-value that falls into the indicated section.

The ordinal scheme for assigning critical ic-values has one great disadvantage: the probability distribution over the critical information will not, in general, be independent of the probability distribution over the non-critical information. This is because, within different members of an optimal constant ratio partition, members of the outcome partition assigned the same number may be different relative sizes. For example, in figure 2.15, most sections numbered 1 take up 1/4 of the space between the dashed lines (that is, 1/4 of the constant ratio partition member to which they belong), but one of them takes up 3/4 of the space. The failure of independence causes three related problems:

1. The critical information about an ic-value would seem, after all, to convey, probabilistically, some information about the member of the optimal constant ratio partition into which the value falls. Thus the ordinal scheme does not fully satisfy the basic definition of a critical ic-value.
2. When the notion of critical-level information is put to work in sections 3.4 and 3.5, I need a definition on which it is plausible to assume that critical-level information is independent of other effective information.
3. Because of the lack of independence, critical-level information assigned according to the ordinal scheme cannot always be *well-tempered* (see definition 3.15). This blocks certain uses of the notion of critical information; see, for example, the end of section 3.B4.

One way around the independence problem is to slice each constant ratio partition member much more finely, so that the slices are of the same thickness but each is wholly inside either a gray or a white area, and to number these slices. On this scheme, however, the critical information will convey low-level information that, according to the basic definition, it ought not to convey, since it locates the full ic-value not just inside a certain outcome partition member, but in a certain slice of that outcome partition member. Still, for some purposes—in particular the purposes of sections 3.4 and 3.5—this is not a disadvantage. In those sections I do not use quite the device suggested here, but I do mingle information at and below the critical level.

For other purposes, it is worthwhile to develop a scheme for assigning critical ic-values that is truer to the basic definition. There is no scheme, as far as I can see, that can be applied entirely generally and that is without disadvantages, but there is one very simple scheme that is almost entirely without disadvantages.

On this new scheme, all members of the outcome partition mapping to zero—all white areas—are assigned the critical ic-value zero, and all members mapping to one—all gray areas—are assigned the critical ic-value one. Figure 2.16 shows this scheme applied to the evolution function from figure 2.15.

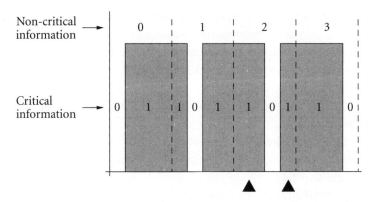

Figure 2.16 The teleological scheme for assigning critical ıc-values. See the caption to figure 2.15 for the interpretation.

Because the scheme labels sections according to the outcome they cause, I call it the *teleological scheme* for assigning critical ıc-values.

There are two disadvantages to the teleological scheme:

1. The critical information about a trial's full ıc-value is sufficient to determine the outcome of the trial. This obscures the fact that the critical information delivered by the teleological scheme is low-level information, but it is not inconsistent with that fact. Nothing in the basic definition of critical information rules out this possibility. The important thing is that the critical information convey no information as to the member of the optimal constant ratio partition in which the full ıc-value falls; the teleological scheme, unlike the ordinal scheme, conforms fully to this requirement.

2. The critical information about a trial's full ıc-value, together with the non-critical effective information, is not always sufficient to determine the trial's effective ıc-value. In figure 2.16, for example, knowing that a full ıc-value has non-critical value 2 and critical value 1 is not enough to distinguish which of the two sections indicated by the black triangles contains the value.

The second disadvantage deserves further comment. Unlike the first, it is inconsistent with the basic definition, which requires that the critical and non-critical information together determine the effective ıc-value. There are, however, two ways to rectify this discrepancy.

First, the scheme for conveying the non-critical information can be amended so that the non-critical information specifies not just into which member of the constant ratio partition an ic-value falls, but also the information needed, in addition to the critical information, to determine into which member of the outcome partition the value falls. It is a consequence of this suggestion that some of the non-critical information will be at a lower level than the critical information.

Second, the formal definition of effective information for microconstant experiments can be revised. Rather than specifying a particular member of the outcome partition, the effective information about a value would have only to specify the member of the constant ratio partition within which the value falls, and whether within that member it falls into a gray or a white section. Thus a trial having a full ic-value falling into either of the indicated sections in figure 2.16 would count as having the same effective ic-value. The main problem with this suggestion is that it results in effective ic-values that are not contiguous. For many purposes, this is of no concern. Note, however, that throughout this study, I assume contiguous effective ic-values unless I say otherwise.

To summarize, there are now three proposals on the table for assigning critical ic-values: the ordinal scheme, the teleological scheme with a revamped scheme for conveying non-critical information, and the teleological scheme with a revamped formal definition of an effective ic-value. These each have advantages and disadvantages, and they are each useful in certain situations. Together, they provide a toolkit that is more than adequate to present all the arguments involving the concept of critical information in this study, and more besides. For the formal apparatus used to provide the scaffolding for the notion of low level information, I refer the reader to sections 3.B2 and 3.B4.

Most of the discussion in this section has assumed implicitly that the non-critical information about a full ic-value is at a higher level than the critical information. Three remarks on this assumption. First, the practical value of the results concerning critical information presented in section 3.4 does indeed depend on the truth of the assumption or, at least, on there being *some* information in the full ic-value above the level of the critical ic-value. Second, the assumption does not always hold. One can easily conceive of a probabilistic experiment in which the lower-level part of the effective ic-value determines the constant ratio partition member U, and the higher-level part the approximate location within U. Third, neither the basic nor the formal definitions of a critical ic-value depend on the truth of the assumption. The practical value of the notion of critical information may depend on the assumption, but the notion itself does not.

Critical IC-values are, as I remarked at the beginning of this section, important because some facts about microconstant probabilities depend only on facts about distributions of critical information. One example of this claim is afforded by the fact that the fundamental microconstancy theorem, concerning the equality of probability and strike ratio, can be proved using the notion of critical level information (see section 3.B4, example 3.5). Another example, perhaps the central fact of chapter three, is the following observation: correlation between the non-critical parts of two microconstant experiments' effective IC-values does not destroy the stochastic independence of trials on those experiments, provided that the critical IC-values are independent of each other (and of the non-critical information). Since the non-critical parts of effective IC-values tend to be at a higher level than the critical parts, this result allows for independence even when there is high-level correlation between IC-values.

2.74 The Multivariable Case

The definitions of effective and critical IC-values given above can easily be modified so as to apply to multivariable experiments as well as to one-variable experiments. But because I have used only one-variable experiments as examples, I should point out a significant difference between the one-variable case and the multivariable case.

In the one-variable case, each value of an IC-variable has a corresponding effective and (if the experiment is microconstant) critical IC-value. Any particular speed for the rolled ball, for example, has an effective and critical counterpart. But in the multivariable case, it makes no sense to talk of the effective and critical IC-values corresponding to the values of individual IC-variables. Consider the coin toss (see section 2.25), an experiment which has two IC-variables, spin speed ω and spin time τ. For a particular value of spin speed, what is the corresponding effective IC-value? The question has no answer, because no part of the spin speed alone is sufficient to determine the outcome of a toss. A value for the spin time is required as well.

The effective IC-values for the coin toss, then, are not intervals of ω or τ, but regions of the two-dimensional space $\omega \times \tau$. (This is implicit in the definition of an effective IC-value given in section 2.71, on which effective IC-values correspond to members of the relevant outcome partition, where an outcome partition for a multivariable evolution function is a partition of the space of all possible *sets* of ic-values.) An experiment with more than one ic-variable, then, will have only one effective ic-variable. Since critical ic-values are defined in terms of constant ratio partitions, the same comments apply. I will sometimes talk about the "effective ic-variables" of a multivariable experiment, however, so as not to convey the false impression that the experiment has only one ic-variable.

The case where an experiment is microconstant and has eliminable ic-variables deserves special comment. Discussion of the details of the case involves some technical notions that I have relegated to the appendix; I refer to the reader to the later part of section 2.B for this discussion. What emerges is that effective-level information about the ic-values of a trial on a microconstant experiment with eliminable ic-variables is high-level information only about the values of the ic-variables in a strike set for the experiment. The effective information specifies the values of eliminated ic-variables *exactly*. This may sound like a limitation, but, for reasons given in section 2.B, it is not. In any case, the reader can, for the most part, safely ignore this fact.

APPENDIX

2.A The Method of Arbitrary Functions

A property akin to microconstancy is the central concept of the mathematical technique known as the method of arbitrary functions (which, oddly, neglects to give this central concept a name). My use of microconstancy to explain probabilistic patterns and insensitivity to initial conditions constitutes an application of the method in a modified form. The purpose of this section is to sketch the historical development of the method and to give the reader some sense of how my techniques differ from or extend the method.

The best known early presentation of the method of arbitrary functions is that of Poincaré (1896). (For further references see von Plato 1983.) Poincaré examines the simple wheel of fortune discussed in section 2.22. He considers the relationship between the final resting place of the wheel and the outcome of a spin, red or black. The final resting place is represented as an angle θ, and the relation between the resting place and the outcome by a function $f(\theta)$ equal to one if θ is in a red section, zero if θ is in a black section. Note that, despite the similarities, θ is not an IC-variable and $f(\theta)$ is not an evolution function, since θ describes not an initial condition but an outcome of a spin on the wheel.

Poincaré observes that as θ goes from 0 to 2π, $f(\theta)$ oscillates quickly between zero and one with a constant ratio of red values of θ to black values of θ. He points out that, for a sufficiently smooth probability distribution over θ, the probability of red will come out as one half. He further notes that, as the number of red and black sections on the wheel increases, the probability distribution over θ need be less and less smooth to give a probability of one half. In the limit, as the number of sections goes to infinity, any probability distribution that has a density will give a probability for red of one half. That is, an *arbitrary density function* will assign red a probability of one half. This, Poincaré concludes, is why the probability of red must be one half.

Poincaré's method differs from mine in two important respects. First, rather than starting with the initial conditions of a spin of the wheel, he starts with the final condition of the wheel, its resting place θ at the end of a trial. Thus the mechanics of the wheel in no way figures in his account. The only physical fact about the wheel that contributes to the value of the probability, on Poincaré's treatment, is the paint scheme.[32]

Second, the explanation of the probability given by Poincaré depends only on properties of $f(\theta)$ *in the limit*. In particular, the actual number of red and black sections is not part of the explanation.

The following example illustrates the limits of Poincaré's analysis. Consider a wheel of fortune with small braking pads under the red sections that slow the wheel whenever a red section passes the pointer. Such a wheel will be characterized by the same $f(\theta)$ as a normal wheel. Thus Poincaré's method will assign a probability of one half to a red outcome on this wheel. But clearly, the probability of red is greater than one half. The neglect of the wheel's mechanics is an error.

Hopf (1934) made the next significant contribution to the method, centering his analysis, too, on the wheel of fortune. Unlike Poincaré, Hopf begins with the initial speed of a spin on the wheel, and examines its relationship to the outcome. That is, Hopf examines the properties of what I call the wheel's evolution function. For this reason, his treatment deals successfully with the brake pads, and is closer to my own approach.

Like Poincaré, however, Hopf emphasizes limiting behavior. (One difference: whereas Poincaré lets the number of sections go to infinity, Hopf lets the time the wheel is allowed to spin go to infinity, by allowing the frictional force slowing the wheel to go to zero.)

Hopf (1934, §9) also derives, again relying on limiting behavior. a restricted version of the independence result I present in section 3.4.

The method of arbitrary functions has been developed more recently by Keller (1986) and Engel (1992). Keller applies the method to a coin toss, using the evolution function described in section 2.25. Engel applies the method to a wide range of different systems, and provides a more general version of Hopf's independence result. Keller and Engel, like Hopf, use a limiting approach, basing the relevant probability on the behavior of the evolution function as some parameter of the mechanism goes to infinity. Engel derives some interesting quantitative convergence results; nevertheless, I will argue, the limiting approach has serious drawbacks.

One might object to the limiting approach on the grounds that it involves parameters approaching infinity that are in reality physically bounded. But this is not, in itself, a powerful objection, since limiting behavior often does provide a good approximation to actual behavior. For just this reason I myself rely, in other parts of this book, on limiting results, the law of large numbers being an obvious example. My reasons for avoiding the limiting approach inherent in the method of arbitrary functions concern specific problems that arise when applying the approach to the questions in this study.

Let me begin by stating the one considerable advantage that the limiting approach has over my approach: the equality of probability and strike ratio may be derived without requiring anything of the ic-density, save that it exist. This is because the equality is determined entirely by the limiting behavior of the evolution function: as some parameter, such as time, increases, the probabilities determined by the different possible initial condition densities all converge on a single value, equal to the strike ratio. My dissatisfaction with the limiting approach stems from the fact that this convergence is neither necessary nor sufficient for the equality of probability and strike ratio in real experiments.

Convergence is not sufficient because actual behavior sometimes does not reflect limiting behavior. For an actual wheel of fortune, there are possible ic-densities, namely, the non-macroperiodic densities, that will induce a probability for red other than one half. The reason is, of course, that the parameters that Poincaré and Hopf allow to go to infinity do not go to infinity at all. The wheel has a moderate number of sections, and it spins for less than a minute.

Is this a problem? As noted above, it is in some ways a part of the charm of the limiting approach. But the ugly details that are charmed away are in some contexts rather important. It is only by attending to the actual values of parameters such as the wheel's spin time and number of sections that one can determine accurately the conditions that must be satisfied by a density in order for the microconstancy of the evolution function to have its significance, and thus for the many results based on microconstancy to go through. For the purposes of my project, which makes some sweeping claims about the explanatory power of microconstancy, it is desirable to state these conditions as clearly as possible. Thus, I need a method that, unlike the limiting approach, takes the real values of relevant physical parameters seriously.

Convergence is not necessary for the equality of probability and strike ratio because there are microconstant experiments in which it is neither relevant nor even possible for parameters to go to infinity. The probability of a rabbit surviving a month cannot be calculated by examining the probability of a rabbit surviving for n days as n goes to infinity. Ignoring the relevant parameter in this case—the number of days in a month—makes the problem more difficult rather than easier. Again, what is required is an approach that bases probabilities on the actual values of a system's physical parameters, as does mine.

A further problem with the limiting approach is that it cannot capture the fact that microconstant probabilities do not depend on the distribution of very low level information about ic-values (see section 2.72). Because this lack of low-level dependence plays a key role in founding the independence properties of enion probabilities (sections 3.7 and 4.53), the method of arbitrary functions, in its standard form, is not a suitable tool for dealing with enion probabilities and their role in EPA. For this and the preceding reasons, I have taken the basic mathematical idea underlying the method and recast it in a new form.

Let me note the several, all too brief, appearances of the method of arbitrary functions in the philosophical literature. Reichenbach (1949, §69) uses the method to explain our ability to infer probabilities from symmetries. He adopts Poincaré's approach, though he is aware of the problem caused by focusing on limiting behavior. (See Strevens (1998) for comments and an alternative account of the inference of probabilities from symmetries based on the notion of microconstancy.) Kneale (1949) also makes use of Poincaré's approach, oddly, to argue that equiprobability is the fundamental notion from which all probabilities are derived. (In fact, the method of arbitrary functions is just as capable of fixing the values of unequal probabilities as it is of fixing the values of equal probabilities, as shown by some of the more complex physical cases discussed by Hopf. Mellor (1971, 139–146) criticizes Kneale's case for equiprobability, but talks as though the fault lies with the method and not with Kneale.) Savage (1973) uses the method to explain intersubjective agreement about the probability of red, in line with his subjectivist approach to such probabilities. Suppes (1973, 1987) proposes using the method to ground a propensity account according to which some probabilities are reducible propensities (at last incorporating Hopf's improvements to Poincaré's method); see also Earman's programmatic comments along the same lines (Earman 1986, §VIII.6). Von Plato (1983) gives a history of the method, summarizes Hopf's work, and makes the same suggestion as Suppes and Earman.

To give the reader some idea of what is new in my treatment: None of the writers cited suggests that the method may be applied very widely, certainly not to the biological enion probabilities that are the subject of chapter four. (The apparently limited scope of the method is surely the reason that it has been confined to the back pages of philosophical books on probability.) With the exception of Reichenbach and Savage, all have focused on limiting behavior. The concepts of microconstancy, macroperiodicity, effective and critical level information, and so on, do not appear in the arbitrary functions literature. Most of the material on independence is new, including all of sections 3.6 and 3.7. Finally, Hopf, Keller, Engel, and others tend to derive probabilities from the detailed physics of the setups they analyze, which has the disadvantage of not making clear what in the physics is essential and what is unimportant in determining the value of a probability (see section 2.24).

2.B More on the Tossed Coin

In this section, I treat the case of the tossed coin (section 2.25) more formally. My aim is not to pursue rigor for its own sake, but rather to provide a useful prelude to the fully general treatment of multivariable experiments presented in section 2.C2, by motivating, using a worked example, the definitions of a restricted evolution function, a restricted ic-density, and the multivariable definition of a constant ratio partition.

I first show how to derive a probability for heads of one half even when the ic-density for the coin in the direction of the spin time τ is not macroperiodic, and I then propose a definition of the constant ratio partition for the coin toss in such a case. Next, I discuss the way in which the notion of an effective ic-value (section 2.71) should be understood for a multivariable experiment such as the coin, and finally, I comment on a way to generalize the approach described here.

In accordance with the formal tenor of the discussion, I assume macroperiodicity and microconstancy in the stringent senses defined in section 2.C. Given more realistic, less demanding assumptions than these, what would be derived is not that the probability of heads is exactly one half, but that it is approximately one half.

As in the main text, I assume that there are only two possible values for the coin's spin time τ, namely, t_1 and t_2. The probability of the outcome heads, abbreviated as e, may then be written as follows:

$$\text{cprob}(e) = \text{cprob}(et_1) + \text{cprob}(et_2)$$

where $\text{cprob}(et_1)$ is the probability that the coin lands heads *and* that the duration time of the spin is t_1, and $\text{cprob}(et_2)$ is the corresponding probability for t_2. (Note that these are not conditional probabilities; the treatment in this section, unlike the treatment in the main text, does not rely on the notion of conditional probability.)

From the two-variable evolution function $h_e(\omega, \tau)$, define the one-variable evolution function $h_e[t_1](\omega)$ as follows:

$$h_e[t_1](\omega) = h_e(\omega, t_1).$$

The one-variable function has as its domain all possible values of ω. (This is what I call in section 2.C2 a restricted evolution function.)

Define the one-variable ic-density $q[t_1](\omega)$ in the same way, so that

$$q[t_1](\omega) = q(\omega, t_1).$$

Note that $q[t_1](\omega)$ is not a probability density, as it does not sum to one, but rather sums to $Q(t_1)$, the probability that t_1 is the duration of the toss. (This is what I call in section 2.C2 a restricted ic-density.) A restricted evolution function and ic-density are constructed for t_2 in exactly the same way.

Before going on, let me recap the two main claims made about the restricted evolution functions for the tossed coin in the main text. First, $h_e[t_1](\omega)$ and $h_e[t_2](\omega)$ are both microconstant. Second, they have the same strike ratio, namely, 1/2. Writing the strike ratio of $h_e[t_1](\omega)$ as $\text{srat}[t_1](e)$ and that of $h_e[t_2](\omega)$ as $\text{srat}[t_2](e)$, then,

$$\text{srat}[t_1](e) = \text{srat}[t_2](e) = \text{srat}(e) = 1/2.$$

Now I proceed with the derivation of the one-half probability of heads. From the definition of complex probability, it follows that

$$\text{cprob}(et_1) = \int_V h_e[t_1](\omega)\, q[t_1](\omega)\, d\omega$$

where V is the set of all possible values of ω.

Because $h_e[t_1](\omega)$ is microconstant, if $q[t_1](\omega)$ is macroperiodic, then

$$\text{cprob}(et_1) = \text{srat}[t_1](e) \int_V q[t_1](\omega)\, d\omega$$
$$= \text{srat}[t_1](e)\, Q(t_1)$$

where $\text{srat}[t_1](e)$ is, as above, the strike ratio of $h_e[t_1](\omega)$, and $Q(t_1)$ is the probability of a spin time of t_1. Applying the same reasoning to $h_e[t_2](\omega)$,

$$\text{cprob}(et_2) = \text{srat}[t_2](e)\, Q(t_2).$$

Since both $\text{srat}[t_1](e)$ and $\text{srat}[t_2](e)$ are equal to $\text{srat}(e)$,

$$\text{cprob}(e) = \text{srat}(e)\, Q(t_1) + \text{srat}(e)\, Q(t_2)$$
$$= \text{srat}(e)(Q(t_1) + Q(t_2))$$
$$= \text{srat}(e)$$
$$= 1/2$$

as desired.

The microconstancy of the coin is due to the microconstancy of the restricted evolution functions $h_e[t_1](\omega)$ and $h_e[t_2](\omega)$. This suggests that the proper conception of the optimal constant ratio partition for the tossed coin would take it to be the union of optimal constant ratio partitions for the two restricted evolution functions. But this is not quite right, because both sets are partitions of ω alone, whereas the optimal constant ratio partition ought to be a partition not just of the range of possible values for ω, but of the range of the possible values for ω and τ (besides which, the proposed "partition" would contain intersecting sets). The obvious fix is to take the Cartesian product of each optimal constant ratio partition with its particular value of τ. Thus if $\mathcal{U}$ is an optimal constant ratio partition for $h_e[t_1](\omega)$ and $\mathcal{V}$ is an optimal constant ratio partition for $h_e[t_2](\omega)$, then an optimal constant ratio partition for the entire experiment is

$$(\mathcal{U} \times \{t_1\}) \cup (\mathcal{V} \times \{t_2\}).$$

Then every possible ordered pair of IC-values (ω, τ) will fall into some member of this partition, as desired (remembering that there are only two possible values for τ). This notion of a constant ratio partition is generalized in definition 2.12.

What is unusual about this partition is that its members are, as it were, infinitely thin in one direction, namely, the direction of the τ axis. This is true not just in the special case under discussion, where there are only two possible duration times for a spin, but in any case where time is an eliminable variable and the optimal constant ratio partition is constructed as suggested above.

One place where this thinness looks odd is in the definition of an effective IC-variable. (The remainder of the discussion assumes that the reader is familiar with the material on effective IC-values in section 2.7.) The effective IC-values of a microconstant experiment are defined so as to correspond to maximal contiguous sets of IC-values within an optimal constant ratio partition member all mapping to the same outcome. If the members of an optimal constant ratio partition are infinitely thin, then so are the sets corresponding to effective IC-values. For example, in the case of the tossed coin, an effective IC-value will correspond to a set of the form

$$\{ (\omega, \tau) : \omega \in [x, y], \tau = t \}$$

where $[x, y]$ is some interval of values of ω and t is either t_1 or t_2.

Suppose that we know the effective IC-value of a given coin toss, and so we know that the full IC-value falls into some particular set of this form. Then we have high-level information about the speed of the toss, but *exact* information about the duration. More generally, effective-level information about a microconstant experiment with eliminable IC-variables will be high-level information about the values of IC-variables in the strike set but complete information about IC-variables not in the strike set. Curious though it may sound, this conclusion in no way compromises the usefulness of the notion of effective information, because values of IC-variables not in the strike set make no difference to the distribution of the outcomes of an experiment, and so may be ignored. (This is particularly clear in the independence result discussed in section 3.45.) For all practical purposes, then, effective-level information may be understood as being simply high-level information about the values of the IC-variables in the strike set.

The approach to multivariable experiments laid out above can be generalized as follows.

What I have done with the tossed coin is, in effect, to look at the properties of the coin's IC-density and its evolution function for heads along lines parallel to the spin speed axis, that is, along all the lines characterized by the following family of the equations:

$$\tau = c$$

where c is a constant. But the mathematical technique I used to derive the equality of the probability and the strike ratio for heads would be equally applicable if the behavior of the IC-density and evolution function were examined along any other set of lines that partitioned the coin toss's domain.

For example, one might choose the lines characterized by the following family of equations:

$$\tau = \omega + c.$$

If the one-variable evolution functions obtained by restricting the coin's full evolution function to each of these lines are all microconstant with a strike ratio of one half (as they are), then, if the IC-density is macroperiodic along the same lines, it follows that the probability for heads is equal to the strike ratio, one half.

This result is of no use in the scenario described above, where τ can take on only two possible values, since no line not parallel to the ω axis belongs to the experiment's domain. But there are many other scenarios where dividing the coin's domain into lines not parallel to either axis might be very useful, a simple example being a case in which the possible values for the spin time τ depend on the spin speed ω of a particular toss (say, if τ is always $\omega - d$ or $\omega + d$ for any toss and a constant d).[33]

The formal treatment of multivariable experiments given in section 2.C2 assumes (confining my attention here to the two IC-variable case) that the lines along which microconstancy and macroperiodicity are assessed are parallel to the axis of one of an experiment's IC-variables. The technique just described can be subsumed under this formal treatment by reparametrizing an experiment so that it has two new IC-variables, one with an axis parallel to and one with an axis perpendicular to the lines into which the experiment's domain is to be divided. The assumption that the lines are parallel to one of the ic-variable axes is not, then, a genuine limitation on the scope of the proofs. Something like a reparametrization of this sort is used in the discussion of statistical physics in chapter four; see note 23 of chapter 4.

2.C Proofs

2.C1 Single-Variable Experiments

I will begin by defining the notions of a probabilistic experiment, a complex probability, a strike ratio, and a constant ratio partition.

Definition 2.1 A *one-variable, deterministic probabilistic experiment* is a triple $\langle q(\zeta), S, H \rangle$ consisting of (a) a probability density $q(\zeta)$, having some appropriate subset of the real numbers as its domain; (b) a non-empty set S, the set of designated outcomes; and (c) a set H containing one function $h_e(\zeta)$ for every member e of S, such that for every e in S,

1. $h_e(\zeta)$ has the same domain as $q(\zeta)$,
2. The range of $h_e(\zeta)$ is $\{0, 1\}$, and
3. The set of values of ζ for which $h_e(\zeta)$ is equal to one is measurable.

The density $q(\zeta)$ is the experiment's *ic-density*; the functions $h_e(\zeta)$ are the experiment's *evolution functions*. The common domain of the ic-density and the evolution functions I simply call the experiment's *domain*.

I require that the domain of a probabilistic experiment exclude all ic-values that are not physically possible. This requirement is necessary because, without it, any experiment with a finite range of possible initial conditions will count as being approximately macroperiodic, since the density will be zero, hence uniform, almost everywhere. Note that the set of designated outcomes does no real work, being simply an index set for the evolution functions.

Definition 2.2 Let $X = \langle q(\zeta), S, H \rangle$ be a one-variable, deterministic probabilistic experiment, and e an outcome in S. The *complex probability* of e on X is

$$\text{cprob}(e) = \int_V h_e(\zeta) q(\zeta) \, d\zeta$$

where V is the domain of X.

Definition 2.3 Let $X = \langle q(\zeta), S, H \rangle$ be a one-variable, deterministic probabilistic experiment, e an outcome in S, and U a measurable subset of the domain of X. Then the *strike ratio* for e over U, written $\text{srat}_U(e)$, is the proportion of U for which $h_e(\zeta) = 1$, that is

$$\text{srat}_U(e) = \int_U h_e(\zeta)\, d\zeta \Big/ M(U)$$

where $M(U)$ is the measure of U. The strike ratio for e over the entire function, written $\text{srat}(e)$, is the strike ratio over the domain of X.

A strike ratio is well defined only for regions of non-zero measure. In what follows, it is assumed that all regions for which strike ratios are assessed have non-zero measure.

Definition 2.4 Let $X = \langle q(\zeta), S, H \rangle$ be a one-variable, deterministic probabilistic experiment, and e an outcome in S. A *constant ratio partition* for e is a partition $\mathcal{U}$ of the domain of X into measurable subsets with non-zero measure such that

1. Every set in $\mathcal{U}$ is contiguous,
2. Every set in $\mathcal{U}$ has the same strike ratio.

I can now give a definition of microconstancy. The evolution function for a designated outcome e is microconstant just in case there is a constant ratio partition for e of which every member is micro-sized.

This is not a formal definition, because it invokes the informal notion of a micro-sized set. Rather than make any explicit references to a standard for what is micro-sized, my strategy in the proofs is to use the ic-density as an implicit standard; thus, a partition is micro-sized just in case the relevant ic-density is macroperiodic relative to that partition. In the statement of theorem 2.2, then, I assume the existence of a constant ratio partition relative to which the ic-density is macroperiodic, without using the terms *micro-sized* or *microconstant* explicitly.

The following result is obvious but important.

Proposition 2.1 Let $X = \langle q(\zeta), S, H \rangle$ be a one-variable, deterministic probabilistic experiment, e an outcome in S, and $\mathcal{U}$ a constant ratio partition for e. Then the strike ratio for e over each member of $\mathcal{U}$ is equal to the strike ratio over *the function as a whole. That is, for any U in U,*

$$\mathrm{srat}_U(e) = \mathrm{srat}(e).$$

Proof. Let V be the domain of X, hence the domain of $h_e(\zeta)$. Let $\mathrm{srat}_U(e) = c$ for all U in $\mathcal{U}$. Then

$$\mathrm{srat}(e) = \int_V h_e(\zeta)\,d\zeta \Big/ M(V)$$

$$= \sum_{U \in \mathcal{U}} \left(\int_U h_e(\zeta)\,d\zeta \right) \Big/ M(V)$$

$$= \sum_{U \in \mathcal{U}} \left(\mathrm{srat}_U(e)\,M(U) \right) \Big/ M(V)$$

$$= c \sum_{U \in \mathcal{U}} M(U) \Big/ M(V)$$

$$= c$$

as desired. ∎

Definition 2.5 A probability density $q(\zeta)$ is *macroperiodic* relative to a partition $\mathcal{U}$ of its domain just in case it is uniform over every set in $\mathcal{U}$. That is, for each U in $\mathcal{U}$, there must exist a constant k_U such that, for every z in U, $q(z) = k_U$.

The formal definition of macroperiodicity is too stringent to be very useful in itself. (It is for this reason that the definition of a macroperiodic density in the main text requires only that the density be *roughly* uniform over every set in the partition.) It is very important, then, that the theorems in this study that are stated using the notion of macroperiodicity hold approximately true when macroperiodicity is only approximate. For theorem 2.2, this is established in approximations 2.2.1 and 2.2.2.

The central theorem of this chapter is, as noted above, here stated without any overt reference to microconstancy. This is due to the relativity of the micro/macro distinction. If what is micro-sized is determined by what is macroperiodic, then it will be seen that any evolution function satisfying the conditions of application of the theorem will count as microconstant. The proof sketch offered in the main text (section 2.23) uses conditional probabilities; here I do things differently in order to postpone the introduction of the notion of conditional probability (for which, see section 3.A).

Theorem 2.2 *If the IC-density $q(\zeta)$ of a one-variable, deterministic probabilistic experiment $X = \langle q(\zeta), S, H \rangle$ is macroperiodic relative to a constant ratio partition for an outcome e in S, then the complex probability of e is equal to the strike ratio for e. That is, $\mathrm{cprob}(e)$ is equal to $\mathrm{srat}(e)$.*

Proof. Suppose that $\mathcal{U}$ is a partition (into measurable sets) of the domain of X. For any U in $\mathcal{U}$, define $\mathrm{cprob}_U(e)$ as the contribution made to $\mathrm{cprob}(e)$ by U, that is, as

$$\mathrm{cprob}_U(e) = \int_U h_e(\zeta) q(\zeta)\, d\zeta.$$

Then the relation between $\mathrm{cprob}_U(e)$ and $\mathrm{cprob}(e)$ is:

$$\mathrm{cprob}(e) = \sum_{U \in \mathcal{U}} \mathrm{cprob}_U(e).$$

Now suppose that $\mathcal{U}$ is the constant ratio partition for e with respect to which $q(\zeta)$ is macroperiodic. For any U in $\mathcal{U}$,

$$\mathrm{cprob}_U(e) = \int_U h_e(\zeta) q(\zeta)\, d\zeta$$

$$= k_U \int_U h_e(\zeta)\, d\zeta \qquad \text{(macroperiodicity)}$$

$$= k_U \, \mathrm{srat}(e) \, M(U) \qquad \text{(proposition 2.1)}$$

where $M(U)$ is the measure of U.

So, writing $Q(U)$ for the probability of U, that is, for $\int_U q(\zeta)\, d\zeta$,

$$\text{cprob}(e) = \sum_{U \in \mathcal{U}} \text{cprob}_U(e)$$

$$= \sum_{U \in \mathcal{U}} k_U \, \text{srat}(e) \, M(U)$$

$$= \text{srat}(e) \sum_{U \in \mathcal{U}} k_U M(U)$$

$$= \text{srat}(e) \sum_{U \in \mathcal{U}} Q(U)$$

$$= \text{srat}(e)$$

as desired. ■

Approximation 2.2.1 The definition of macroperiodicity can be relaxed so as to include functions that are approximately, rather than exactly, uniform over each member of a partition. The definition of a constant ratio partition (and hence of microconstancy) can be relaxed so as to include partitions in which the strike ratios of the members of the partition are approximately, rather than exactly, equal. The correspondingly relaxed theorem asserts only the approximate equality of the probability and the strike ratio. The approximate equality is one of proportion; that is, the difference between the strike ratio and the probability is always a small proportion of the whole. How small? That is determined by the deviation from uniformity or equality, respectively. Note that there will be some tendency for deviations to cancel one another out, so that the correspondence of strike ratio and probability may be better than one would expect.

Approximation 2.2.2 The definition of macroperiodicity can be relaxed so as to include functions that are uniform over almost all, but not all, members of the partition. The definition of a constant ratio partition (and hence of microconstancy) can be relaxed so as to include constant ratio partitions in which the strike ratios of the members of the partition are almost all, but not all, equal. In both cases, call the exceptional members of the partition **bad regions**. The effect of bad regions on the equality of the probability and the strike ratio will depend on the probability of the bad regions themselves, that is, on the proportion of the ic-density that spans the bad regions. The smaller the probability of the bad regions, the smaller their disruptive influence on the equality. Thus the probability will approximately equal the strike ratio provided that the bad regions are improbable, that is, provided that there is only a very low probability that an ic-value will fall into a bad region.

2.C2 Multivariable Experiments

I now prove a version of theorem 2.2 for multivariable experiments. I strongly recommend that the reader, before continuing, look at the treatment of the tossed coin in section 2.B, which will help to make clear the purpose of the rather abstract definitions that follow.

I begin by defining the notions of a probabilistic experiment and a complex probability for the multivariable case.

Definition 2.6 A *k-variable, deterministic probabilistic experiment* is a triple $\langle q(\zeta_1, \ldots, \zeta_k), S, H \rangle$ consisting of (a) a joint probability density $q(\zeta_1, \ldots, \zeta_k)$; (b) a non-empty set S, the set of designated outcomes; and (c) a set H containing one function $h_e(\zeta_1, \ldots, \zeta_k)$ for every member e of S, such that, for every e in S, conditions (1) to (3) of definition 2.1 hold. As in the one variable case, the common domain of the ic-density and the evolution functions is called the *domain* of the experiment.

Again, I require that the domain of a probabilistic experiment exclude all impossible ic-values.

Definition 2.7 Let $X = \langle q(\zeta_1, \ldots, \zeta_k), S, H \rangle$ be a k-variable, deterministic probabilistic experiment, and e an outcome in S. The *complex probability* of e on X is

$$\text{cprob}(e) = \int \cdots \int_V h_e(\zeta_1, \ldots, \zeta_k) \, q(\zeta_1, \ldots, \zeta_k) \, d\zeta_1 \ldots d\zeta_k$$

where V is the domain of X.

I now set up concepts of microconstancy and macroperiodicity that play a role in proving the central theorem, theorem 2.3. In what follows, I divide the ic-variables of a multivariable experiment into two sets. I assign an arbitrary set of values to the ic-variables in the first set, and examine the behavior of the evolution function and ic-density over the second set. To this end, I introduce the following notation.

Notation. Let $f(\zeta_1, \ldots, \zeta_k)$ be a function with k arguments. Assign the values $a_{j+1}, \ldots, a_k$ to the arguments $\zeta_{j+1}, \ldots, \zeta_k$, so that $\zeta_{j+1} = a_{j+1}$ and so on. Then $f[a_{j+1}, \ldots, a_k]$ is a function with j arguments defined as follows:

$$f[a_{j+1}, \ldots, a_k](\zeta_1, \ldots, \zeta_j) = f(\zeta_1, \ldots, \zeta_j, a_{j+1}, \ldots, a_k).$$

That is, $f[a_{j+1}, \ldots, a_k]$ is just f with $\zeta_{j+1}, \ldots, \zeta_k$ held constant. I call $f[a_{j+1}, \ldots, a_k]$ a *restricted* version of f, since $f[a_{j+1}, \ldots, a_k]$ is, in a sense, f restricted to that part of f's domain in which $\zeta_{j+1} = a_{j+1}$ and so on. But only in a sense: it is important to remember that a restricted function has fewer arguments than the original.

In order that the values $a_{j+1}, \ldots, a_k$ need not be spelled out every time a restricted function is mentioned, I will write them as an ordered set a, so that $f[a](\zeta_1, \ldots, \zeta_j)$ is shorthand for $f[a_{j+1}, \ldots, a_k](\zeta_1, \ldots, \zeta_j)$. I call the ordered set a a *restriction* of $\zeta_{j+1}, \ldots, \zeta_k$. I also sometimes abbreviate $\zeta_1, \ldots, \zeta_j$ as ζ, allowing the relatively compact notation $f[a](\zeta)$ for $f[a_{j+1}, \ldots, a_k](\zeta_1, \ldots, \zeta_j)$.

The notation is now put to use in defining restricted versions of an experiment's evolution functions and ic-density.

Definition 2.8 Let $X = \langle q(\zeta_1, \ldots, \zeta_k), S, H \rangle$ be a k-variable, deterministic probabilistic experiment. A *restricted ic-density* of X over a subset $\zeta_1, \ldots, \zeta_j$ of X's ic-variables is X's ic-density restricted by assigning fixed values to $\zeta_{j+1}, \ldots, \zeta_k$. A *restricted evolution function* of X over $\zeta_1, \ldots, \zeta_j$ is a similarly restricted evolution function of X. A restricted ic-density is written $q[a](\zeta_1, \ldots, \zeta_j)$ and a restricted evolution function for e is written $h_e[a](\zeta_1, \ldots, \zeta_j)$, where a is the restriction, that is, the set of fixed values assigned to $\zeta_{j+1}, \ldots, \zeta_k$.

That the arguments of a restricted evolution function or ic-density are the first j ic-variables of the experiment does not involve any real loss of generality, as the ic-variables may be reordered as necessary.

The goal of the next few definitions is to work towards notions of microconstancy and macroperiodicity for a multivariable experiment that are relative to a subset of the experiment's ic-variables. The notion of microconstancy will, as in the one variable case, remain informal. I begin by generalizing the notions of a strike ratio and of a constant ratio partition to functions of more than one argument, and in particular, to restricted evolution functions.

Definition 2.9 Let $X = \langle q(\zeta_1, \ldots, \zeta_k), S, H \rangle$ be a k-variable, deterministic probabilistic experiment, and e an outcome in S. The *strike ratio* for a restricted evolution function $h_e[a](\zeta_1, \ldots, \zeta_j)$ of X over a measurable subset U of the domain of $h_e[a]$ is

$$\int \cdots \int_U h_e[a](\zeta_1, \ldots, \zeta_j) \, d\zeta_1 \ldots d\zeta_j \Big/ M(U)$$

where $M(U)$ is the measure of U. The strike ratio for the entire restriction a, written $\mathrm{srat}[a](e)$, is the strike ratio over the entire domain of $h_e[a]$.

As with the one variable definition of a strike ratio, it is assumed that U has non-zero measure. Note that this measure is relative to $\zeta_1, \ldots, \zeta_j$, not to $\zeta_1, \ldots, \zeta_k$, else $M(U)$ would, in non-exceptional cases, be zero. That is, $M(U) = \int_U d\zeta_1 \ldots d\zeta_j$.

Definition 2.10 Let $X = \langle q(\zeta_1, \ldots, \zeta_k), S, H \rangle$ be a k-variable, deterministic probabilistic experiment, and e an outcome in S. A *constant ratio partition* for any restricted evolution function $h_e[a](\zeta)$ of X is a partition $\mathcal{U}$ of the domain of $h_e[a]$ into measurable subsets with non-zero measure such that

1. Every set in $\mathcal{U}$ is contiguous,
2. Every set in $\mathcal{U}$ has the same strike ratio.

This is identical to the definition of a constant ratio partition for a one-variable evolution function.

A definition of a strike ratio and of a constant ratio partition for a multivariable experiment's full evolution functions may be obtained by letting $j = k$ in definition 2.9 and by letting $\zeta = \zeta_1, \ldots, \zeta_k$ in definition 2.10. But I offer a somewhat different definition below, modeled on the constant ratio partition suggested for the tossed coin in section 2.B.

Definition 2.11 Let $X = \langle q(\zeta_1, \ldots, \zeta_k), S, H \rangle$ be a k-variable, deterministic probabilistic experiment, and e an outcome in S. Then X has a *constant strike* *ratio* over a subset of X's ic-variables ζ just in case for every restriction a, the strike ratio of the restricted evolution function $h_e[a](\zeta)$ is the same. This ratio, when it exists, is called X's strike ratio for e over ζ, written $\mathrm{srat}[\zeta](e)$.

A multivariable experiment is microconstant over a subset of its ic-variables ζ just in case (a) it has a constant strike ratio over ζ, and (b) for every restriction a, $h_e[a](\zeta)$ has a micro-sized constant ratio partition, that is, for every restriction a, $h_e[a](\zeta)$ is microconstant. (As in the one-variable case, there is no formal definition of microconstancy.) When I say that a multivariable experiment is microconstant, I mean that it is microconstant over some subset of its ic-variables.

This definition of microconstancy, and my earlier treatment of the tossed coin (section 2.B), suggest a conception of a constant ratio partition different from that presented in definition 2.10. On this new conception, a constant ratio partition for an entire experiment is, roughly, the union of constant ratio partitions for each of its restricted evolution functions. The definition is, of course, relative to that subset of the experiment's ic-variables with respect to which the evolution functions are restricted.

The partition is defined so as to be a partition of the full space of ic-variables for the experiment; thus, the members of the partition explicitly represent with respect to which restriction they form a constant ratio partition for a particular restricted evolution function. The idea, then, is to find a constant ratio partition for each restriction a, to take the Cartesian product of this partition with its a, and to put together the sets obtained in this way to form one big partition.

Definition 2.12 Let $X = \langle q(\zeta_1, \ldots, \zeta_k), S, H \rangle$ be a k-variable, deterministic probabilistic experiment, and e an outcome in S. Suppose that X has a constant strike ratio over a subset ζ of its ic-variables. Then a *constant ratio partition* for e on X relative to ζ is a set $\mathcal{U}$ constructed as follows:

$$\mathcal{U} = \bigcup_a (\mathcal{U}_a \times a)$$

where $\mathcal{U}_a$ is a constant ratio partition for the restricted evolution function $h_e[a](\zeta)$, and where the union is taken for every restriction a of the ic-variables not in ζ.

An experiment is defined to have a constant ratio partition on ζ, note, only if there is a constant strike ratio over ζ.

Definition 2.13 If an experiment has a constant strike ratio for an outcome e over a subset of X's ic-variables ζ, the set ζ is said to be a *strike set* for e on X.

The definition of macroperiodicity is now extended in the obvious way to the multivariable case.

Definition 2.14 A multivariable probability density $q(\zeta_1, \ldots, \zeta_k)$ is *macroperiodic* relative to a partition $\mathcal{U}$ of its domain just in case it is uniform over every set in $\mathcal{U}$.

Theorem 2.3 *Let $X = \langle q(\zeta_1, \ldots, \zeta_k), S, H \rangle$ be a k-variable, deterministic probabilistic experiment, and e an outcome in S. If for some subset $\zeta = \zeta_1, \ldots, \zeta_j$ of X's ic-variables,*

1. *ζ is a strike set for X, that is, X has a constant strike ratio srat$[\zeta](e)$ for e over ζ, and*
2. *For every restriction a of the variables not in ζ, there exists a constant ratio partition of $h_e[a](\zeta)$ over which $q[a](\zeta)$ is macroperiodic,*

then the complex probability of e is equal to the strike ratio for e over ζ. That is, cprob(e) is equal to srat$[\zeta](e)$.

Proof. The proof uses the following result: provided the conditions stated in the theorem hold,

$$\int \cdots \int h_e[a](\zeta_1, \ldots, \zeta_j) \, q[a](\zeta_1, \ldots, \zeta_j) \, d\zeta_1 \ldots d\zeta_j$$

$$= \mathrm{srat}[\zeta](e) \int \cdots \int q[a](\zeta_1, \ldots, \zeta_j) \, d\zeta_1 \ldots d\zeta_j.$$

I omit the argument for this claim, as it is exactly analogous to the proof of theorem 2.2.

To simplify the expressions, I have omitted the bounds on the integrals; they should be understood as definite integrals over the entire domain of the relevant functions.

$$\text{cprob}(e) = \int \cdots \int h_e(\zeta_1, \ldots, \zeta_k) \, q(\zeta_1, \ldots, \zeta_k) \, d\zeta_1 \ldots d\zeta_k$$

$$= \int \left(\int \cdots \int h_e[a](\zeta_1, \ldots, \zeta_j) \, q[a](\zeta_1, \ldots, \zeta_j) \, d\zeta_1 \ldots d\zeta_j \right) da$$

$$= \int \left(\text{srat}[\zeta](e) \int \cdots \int q[a](\zeta_1, \ldots, \zeta_j) \, d\zeta_1 \ldots d\zeta_j \right) da$$

$$= \text{srat}[\zeta](e) \int \cdots \int q[a](\zeta_1, \ldots, \zeta_j) \, d\zeta_1 \ldots d\zeta_j \, da$$

$$= \text{srat}[\zeta](e) \int \cdots \int q(\zeta_1, \ldots, \zeta_k) \, d\zeta_1 \ldots d\zeta_k$$

$$= \text{srat}[\zeta](e)$$

as desired. ∎

Approximation 2.3.1 With respect to approximation, there is nothing of substance to add to the comments made concerning theorem 2.2 (approximations 2.2.1 and 2.2.2). In the multivariable case, microconstancy and macroperiodicity can be approximate in the same sorts of ways, with the same sorts of consequences.

The next proposition is, in a way, a multivariable version of proposition 2.1.

Proposition 2.4 *If ζ is a strike set for a multivariable experiment X, then* $\text{srat}[\zeta](e)$ *is equal to* $\text{srat}(e)$, *the strike ratio over the full evolution function for X.*

Proof. Let $\zeta_1, \ldots, \zeta_k$ be X's ic-variables. Let V be the domain of X. Let A be the set of all possible restrictions a of the ic-variables not in ζ, and C_a the set of all possible values for ζ given a particular restriction a, that is, the domain of $h_e[a](\zeta)$. Then

$$\mathrm{srat}(e) = \int \cdots \int_V h_e(\zeta_1, \ldots, \zeta_k) \, d\zeta_1 \ldots d\zeta_k \Big/ M(V)$$

$$= \int_A \int_{C_a} h_e[a](\zeta) \, d\zeta \, da \Big/ M(V)$$

$$= \int_A \mathrm{srat}[\zeta](e) \, M(C_a) \, da \Big/ M(V)$$

$$= \mathrm{srat}[\zeta](e) \int_A M(C_a) \, da \Big/ M(V)$$

$$= \mathrm{srat}[\zeta](e)$$

as desired. ■

Corollary 2.5 *If ξ and η are each strike sets for a multivariable experiment X, then* $\mathrm{srat}[\xi](e)$ *is equal to* $\mathrm{srat}[\eta](e)$.

Proof. By proposition 2.4,

$$\mathrm{srat}[\xi](e) = \mathrm{srat}(e) = \mathrm{srat}[\eta](e)$$

as desired. ■

3

The Independence
of Complex Probabilities

The importance of stochastic independence was made clear in section 1.2: enion probability analysis can account for the simple behavior of complex systems only if enion probabilities are largely or completely independent of one another. The independence requirement is the second part of what I call the probabilistic supercondition. Later (section 4.4), it will emerge that an understanding of independence is crucial not only for establishing the independence of enion probabilities but also for establishing their microconstancy, hence, for establishing their satisfaction of the first part of the supercondition. An understanding of stochastic independence, then, lies at the heart of the explanation of complex systems' simple behavior.

This chapter examines the conditions under which outcomes produced by complex probabilistic experiments, and in particular, by microconstant experiments, are stochastically independent. It turns out that there is no single set of conditions that is necessary and sufficient for independence. As expected, causal independence is more or less sufficient for stochastic independence (section 3.3), but there are cases where it is not necessary (sections 3.4, 3.6, and 3.7). Because there is very little causal independence in complex systems, the conditions under which stochastic independence exists despite causal interaction are of great interest for the vindication and explanation of EPA.

The main goal of what follows is to establish some very general results about stochastic independence. But the best way to understand independence in any particular type of complex system will always involve some tricks and techniques that are specific to that type of system. Thus, almost as important as proving general results about independence is developing an understanding of the kinds of properties of mechanisms that underlie independence. This chapter, especially section 3.7, is therefore rather longer than the proofs and arguments strictly require; as well as offering a better appreciation of the reasons that my conclusions hold, the additional material will, I hope, help the reader to see both how the results might be generalized, and how they might be tailored to specific kinds of systems.

3.1 Stochastic Independence and Selection Rules

Two events a and b are stochastically independent just in case the probability of a is unaffected by conditionalizing on b, that is, just in case

$P(a|b) = P(a).$

(assuming that $P(b) > 0$). This relation holds just in case

$P(ab) = P(a)P(b)$

where $P(ab)$ is the probability of both a and b occurring. To avoid certain complications involved in defining conditional probability (see section 3.A), I will use this latter formulation as my principal definition of stochastic independence.

Independence between three or more events can be defined in a similar way, although there is something of a combinatorial explosion in spelling out the conditions. For example, the events a, b, and c are independent just in case

$P(ab) = P(a)P(b)$
$P(ac) = P(a)P(c)$
$P(bc) = P(b)P(c)$
$P(abc) = P(a)P(b)P(c).$

To keep things simple, in this chapter I will consider only the independence of pairs of events. It will be quite clear how to extend the treatment to larger sets of events.

I have begun by speaking of independence as a relation between events, but for the reasons given in note 11 of chapter 1, it is just as natural, and perhaps better, to regard independence as a relation between probabilities. This introduces a complication, however (which is in fact unavoidable on any way of talking). Complex probabilities are attached to types of outcomes on types of setups, rather than to particular outcomes of particular trials, but independence cannot be, in the first instance, a relation holding between the probabilities of outcomes of experiment types, at least as experiment types are normally individuated. This is because, given two experiment types X and Y, independence may hold for the outcomes of one pair of trials on X and Y but fail for another.

To illustrate this point, consider a die-throwing machine that, when triggered, imparts the same randomly selected velocity, spin, and so on to two physically identical dice, one red and one green. On any given trial, the two dice will yield the same outcome, but repeated trials will yield the usual probabilistic patterns.

Clearly, the outcomes of simultaneous throws of the red and green dice are not independent. The probability of getting two sixes, for example, is 1/6, not 1/36, and the probability of getting a five and a six is zero. But the outcomes of non-simultaneous throws are independent. The probability that the red die produces a five this throw and the green die produces a six in the next throw, for example, is 1/36.

Although, for a die-throwing machine of the sort just described, independence does not hold for all pairs of trials, it does hold in a systematic way at a level above that of individual pairs of trials. To characterize these sorts of facts about independence, I make use of the notion of a **selection rule**, or *selection* for short. A selection rule picks out certain combinations of trials from some group of experiments. An example is the rule that picks out simultaneous throws of the red and green dice by the same machine, which I will call $s^\dagger$; another is the rule that picks pairs of throws of the red and green dice that are not simultaneous, which I will call s'. Using the notion of a selection rule, it is possible to think of stochastic independence as a relation between probabilities attached to experiment types, provided that the relation is always relativized to a selection rule. For example, the probabilities of outcomes on the red and green dice are independent with respect to s', and not independent with respect to $s^\dagger$.

In many cases, a set of trials is independent relative to every selection or, at least, to every selection that might reasonably figure in the scientific study of some setup or system.[1] Examples include sets of Bernoulli trials, such as repeated tosses of a normal coin. When dealing with such a set, the relativization to a selection is not necessary.

3.2 Probabilities of Composite Events

The definition of independence refers to the probability that two events both occur. I next define complex probabilities of this sort. Both in philosophy and in probability mathematics, it is standard to conceive of the occurrence of two events as itself constituting a third event. I follow this practice, calling such events **composite events**. The only real complication in the definition of probabilities of composite events is the need to incorporate reference to a selection rule.

Consider the claim that the two outcome types e and f, produced by experiments X and Y respectively, are stochastically independent with respect to a selection s, that is, that

$$\text{cprob}(ef) = \text{cprob}(e)\,\text{cprob}(f)$$

for all pairs of trials that satisfy s. I will write this as

$$\text{cprob}(ef_s) = \text{cprob}(e)\,\text{cprob}(f).$$

The term ef_s represents a composite event type. The type is instantiated by any two events of types e and f produced by trials that satisfy s.

In order to define a probability such as $\text{cprob}(ef_s)$ with the least possible additional apparatus, I conceive of the composite event ef_s as the outcome of a single probabilistic experiment. A trial on this experiment, which I will call the **composite experiment** XY_s, consists of a trial on X and a trial on Y, such that the two trials satisfy s (see also definition 3.1). In effect, s is built into the experiment type itself. It is allowed that X is the same experiment type as Y, and that e is the same outcome type as f.

Suppose that X and Y have just one ic-variable each, respectively ξ and η. (Throughout this chapter, I assume that all experiments have only one ic-variable. Generalization to multivariable experiments is straightforward.) Associated with the composite experiment XY_s is an evolution function $h^s_{ef}(\xi, \eta)$ and an ic-density $q^s(\xi, \eta)$. The ic-density is relativized to s because the distribution of pairs of ic-values may vary with the selection rule used to pick out the pairs, as demonstrated by the example of the red and green dice given above. The evolution function is also relativized to s; the extent to which this is necessary will be discussed shortly.

The value of $\text{cprob}(ef_s)$ is defined just as one would expect (see also definition 3.1):

$$\text{cprob}(ef_s) = \iint_V h^s_{ef}(\xi, \eta)\, q^s(\xi, \eta)\, d\xi\, d\eta$$

where V is the set of all possible pairs of values of ξ and η. (Note that s may play a role in determining what pairs of values are possible.) In the remainder of this section, I define $h^s_{ef}(\xi, \eta)$ and $q^s(\xi, \eta)$.

The evolution function $h^s_{ef}(\xi, \eta)$ is defined in the obvious way. A deterministic evolution function takes the value one for pairs of values of ξ and η that produce both e and f outcomes on trials satisfying s, and zero otherwise. An explicitly probabilistic evolution function (defined in section 2.61) takes the value of the probability that a pair of values of ξ and η produces both e and f outcomes.

When the evolution functions of X and Y are deterministic, whether or not e occurs is completely determined by the value of ξ, and whether or not f occurs is completely determined by the value of η, from which it follows that $h^s_{ef}(\xi, \eta) = 1$ just in case $h_e(\xi) = 1$ and $h_f(\eta) = 1$. Equivalently,

$$h^s_{ef}(\xi, \eta) = h_e(\xi)h_f(\eta).$$

Thus $h^s_{ef}(\xi, \eta)$ does not depend on s and so the relativization of the evolution function to the selection may be dropped.

This is true even if X and Y have mechanisms that causally interact. The reason is that the physics of the interaction is already built into the evolution functions for the individual trials. There is the possibility of some confusion here. Suppose that the mechanisms of two coin tosses are not independent, because one coin collides with the other while both are spinning. Then the evolution function for the composite experiment will, as I have claimed, be the product of the evolution functions for the individual coins. But it would be a mistake to think that the evolution functions for the individual coins in this case are the evolution functions for individual coins that do not collide. They are, rather, evolution functions for a coin that collides with another coin. Each such evolution function determines the outcome, heads or tails, of one of the coins, given the spin speeds and times of both coins. Thus each evolution function takes as its IC-values the relevant facts about both coins. The evolution functions for the individual coins in the colliding case will therefore be quite different from the evolution functions for individual coins that do not collide. The details are given in section 3.6, where the case of the colliding coins is discussed at length. Building the effects of the interaction into the evolution functions for the individual experiments makes the physics no easier, but it does make the treatment of independence more straightforward, by simplifying the relationship between the evolution function for the composite experiment and the evolution functions for the individual experiments.

When the evolution functions of X and Y are explicitly probabilistic, then provided that both mechanisms' nomic probabilities are independent, $h^s_{ef}(\xi, \eta) = h_e(\xi)h_f(\eta)$, as in the deterministic case. The distinction between the probabilistic and the deterministic cases may be collapsed by representing the effects of the nomic probabilities as additional ic-variables of X and Y, that is, by representing the composite mechanism using an implicitly rather than an explicitly probabilistic evolution function (see section 2.6). For an implicitly probabilistic evolution function, the relation between the composite and individual evolution functions holds without the proviso. (The independence or otherwise of the nomic probabilities is now represented as a fact about the composite experiment's joint ic-density, rather than as a fact about its evolution function.) For the purposes of this chapter, I propose to represent all probabilistic mechanisms using implicit evolution functions. Then it can simply be said that, regardless of the selection chosen, the evolution function for the composite experiment is the product of the evolution functions for the individual experiments. From this point on I write the composite evolution function $h_{ef}(\xi, \eta)$, without any reference to the selection.

The ic-density $q^s(\xi, \eta)$ represents the distribution of pairs of ic-values of ξ and η that feed composite experiments of type XY_s, that is, the distribution of pairs of ic-values that feed pairs of trials that satisfy s. The interpretation of this distribution will vary with the nature of the distributions over ξ and η. When the ic-densities $q(\xi)$ and $q(\eta)$ are simple, complex, or frequency-based, I suggest the following scheme for interpreting $q^s(\xi, \eta)$.

If $q(\xi)$ and $q(\eta)$ are simple probability densities, then I take it that $q^s(\xi, \eta)$ can be read off from the fundamental laws of nature.

If $q(\xi)$ and $q(\eta)$ are frequency-based densities, then $q^s(\xi, \eta)$ is simply the frequency-based density of pairs of values of ξ and η that feed pairs of trials that satisfy s.

If $q(\xi)$ and $q(\eta)$ are complex probability densities, then $q^s(\xi, \eta)$ is to be interpreted as the relevant complex probability density. The relevant complex probability density will depend on, first, the evolution functions for the experiments that produce values of ξ and η, and second, a joint density over the ic-variables of these experiments. This joint ic-density will itself be relativized to a selection; it will be the distribution over sets of ic-values that give rise to pairs of values of ξ and η that in turn feed trials that satisfy s. The selection is in this way passed up the probabilistic network, so that a complex probability distribution relativized to a selection s has as its foundation a basic joint distribution relativized to what might be called the ancestor of s, in the sense that the selection for the basic density picks out sets of IC-values that are destined, after having traveled down the network, in whatever way, to feed pairs of trials that satisfy s.

If $q(\xi)$ and $q(\eta)$ are different kinds of densities then $q^s(\xi, \eta)$ is constructed from the relevant conditional probabilities, as discussed in section 2.34.

3.3 Causal Independence

The goal of this section is to show that the causal independence of two probabilistic experiments is more or less sufficient for the stochastic independence of the outcomes produced by those experiments. The section has three parts. In the first, I state a condition sufficient for stochastic independence. In the second, I argue that experiments that are, intuitively, causally independent will tend to satisfy this condition. Finally, I propose a formal definition of causal independence, and discuss its merits.

The larger significance of the discussion is, first, that it partly vindicates, in the case of complex probabilities, the intuition that causal independence is sufficient for stochastic independence; second, that it introduces a notion, causal independence, and a style of argument, that will play important roles in the remainder of the discussion of independence; and third, that it emerges that causal independence cannot *explain* stochastic independence. This third consequence of the discussion—a consequence that arises, in a sense, from the weakness of the connection established in what follows between causal independence and stochastic independence—will, I hope, stimulate the reader's appetite for the far more interesting results concerning stochastic independence that I present in the remainder of the chapter.

3.31 A Sufficient Condition for Stochastic Independence

The sufficient condition for stochastic independence is as follows. The outcomes of two experiments are stochastically independent relative to a selection s if the basic IC-densities that found the probabilistic network in which the experiments are embedded are stochastically independent relative to s (in a sense to be made clear shortly).

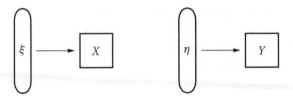

Figure 3.1 Independence in two simple networks.

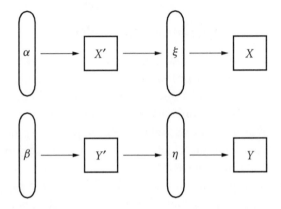

Figure 3.2 Independence in two slightly less simple networks.

To understand why the condition is sufficient for independence, consider first the simple case in which the IC-densities of the two experiments X and Y are themselves basic densities, so that the networks in which X and Y are embedded are the simple networks shown in figure 3.1. It is easy to show that the outcomes of trials on X and Y will be independent relative to a selection s if the IC-densities for X and Y are independent relative to s. (Remember that s in this context selects pairs of IC-values that feed pairs of trials satisfying s.) This observation is formalized as theorem 3.1 in section 3.B1.

Now consider the slightly more complicated case in which the IC-values for X and Y are generated by complex probabilistic experiments that themselves have basic IC-densities, as in figure 3.2. By the reasoning above, the outcomes of trials on X and Y will be independent relative to a selection s if the IC-densities for X and Y are independent relative to s, that is, if values of ξ and η generated by X' and Y' are independent relative to s. By the same style of argument, values of ξ and η are independent relative to s if the two basic

IC-densities, $q(\alpha)$ and $q(\beta)$, are independent relative to s. (In this context, s picks out pairs of values of α and β that feed pairs of trials that produce pairs of values of ξ and η that in turn feed pairs of trials picked out by s.) By iterating this argument, one reaches the more general conclusion that the stochastic independence of a network's basic densities relative to a selection s is a sufficient condition for the independence relative to s of any two trials in the network, as desired.

3.32 Causal Independence and the Sufficient Condition

If two experiments are causally independent, I now argue, then the sufficient condition for stochastic independence stated in the last section will tend to hold. Thus, causally independent experiments will tend to produce stochastically independent outcomes. I will begin by discussing the circumstances under which basic IC-values are independent; causal independence will be explicitly introduced only near the end of the section.

Suppose that you pick out, according to some selection procedure,[2] two IC-values from basic IC-densities. Ought you to expect the probability distribution over choices of pairs such as this to exhibit independence? That is, ought the probability of getting a particular pair to be equal to the probability of getting one of the pair multiplied by the probability of getting the other? There are three cases to consider:

1. Suppose that the values are chosen from two different basic IC-variables. Then, because the IC-variables are basic, the two values have no common causal history,[3] and so, it would seem, ought not to be correlated. Thus it is reasonable to expect independence. (I return to the assumption behind this line of reasoning below.)

2. Suppose that both IC-values are values of the same basic IC-variable, for example, that you choose the spin speeds that a certain croupier imparts to a certain wheel of fortune on two different occasions. (Wheel speed is probably not a basic IC-variable, but never mind.) Then, although the values are values of the same type, they are distinct physical property instances. Because the IC-variable is basic, distinct instances have no common causal history. Thus it would seem that they ought not to be correlated, and so that it is reasonable to expect independence.

3. Suppose that you choose the same IC-value twice, for example, that you choose the spin speed that a certain croupier imparts to a certain wheel

of fortune on one occasion, and then chooses the spin speed imparted on that occasion again. Then the two values chosen must, of course, be equal, and so they will not, except in the degenerate case where there is only one possible spin speed, be independent.

I conclude that two basic IC-values are independent just in case they are distinct, and so that two sets of basic IC-values are stochastically independent just in case they do not overlap.

It follows that two outcomes are independent if their genesis can be traced up the probabilistic network to wholly distinct sets of basic IC-values. To make this condition more succinct, define the *ancestry* of an outcome as the following set of IC-values: the IC-values for the trial that produced the outcome, the IC-values for the trials that produced those IC-values, and so on, up to and including the basic IC-values that lie at the beginning of the process. Then a sufficient condition for the stochastic independence of two outcomes is that they have non-overlapping ancestries.[4]

Two experiments are independent with respect to a selection, then, if the ancestries of pairs of outcomes chosen by the selection never overlap. Consider an example, the case of the die-tossing mechanism described in section 3.1. Recall that the mechanism tosses two dice at a time, in such a way that the dice have identical IC-values. Relative to a selection that picks out simultaneous throws of the dice, the sufficient condition for independence does not hold, because the outcomes of simultaneous throws can each be traced back to the same initial condition of the throwing mechanism. At this point, their ancestries overlap. Relative to a selection that picks out non-simultaneous throws, however, the sufficient condition will presumably be satisfied, since the ancestries of each of the throws will include only initial conditions of the throwing mechanism on different occasions. (Of course, if these distinct initial conditions were effects of a single earlier initial condition, the ancestries would, after all, overlap.)

The causal independence of two trials is sufficient for their outcomes to have non-overlapping ancestries, and thus for their stochastic independence. The argument is made by showing that trials whose outcomes have overlapping ancestries are causally dependent. I assume that the following is a reasonably accurate definition of causal independence: two trials are causally independent just in case there are no common causal influences on their outcomes. Now, any IC-value in the ancestry of an outcome is a causal influence on the outcome, by the very nature of what it is to be an IC-value. If the ancestries of the outcomes of two trials overlap, then, there is a common causal influence on the outcomes, and so the trials are not causally independent. Causal independence, then, implies non-overlapping ancestries and so stochastic independence. Thus if a selection rule picks out pairs of trials on two experiments in such a way that the trials are always causally independent, then the experiments are stochastically independent with respect to that selection.

The argument presented in this section makes one notable assumption: that because different basic ic-values are causally independent, they are stochastically independent. To have to make such an assumption is rather disappointing in two respects.

First, it is a rather strong assumption. At least in the case where basic ic-densities are frequency-based, it is not so hard to imagine the stochastic independence of two basic ic-densities relative to some scientifically important selection rule systematically failing due to a string of coincidences. The assumption can in fact be weakened: as demonstrated by theorem 3.2, independence of outcomes requires only independence of effective ic-values, not of full ic-values. But although this result will prove very important later, it is not helpful here: if coincidence introduces a correlation into a joint distribution of basic ic-values, it seems likely to do so at the level of the effective information concerning those values, and so to correlate the effective ic-values. (The reasons for this are discussed in section 3.5.)

Second, and more importantly, the assumption behind the argument that drives this section is of exactly the same sort as the argument's conclusion, that causal independence is a sufficient condition for stochastic independence. As a consequence, the argument at best provides a rather weak explanation of the connection between causal independence and stochastic independence. It shows how the stochastic independence of a set of basic ic-values is translated into stochastically independent outcomes, but it does not show how new stochastic independence may be created.

Perhaps this is as it should be. Perhaps the connection between causal independence and stochastic independence is a negative one: causal dependence of certain sorts can destroy stochastic independence, but the source of stochastic independence is something altogether different. The results presented in section 3.4 go some way towards bolstering this view. If the view is correct, then the real lesson of this section is that what destroys stochastic independence is the sharing of ic-values (but not all such sharing, as I will show in sections 3.6 and 3.7).

3.33 Causal Independence Defined

Let me pause to suggest the following formal definition of **causal independence**: two experiments are causally independent relative to a selection *s* just in case *s* never selects pairs of trials with overlapping ancestries. Two experiments are not just relatively, but absolutely causally independent, if they are causally independent relative to all selections.[5]

If this definition does not quite capture the intuitive notion of causal independence, it is because the definition counts as causally independent two experiments, one of which influences the other in a way that does not vary with its ic-values. Suppose, for example, that two massive wheels of fortune exert a constant gravitational influence on one another. The influence does not change with the speed of the wheels, as the distribution of mass within a wheel remains the same as it rotates.[6] Thus, regardless of how fast one wheel rotates, the outcome on the other wheel will be the same, but if the one were not present at all, the outcome on the other would perhaps be different.

Ought trials on two such wheels to count as causally independent? If not, my suggested definition is deficient; if so, my definition captures an interesting subtlety in our notion of causal independence. I leave the decision to the reader. I do assert, however, that my definition of causal independence captures that part of our intuitive notion that is of interest in the study of stochastic independence; this is, of course, my reason for presenting it here.

3.4 Microconstancy and Independence

3.41 Microconstancy in Composite Experiments

The results presented in the last section apply to all complex probabilistic experiments. In this section I show that there is a more interesting, because considerably weaker, sufficient condition for the stochastic independence of outcomes of microconstant experiments (the analog, in my approach, of a result first stated by Hopf within the framework of the method of arbitrary functions, for which see section 2.A). For the reasons given in section 3.2, in what follows I assume that all evolution functions are either deterministic or implicitly probabilistic, so that all evolution functions map only to zero and one.

The theoretical backbone of this section is the following fact: the composite experiment for two or more microconstant experiments is itself microconstant, with a strike ratio for any composite event equal to the product of the strike ratios for the individual events (theorems 3.3 and 3.4). In symbols,

$$\mathrm{srat}(ef) = \mathrm{srat}(e)\,\mathrm{srat}(f)$$

an equality which has obvious implications for the independence of e and f. These implications are explored in the next two sections (sections 3.42 and 3.43); in this initial section, I explain the microconstancy result.

The result is, perhaps, best appreciated graphically. Figure 3.3 shows the microconstant evolution functions for outcomes on a pair of one-variable probabilistic experiments (top), and the way that these two evolution functions are

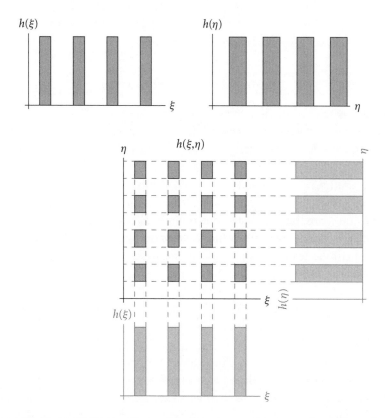

Figure 3.3 Two individual evolution functions $h(\xi)$ and $h(\eta)$ *(top)* and their relation to the corresponding composite evolution function $h(\xi, \eta)$ *(bottom)*. From left to right, that is, along the ξ axis, the composite evolution function oscillates in step with $h(\xi)$, from bottom to top, that is, along the η axis, in step with $h(\eta)$. The strike ratios for the individual evolution functions are, respectively, 1/3 and 1/2; that of the composite function is, as theorem 3.3 requires, 1/6.

put together to make the evolution function for the corresponding composite outcome on the composite experiment (bottom). The composite evolution function simply inherits the constant ratio structure of the two individual evolution functions, putting them together into a two-dimensional function that has the constant ratio structure of one evolution function in one direction, and that of the other in the other direction.

Note that the optimal constant ratio partition for the composite outcome is the set containing the Cartesian products of all members of the optimal constant ratio partition for one of the individual outcomes with all the members of the optimal constant ratio partition for the other. That is, if $\mathcal{U}$ is an optimal constant ratio partition for e and $\mathcal{V}$ is an optimal constant ratio partition for f, then an optimal constant ratio partition for the composite outcome ef is $\{\, U \times V : U \in \mathcal{U}, V \in \mathcal{V} \,\}$. This is in general true.

3.42 Weak Independence

I showed in the last section that a composite experiment made up of two microconstant experiments X and Y is itself microconstant, with a strike ratio equal to the product of the strike ratios for X and Y. In symbols,

$$\mathrm{srat}(ef) = \mathrm{srat}(e)\,\mathrm{srat}(f)$$

where e and f are the designated outcomes of X and Y. Clearly, e and f are stochastically independent if (a) the ic-densities for X and Y are macroperiodic, and (b) the ic-density for the composite experiment is macroperiodic, since, if these conditions obtain, then the probabilities are equal to the strike ratios, so that

$$\mathrm{cprob}(ef) = \mathrm{cprob}(e)\,\mathrm{cprob}(f)$$

as required for independence. Throughout the remainder of this section, I assume that I am dealing with microconstant experiments that have macroperiodic ic-densities. Under these circumstances, then, a sufficient condition for stochastic independence is the macroperiodicity of the composite ic-density.

Let me give this sufficient condition a name. If the joint ic-density, relative to some selection, of two or more macroperiodically distributed ic-variables is itself macroperiodic, I will say that the ic-variables are **weakly independent** relative to that selection. Note that it is a presupposition of the definition that weakly independent IC-variables are macroperiodic; the definition does not apply to variables that are not distributed macroperiodically. The justification for the term *weak independence* is given later in this section.

If two macroperiodic IC-variables are stochastically independent relative to a selection, the composite density must be macroperiodic. But the converse does not hold. Consequently, weak independence is a strictly weaker condition than the sufficient condition for independence stated in section 3.3, stochastic independence of the IC-variables (remembering that this claim is confined to those cases where IC-variables are macroperiodic).

The purpose of this and the next section is to show that weak independence is a useful and illuminating weakening of the independence requirement and, in particular, that it enhances our understanding of the sources of stochastic independence. This section provides a more informal treatment; the next is more formal.

Two variables are independent just in case there is no correlation between the values of one and the values of the other. Two variables are weakly independent, I will show, just in case there is at most a loose correlation between the values of one and the values of the other, in a certain sense of *loose*. All claims about independence and correlation in what follows are relative to a selection, but for economy's sake, I usually omit reference to the selection.

Consider two IC-variables ξ and η. For the two variables to have no correlation at all is for values of the variables to be stochastically independent, so that

$$q(\xi \mid \eta = y) \;=\; q(\xi)$$

and vice versa. For the two variables to be as tightly correlated as possible is for the value of one to determine the other, so that the probability distribution $Q(\xi \mid \eta = y)$ is equal to one for exactly one value x of ξ, and zero for all other values or sets of values not including x. (Such a distribution has no density.) A loose correlation sits somewhere between these two extremes. What I propose to do is to examine the behavior of the conditional density at different levels of correlation. The discussion appeals to our natural expectations about correlations, tight and loose; the aim is to give the reader an intuitive sense of what I mean by a *loose correlation*. The characterization of looseness is derived from a certain conception of the natural way to loosen a correlation. In what follows, I rely at first on the intuitive appeal of this conception. Later in the discussion, the loosening is characterized at a more theoretical level; then, in the next section, the nature of the loosening is fully explained. Those readers who desire precision at once may wish to move on to the next section (section 3.43).

Suppose that ξ and η are almost, but not quite, as tightly correlated as possible. Then a value of η will determine an almost exact value for ξ. I interpret this to mean that the conditional density $q(\xi \mid \eta = y)$ is extremely highly peaked around some particular value for ξ. Now relax the correlation bit by bit. The peak will become less and less sharp. Additional peaks may spring up in other places, where the density was previously zero, but they will also be less sharp than the original peak. At any point, the degree to which the correlation has been relaxed determines the maximum sharpness of the peaks. When the correlation is fully relaxed—that is, when the correlation vanishes—the conditional density is equal to the unconditional density $q(\xi)$. On the assumption that the unconditional density is macroperiodic, the conditional density in a micro-sized locale of the original peak will have gone from highly peaked to flat. Any other peaks that formed partway through the transition will also have been flattened.

There exists a point in this transition from a tight to a non-existent correlation at which the maximum sharpness of a peak becomes just sufficiently blunt that any micro-sized portion of such a peak is approximately flat, though not necessarily equal to the unconditional density. At this point, the conditional density, though it has not yet converged with the unconditional density, has become macroperiodic. Assuming that this has also happened to the conditional density for η, it follows that the joint ic-density has also become macroperiodic (proposition 3.5). At this level of correlation, then, the weak independence condition comes to hold. This is my reason for saying that weak independence corresponds to a degree of correlation partway between tight and non-existent.

To what extent are weakly independent ic-variables still correlated? Although the conditional ic-density is, like the unconditional ic-density, flat over any micro-sized locale, it need not be equal to the unconditional density. Thus the probability that a pair of values of ξ and η falls into a given micro-sized locale U need not equal the product of the probability that ξ falls into U with the probability that η falls into U. Where there is an absence of correlation is within each locale: the probability that ξ falls in one part of U rather than another is independent of the position of η within U.

Since the members of an optimal constant ratio partition are micro-sized locales, weak independence implies that, while on the one hand, the member of such a partition into which ξ falls may be correlated with the member of the partition into which η falls, on the other hand, given that ξ and η fall into a certain member U, their positions within U are uncorrelated. Recall that the information that determines into which member of an optimal constant ratio partition ξ falls is the *non-critical* part of the effective information about ξ, while the information that determines approximately where ξ falls in that constant ratio partition member is the *critical information* about ξ (section 2.73). It would seem, then, that there is a close relationship between the weak independence of two ic-variables and the independence of the critical level information about the two variables. The next section confirms this claim.

Before continuing on to a more formal treatment of weak independence, let me summarize the results presented so far. I have shown that the outcomes of two or more microconstant experiments will be stochastically independent even if the experiments' ic-variables are loosely correlated, provided that they are at most loosely correlated. Microconstant experiments that take loosely correlated ic-values as input, then, are capable of producing completely uncorrelated output, or in other words, microconstant mechanisms have the power to break down loose correlations, and thus to create stochastic independence where there was none before.

This result is significant in two ways. First, it shows that microconstant probabilities may be stochastically independent even if the experiments to which they are attached are not causally independent, provided that the causal dependence induces no more than a loose correlation between the experiments' ic-values. Second, it shows that, even if basic ic-values are not independent, outcomes in networks based on those variables may be independent, because microconstant mechanisms in the network will break down any loose correlations that come their way.

3.43 *Infracritical Independence*

The last section developed the notion of weak independence, important because the weak independence of two or more microconstant experiments' macroperiodically distributed ic-values is sufficient for the independence of the experiments' outcomes. In this section I pursue the suggestion, advanced in the last section, that there is a connection between weak independence and the independence of critical level information.

The connection turns out to be a strong one: if two ic-densities are macroperiodic, then they are weakly independent relative to a selection just in case they are independent at the critical level and below relative to that selection.[7] The part of this result that is more important for my purposes, that independence at the critical level and below entails weak independence, is proved in section 3.B3, theorem 3.12. The other part, that weak independence entails independence at the critical level and below, is proved in section 3.B3, proposition 3.13.

To save words, when two ic-variables are independent at the critical level and below, I say that they have **infracritical independence**. I will begin by making the notion of infracritical independence more precise. (For the formal version, see definition 3.12.)

Recall from section 2.73 that critical-level information about an ic-value determines approximately where in a given member of an optimal constant ratio partition the ic-value falls. **Infracritical information** determines *exactly* where in a given constant ratio partition member the ic-value falls. The infracritical information about an ic-value includes, as the name does not quite convey, the critical information about that value. Thus the information in a full ic-value can be divided into two parts:

1. The non-critical part of the effective information, which determines into which member U of an optimal constant ratio partition the value falls, and
2. The infracritical information, which, given that the value falls into U, determines exactly where in U the value falls.

Whereas effective and critical ic-values are discrete, infracritical ic-values make up a continuum. A formal scheme for representing such information, extrapolated from the formal scheme for representing critical information presented in section 2.73, is developed in section 3.B2; see especially definition 3.4.

Given this formal scheme, infracritical independence is defined as follows (definition 3.12): two ic-variables have infracritical independence relative to a selection s just in case (a) infracritical information about one ic-variable is independent of infracritical information about the other, relative to s, and (b) infracritical information about each variable is independent of any higher-level information about either variable, again relative to s. In short, the infracritical independence of two ic-variables requires that infracritical information about each variable be uncorrelated with any other information about either variable. Because of clause (b), this is perhaps a little stronger than what is informally suggested by the expression *infracritical independence*.

I make four remarks about the notions of infracritical information and infracritical independence.

First, because the critical level is determined by a particular outcome and experiment—more specifically, by a particular optimal constant ratio partition—infracritical independence holds relative to an experiment and outcome. Explicit reference to this relativization is, like explicit reference to the relativization of independence claims to a selection, suppressed in most of the following discussion.

Second, since infracritical independence is relativized to an optimal constant ratio partition, and is equivalent to weak independence, weak independence is also relativized to an optimal constant ratio partition. This was implicit in the use of the notion of macroperiodicity in the definition of weak independence, since macroperiodicity is always relative to a partition, normally an optimal constant ratio partition (section 2.23).

Third, the infracritical information about an ic-value may be divided into two parts: the critical information and the information below the critical level. The latter information is not part of the effective information about an ic-value, and so makes no difference to the outcome of the ensuing trial. Yet infracritical independence requires that this information satisfy certain conditions. Why impose conditions on information that can make no difference to outcomes?

There is, indeed, no good reason, from a purely theoretical point of view, to impose conditions on information below the critical level; it is enough to impose the conditions on the critical information. It can be shown that independence of the critical-level information concerning the ic-values of two microconstant experiments is sufficient for the independence of their outcomes, provided that the ic-densities of the experiments are macroperiodic.[8] Also, critical-level independence can be shown (when all individual ic-densities are macroperiodic) to be equivalent to a notion of weak independence defined not in terms of the macroperiodicity of the joint ic-density, but in terms of the macroperiodicity of the joint distribution of effective ic-values, using the weaker notion of macroperiodicity available for such distributions defined in section 2.72.

The reason for requiring infracritical independence rather than critical level independence is a practical one: it is easier to deal with the mathematics of infracritical independence. The most important, though not the only, reason for this is the difficulty, when developing a formal system for representing critical ic-values, of devising a numbering system that does not inadvertently convey information above the critical level (see the discussion in section 2.73). Dealing with infracritical information, rather than critical information, has the advantage that it is not necessary to presuppose *any* particular scheme for assigning critical-level information. This is because the differences between different schemes, such as the ordinal and the teleological schemes discussed in section 2.73, concern the lower bound of the critical information, not the upper bound. There is, by definition, no lower bound on the infracritical information.

Fourth, it is now possible to explain more exactly what is going on in the transition from tight to loose correlation in section 3.42. When two variables are as tightly correlated as possible, information about two values of those variables is tightly correlated at every level, no matter how low the level. When two variables are very tightly correlated, but not as tightly correlated as possible, information about two values of those variables is correlated at every level except very low levels. That is, there exists a level of information above which all information is correlated, but below which all information is uncorrelated. The transition from tight to loose correlation, as presented in section 3.42, proceeds by moving the border between correlated and uncorrelated information ever higher. That is, the correlation is loosened by making ever higher levels of previously correlated information uncorrelated.

Let me conclude this section by assessing the significance of the results I have presented. This allows me to refine the conclusion to section 3.42 using the more precise terms afforded by the notion of infracritical information.

The IC-variables of two or more microconstant experiments with macroperiodic IC-densities may be correlated above the critical level, yet provided that the variables have infracritical independence, the outcomes of those experiments will be stochastically independent. Thus microconstant experiments have the power to break down correlations above the critical level.

What counts as the critical level is determined by the size of an evolution function's optimal constant ratio partition, that is, its CRI. The smaller the CRI, the tighter the correlation that counts as being above the critical level, and hence the tighter the correlation that can be broken down. Microconstant experiments that have evolution functions with very small CRIs can break down almost any correlation.

As a result, microconstant experiments can produce independence where there was no independence before. Whereas non-microconstant probabilistic experiments can only pass on independence that already exists in the basic IC-variables, microconstant experiments can *create* independence. But I should not overstate the case. Microconstant mechanisms do not so much create independence from nothing, as create the conditions under which independence can spread from the infracritical level to higher levels. A certain amount of low-level independence must be present as a seed, but given such a seed, microconstancy provides the mechanism by which independence may proliferate.

The microconstant mechanisms' contribution to the proliferation of independence can be divided into two parts. First, microconstancy can create completely independent ic-variables from correlated basic ic-variables. In this way, microconstancy creates entirely new independence above the critical level. Second, microconstancy can restore independence where some kind of causal interaction has induced a correlation above the critical level. In this way, microconstancy regenerates independence above the critical level. For both of these reasons, I suggest, microconstancy has a crucial—perhaps *the* crucial—role to play in explaining the stochastic independence that exists in the physical quantities around us.

3.44 True Probability

To benefit from the correlation-destroying power of microconstancy, an experiment need not be microconstant itself. It is sufficient that, somewhere in the probabilistic network between any correlated joint density and the experiment itself, sit microconstant experiments placed so as to break down the correlation.

For this reason, the result stated above is easily extended to true probabilities. A sufficient condition for the independence of the outcomes of any two true probabilistic experiments relative to a selection s is the infracritical independence of the relevant basic ic-densities relative to s, or equivalently, the weak independence of those same basic ic-densities relative to s.

3.45 Strike Sets and Stochastic Independence

I now investigate a different way to weaken the sufficient condition for the independence of outcomes of microconstant experiments, taking advantage of the notion of an eliminable ic-variable, defined in section 2.25. This weakening depends on the result stated in section 3.41, that composites of microconstant experiments are themselves microconstant, but it is not directly related to the weakening developed in sections 3.42 and 3.43. At the end of the present section, I will combine the two kinds of weakening into a single set of sufficient conditions for stochastic independence in microconstant experiments.

In section 2.23 (see also theorem 2.3), I showed that, in order for the probability of an outcome of a multivariable microconstant experiment to equal its strike ratio, it is sufficient that the experiment's ic-density be macroperiodic in the direction of a strike set of its ic-variables—it need not be macroperiodic in all directions. (A strike set of ic-variables for an experiment, recall, is a set in favor of which all other variables in the experiment can be eliminated.)

This suggests that a sufficient condition for independence in microconstant experiments might be stated in terms that refer only to strike sets of ic-variables for the experiments. In what follows, I show that such a condition can be stated, a condition that has as a consequence that two microconstant experiments may be independent even given as much correlation as you like between ic-variables not in the experiments' strike sets.

The key to seeing that this is so is the following, not unexpected, result: ic-variables that are eliminable for a given experiment are also eliminable for any composite experiments of which that experiment is a part. Consequently, the union of two strike sets for any two microconstant experiments is a strike set for the corresponding composite experiment (theorem 3.4). In order to get independence, then, it is not necessary that the composite experiment's ic-density be macroperiodic over all the ic-variables in the individual experiments, but only over strike sets of the individual experiments' ic-variables. Spelling out the details: a sufficient condition for the stochastic independence of the outcomes of two microconstant experiments with macroperiodic ic-densities is the macroperiodicity of the corresponding composite experiment's ic-density over strike sets of ic-variables for the two individual experiments.

The sufficient condition will hold if the ic-variables in a strike set for one experiment are stochastically independent of all the ic-variables for the other experiment, and vice versa. This is weaker than the result stated in section 3.31, because it allows any amount of correlation between the ic-variables not in the strike sets.

The sufficient condition may, of course, be further weakened by applying the results of sections 3.42 and 3.43, that is, by substituting *weakly independent* for *independent*. The stochastic independence of the outcomes of two or more microconstant experiments is not compromised, then, by either

1. Any degree of correlation between ic-variables not in the strike sets for the experiments, or

2. Correlation above the critical level between ic-variables, some or all of which are in the strike sets.

When one or both experiments have more than one strike set, what is required is only that there be some choice of strike sets for which these conditions both hold.

In the remainder of this chapter, I will seek to weaken even this condition, showing that, if the circumstances are right, stochastic independence is possible even when all ic-values are as tightly correlated as you like.

3.5 The Probabilistic Patterns Explained

Microconstancy's explanation of the Bernoulli patterns can now be completed. In note 11 of chapter 2, I remarked that the properties of microconstancy described in section 2.23 did not fully account for the short-term disorder in the probabilistic patterns characteristic of Bernoulli processes. It was clear that microconstancy tended to create short-term disorder, but it was not clear that microconstancy created exactly the kind of short-term disorder found in the probabilistic patterns.

The mark of the probabilistic patterns is a total lack of correlation between the outcomes of small groups of trials in a given locale, aside from that correlation entailed by the existence of a long-run frequency. The notion of stochastic independence captures this lack of correlation exactly: two outcomes a and b are stochastically independent just in case the probability of a is unaffected by conditionalizing on b. Informally, whether or not b occurs makes no difference to whether or not a occurs. If microconstancy explains stochastic independence, then it explains the short-term disorder of the probabilistic patterns.

The results above show that microconstancy does tend to create stochastic independence. Given ic-values that are loosely correlated but have infracritical independence, a microconstant mechanism will break down the correlation, producing completely independent outcomes. And the more microconstancy the better: the smaller the CRI of an evolution function, the tighter the correlation that can be broken down.

This correlation-destroying ability is significant in large part because our world contains many loose correlations, but few very tight, that is, infracritical, correlations. Why so few tight correlations? The perturbation argument (section 2.53) provides an answer. Low-level correlations are fragile. Small perturbations, because they act on only one or a few of a given set of correlated values, will tend to destroy many more low-level correlations than they create. Large perturbations may preserve low-level correlations, and they may create high-level correlations, but they do not have the delicacy to create new low-level correlations.

The perturbation argument shows, then, that low-level correlations, unlike high-level correlations, are on balance far more easily destroyed than created. For this reason, we live in a world where most correlations are at a relatively high level. In such a world, microconstancy is a prime source of stochastic independence.

The reappearance of the perturbation argument should be no surprise: in section 3.43, I showed that macroperiodicity and infracritical independence are more or less the same thing. The perturbation argument establishes the prevalence of either, by showing how perturbations cause uniformity over micro-sized regions of any density, including joint densities.

Microconstancy can explain the Bernoulli patterns, then, but what of Gaussian patterns, Poisson patterns, Markov patterns, and so on (section 1.32)? One interesting possibility is that these other probabilistic patterns are all produced by appropriate arrangements of Bernoulli trials. Poisson patterns, for example, might be the result of a sequence of very short Bernoulli trials, each with a very low probability of producing the designated outcome, a situation which can be shown to induce a probability distribution approximating the Poisson distribution. If all the probabilistic patterns could be explained along these lines, and all the underlying Bernoulli patterns were explained by microconstancy, then microconstancy would explain every probabilistic pattern we see around us. Another possibility, perhaps more plausible, is that some instances of non-Bernoulli probabilistic patterns will be explained, not by microconstancy, but by other properties of evolution functions much like microconstancy.

I commented in section 2.24 that microconstancy seems to explain both short-term disorder and long-term statistical order. This fact has a parallel in the mathematical structure of probability, where stochastic independence first manifests itself as short-term disorder or lack of correlation, and then, by way of the laws of large numbers and other results, plays a pivotal role in the demonstration of long-term statistical order. The study of probability is, in this sense, the study of the mathematics of processes that create order from a certain kind of disorder. I offer microconstant processes as a paradigm— perhaps *the* paradigm.

3.6 Causally Coupled Experiments

3.61 Causal Coupling

This section investigates the circumstances under which events generated by causally coupled mechanisms are stochastically independent. I will show that, although causal coupling makes a difference to the outcomes of individual trials, it does not always make a difference to the distribution of outcomes over a large number of trials, in roughly the same way that, although shuffling a pack of cards likely makes a difference to the suit of the card that will be drawn next, it makes no difference to the proportion of cards in the pack that have a given suit. Because facts about complex probabilities, as I have defined them, are facts not about individual outcomes but about distributions of outcomes, a causal coupling that does not alter the relevant distributions does not destroy stochastic independence.

If the mechanisms of trials on two experiments are causally coupled, then at least some aspects of the initial conditions for one trial will affect the outcome of the other. Thus, at least one of the ic-values for one trial will also be an ic-value for the other. (The sole exception, of no practical interest, is described in the passage in section 3.33 concerning massive wheels of fortune.)

For example, if two dice, red and green, are attached by a rubber band and are then tossed together, the ic-values that determine the outcome of the toss of the red die will include not only the initial speed, spin, and so on of the red die but also the speed, spin, and so on of the green die, and vice versa. It follows that the ic-values that determine the outcome of the toss of the red die will be identical to the ic-values that determine the outcome of the toss of the green die: they will be the initial speeds, spin, and so on of both dice.

In a case such as this, it is not possible to take advantage of the independence results developed in sections 3.3 and 3.4, since those results require that some ic-values for the two trials be at most loosely correlated. Since the ic-values for two coupled trials are identical, they are as tightly correlated as they could be. If the trials are stochastically independent, then, a new technique must be developed to explain the independence. That is my goal in what follows.

Causal couplings can be crudely divided into two classes, long-term couplings and short-term couplings. Two mechanisms are coupled in the short term if they interact just once in the course of a trial. They are coupled in the long term if they continue to interact over the course of an entire trial. (*Long* and *short*, then, are relative to the length of a trial.) The two dice joined by a rubber band are coupled in the long term, while two dice that collide in the course of their otherwise independent tumblings are coupled in the short term.

In complex systems, causal coupling between enions tends to be short term: a rabbit and a fox, or two gas molecules, or a buyer and a seller in a free market, meet, interact, then go their separate ways. For this reason, I focus my attention exclusively on short-term coupling. The aim is to derive a condition on short-term couplings under which the stochastic independence of outcomes will be uncompromised by the coupling. A limited version of this condition is stated and justified in section 3.64; a much more general condition is the subject of section 3.65. Section 3.66 sketches an alternative path to the same result. I hope that the alternative treatment will deepen the reader's understanding of the connections between the different independence results stated in this chapter, but it is strictly optional. The possibility of extending my results to the case of long-term coupling is discussed in website section 3.6A.

3.62 Two Tossed Coins

To examine short-term coupling, I will use the example of two tossed coins that collide in midair. As promised at the beginning of this section, I will show that an interaction—in this case, the coins' collision—that has a great effect on the outcomes of individual tosses, may have no net effect on the statistics of the outcomes of the tosses, and so no net effect on the joint complex probability distribution of the tosses.

Before I continue, let me say something more, at a very general level, about how it could be that individual outcomes are affected by a coupling but the statistics concerning those outcomes are not.

Suppose that I am given a large set of initial conditions for tosses of a pair of coins. The joint frequency-based density over these initial conditions is, I will assume, macroperiodic.

I work my way through the initial conditions, calculating the pairs of outcomes that would be produced were the coins tossed in the way specified by each pair of conditions, without any collisions occurring. For each such pair, there are four possible outcomes: heads-heads, tails-tails, heads-tails, and tails-heads. I plot the distribution of these four kinds of outcomes. Because the joint density is macroperiodic, I will obtain just the distribution one would expect given the causal independence of the tosses: 1/4 of the outcomes will fall into each outcome class.

Very roughly, what I will show in this section is that, if I use the same set of initial conditions to determine instead the pairs of outcomes that would be produced were the coins repeatedly tossed in such a way that they did collide, then given certain assumptions about the collision, I will get the same statistics: 1/4 of the outcomes will fall into each outcome class. Close up, the pattern of outcomes will look quite different from that obtained in the non-colliding case, but it will obey the same statistics. Introducing the collision shifts around individual outcomes, then, but in such a way as to preserve the shape of the distribution.

Now to the elaboration and the vindication of this claim.

I make the following assumptions about tossed coins throughout this section. First, I assume that an individual coin toss has just one IC-variable. When I introduced the coin toss experiment in section 2.25, the experiment had two IC-variables, spin speed and spin duration. In this section, I assume that the spin duration is held constant, so that individual coins take spin speed as their sole IC-variable. I will represent spin speed using the letter ζ to remind the reader of the difference. Second, I assume that the spin speeds of any two coins tossed at the same time are independent. Third, I assume that the spin speed IC-density for any toss, and thus, given independence, the joint density for any pair of tosses, is macroperiodic.

Consider the composite experiment where two coins are tossed and do not collide. Let the designated outcome be the event that the two tosses land same side up, that is, that either both coins land heads or both coins land tails. (This is purely for the sake of the visual presentation.) The evolution function for the designated outcome is

$$h(\zeta_1, \zeta_2) = \begin{cases} 1, & \text{if } h(\zeta_1) = h(\zeta_2); \\ 0, & \text{otherwise.} \end{cases}$$

where ζ_1 and ζ_2 are the initial spin speeds of the two coins. This evolution function is graphed in figure 3.4. It will be useful in this section to have a name for such a graph, conceived of as a mathematical object in its own right: I will call it the **outcome map** for the designated outcome. Note that the outcome map shows the generic structure of any evolution function for composite mechanisms; for the analysis of this structure, see figure 3.3.

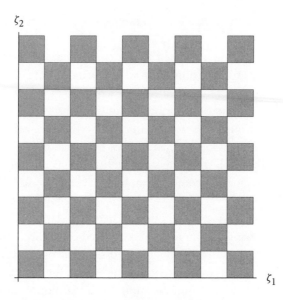

Figure 3.4 Outcome map for non-colliding coins. Gray areas indicate pairs of spin speeds ζ_1 and ζ_2 that yield tosses with the same outcome, either both heads or both tails.

3.63 A Linear Collision

Now suppose that the two tossed coins collide. I will make the following simplifying assumptions about the collision:

1. The only effect of the collision is to reduce the spin speed of each coin. The time spent spinning is not affected.
2. The collision occurs immediately after the coins are launched.
3. The amount by which one coin's spin speed is reduced is some fraction of the speed of the other coin.

Assumption (1) is made to simplify the exposition. Assumption (2) can be relaxed in the case of the coin, and for experiments much like the coin toss, but not for all microconstant experiments. It represents a real limitation on the result stated below. But when the result is applied in the course of the argument for EPA in section 4.54, the limitation will not matter, as things will be arranged so that the coupling occurs at the very beginning of the relevant experiment. This issue is further discussed in website section 3.6B. Assumption (3), also very strong, is replaced with a much weaker assumption in section 3.65.

Given the simplifying assumptions, the spin speeds after the collision ζ_1' and ζ_2' may be written as functions of the spin speeds before the collision ζ_1 and ζ_2 as follows:

$$\zeta_1' = \zeta_1 - \zeta_2/a$$
$$\zeta_2' = \zeta_2 - \zeta_1/b$$

for a and b greater than or equal to one. Call this the simple collision transformation.

The question I wish to ask is: what effect will the simple collision transformation have on the outcome map? Let $h'(\cdot)$ be the evolution function for the colliding coins. Because the collision occurs immediately after the coins are launched, the transformed speeds bear the same relation to the outcomes as the original speeds bear to the outcomes in the case where the coins do not collide. That is,

$$h'(\zeta_1, \zeta_2) = h(\zeta_1', \zeta_2').$$

Thus $h'(\cdot)$ is the composition of $h(\cdot)$ with the simple collision transformation. It follows that the outcome map for $h'(\zeta_1, \zeta_2)$ will be the outcome map for $h(\zeta_1, \zeta_2)$ transformed by the *inverse* of the simple collision transformation.[9] The transformed outcome map—the outcome map for the colliding case—is shown in figure 3.5 (for $a = 3$, $b = 4$).

As can be seen from the transformed outcome map, the evolution function for the colliding coins is microconstant with the same strike ratio—one half—as the evolution function for the non-colliding coins. The probability of two colliding coins' yielding the same outcomes, then, is one half, just as it is in the non-colliding case, and just as it must be for the tosses to count as stochastically independent. (Here I am, of course, making use of the assumption that the joint ic-density is macroperiodic.)[10]

This result generalizes: the evolution function for the colliding coins will be microconstant with strike ratio one half for almost any values of a and b. The reason for this, intuitively, is as follows. When a simple collision transformation is applied to an outcome map, the transformed outcome map has the appearance of the original map viewed from a new angle and distance. The parameters a and b determine which angle and which distance. But whichever point a chessboard is viewed from, it will have a microconstant pattern, with equal proportions of black and white.

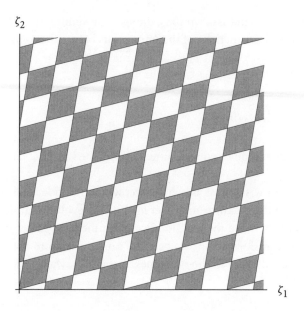

Figure 3.5 Outcome map for colliding coins: linear coupling.

There are two exceptions. First, if a chessboard is viewed exactly side-on, the pattern disappears. Given the constraints imposed by the simple collision transformations, this corresponds to the case where $a = b = 1$, so that $\zeta_1' = -\zeta_2'$.[11] The absolute dependence of ζ_1' on ζ_2' thus created does indeed undermine independence, at least in the case of the coins. But the simple collision transformation corresponding to these values of a and b constitutes the very thinnest slice of all possible transformations.

Second, if a chessboard is viewed very close up, black and white will alternate only over relatively large distances. An evolution function with this aspect will have a non-trivial constant ratio partition, but the members of the partition will be macro-sized; hence, the function will lack microconstancy, and the collision will probably destroy stochastic independence. An overly close viewing of the board corresponds to the case where a and b in the simple collision transformation are near one.

3.64 Linear Coupling

In what follows I will express all this more formally and generalize to other transformations of the IC-variables. My treatment makes use of some basic linear algebra, but the reader who is unfamiliar with this subject should still be able to follow along, if not to see why every claim is true.

The simple collision transformations are a subfamily of the linear transformations, which are transformations of the form

$$\zeta_1' = a\zeta_1 + b\zeta_2$$
$$\zeta_2' = c\zeta_2 + d\zeta_1$$

where the coefficients can be positive or negative. Performing a linear transformation on the variables has the same effect as performing the inverse of that transformation on the outcome map. All invertible linear transformations have inverses which are also linear transformations. For reasons to be explained below, a linear transformation of an outcome map will preserve the microconstancy and strike ratio of the map, unless it makes micro-sized intervals macro-sized. Thus all linear transformations of the colliding coins' IC-variables preserve the microconstancy and the strike ratio of the relevant composite evolution functions, with two exceptions, the generalized cases of the two exceptions described above:

1. Microconstancy and strike ratio may not be preserved when an absolute dependence of the form $\zeta_1' = k\zeta_2'$ is established between the two IC-variables.[12] This can only happen when the linear transformation of the variables effected by the coupling is non-invertible.[13]
2. Transformations which enlarge the outcome map will tend to transform micro-sized regions in a constant ratio partition into macro-sized regions. The map is enlarged by a linear transformation that has a determinant with magnitude greater than one, thus by a linear transformation of the IC-variables that has a determinant of less than one (since the determinant of an inverse transformation is the reciprocal of the determinant of the original transformation).[14] The nearer zero the determinant of the transformation of the IC-variables, the more the map is enlarged.

Because non-invertible transformations have determinants equal to zero, both exceptions occur when the magnitude of the determinant of the IC-variable transformation is too much less than one.

In summary, with the exceptions stated above, a collision between tossed coins that effects a linear transformation of the spin speeds will create a new evolution function $h'(\zeta_1, \zeta_2)$ that, like the original evolution function, is microconstant and has a strike ratio of one half. The microconstancy and strike ratios of other composite evolution functions for the coins are likewise preserved. For example, the evolution function for the composite outcome tails-tails will be, in the colliding as in the non-colliding case, microconstant with strike ratio 1/4.

These conclusions supply the foundation of the following independence result: a collision between two tossed coins will not affect the stochastic independence of the outcomes of the tosses if

1. The joint and individual ic-densities for the tosses are macroperiodic,[15]
2. The transformations of the ic-variables effected by the collision are linear, and
3. The determinant of the transformation has a magnitude not too much less than one.

Just how much less than one the magnitude of the determinant can go depends on the macroperiodicity of the joint ic-density: the more macroperiodic the ic-density, the more the outcome map can be blown up without making its optimal constant ratio partition macro-sized relative to the ic-density. Thus, the more macroperiodic the joint ic-density, the closer to zero the determinant can be.

In the remainder of this section, I explain, as promised, why a linear transformation of an outcome map will (with the exceptions specified above) preserve the microconstancy and the strike ratio of the map.

The preservation of microconstancy and strike ratio is entailed by two properties of invertible linear transformations: (a) their continuity and (b) the fact that they have what I call the *strike ratio preservation property*. A transformation has the strike ratio preservation property if it preserves the ratio of gray to white in any area of an outcome map. More exactly, a transformation has the property just in case any (measurable) region in the original outcome map contains the same ratio of gray to white as the image of that region in the transformed map.

Continuity and strike ratio preservation are sufficient for microconstancy preservation for the following reason. Suppose that Y is the image of an outcome map X under an invertible linear transformation. Consider a constant ratio partition of X. By the strike ratio preservation property, the image in Y of any member of the partition is a set with the same strike ratio, and by continuity, the image is itself a contiguous region. From these two facts it then follows that the image of an optimal constant ratio partition for X is a constant ratio partition for Y, with the same strike ratio.[16] Provided that the transformation does not enlarge regions too much (that is, provided that its determinant is not too close to zero), the members of the partition of Y are micro-sized.

Why do invertible linear transformations have the strike ratio preservation property? That they do follows from a well-known property of linear transformations, lemma 3.7 of section 3.B2:

> An invertible linear transformation enlarges or diminishes the size of any measurable region by the same amount. That is, for any invertible linear transformation there exists a constant a, such that for any region R, the ratio of the size of the image of R to the size of R is equal to a.

The strike ratio preservation property follows from the lemma almost immediately: the lemma entails that a linear transformation changes the gray and the white areas of any region by the same proportion, thus the ratio of one to the other remains unchanged.

3.65 Non-Linear Coupling

Linear couplings are not particularly common, and so the independence result stated in the last section is rather narrow. This section broadens the result.

LINEAR PLUS CONSTANT TRANSFORMATIONS

A first attempt at broadening might look for other transformations with the strike ratio preservation property. There is one important class of these: the linear transformations with a constant added, that is, transformations of the form:

$$\zeta_1' = a\zeta_1 + b\zeta_2 + c.$$

(The inverses have the same form.) But there are very few others.

A NON-LINEAR COUPLING THAT PRESERVES INDEPENDENCE

In general, non-linear transformations do not have the strike ratio preservation property. This does not mean, however, that non-linear transformations cannot preserve the microconstancy and strike ratio of an outcome map, at

ζ_2

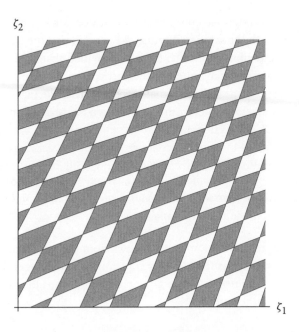

ζ_1

Figure 3.6 Outcome map for colliding coins: non-linear coupling.

least approximately. To see why, consider the following transformation of coin spin speeds:

$$\zeta_1' = \zeta_1 - \zeta_2^2/5$$
$$\zeta_2' = \zeta_2 - \zeta_1^2/7.$$

A part of the transformed outcome map is pictured in figure 3.6.

The transformation of the outcome map, which is the inverse of the transformation of the IC-variables, clearly preserves, approximately, the evolution function's microconstancy and its strike ratio of 1/2 (except for very low values of ζ_1 and ζ_2, not shown in the figure). A non-linear transformation of this sort, then, will preserve stochastic independence.

Let me now introduce two abbreviations: I will call the transformation of the IC-variables effected by a causal coupling the *IC-variable transformation*, and the transformation of the outcome map, its inverse, the *outcome transformation*.

Why does the non-linear outcome transformation described above preserve the evolution function's microconstancy and strike ratio? The answer is that, although the transformation does not preserve the strike ratio over every region, as required by the strike ratio preservation property, it does preserve the strike ratio, roughly, over small, contiguous regions. In particular, it preserves the strike ratio for any member of a micro-sized constant ratio partition. This, together with continuity, ensures that the transformation of a micro-sized constant ratio partition for the non-colliding evolution function is a constant ratio partition for the colliding evolution function, with the same strike ratio. Provided that the transformed constant ratio partition is still micro-sized—provided that the outcome transformation does not enlarge any members of the constant ratio partition too much—this is sufficient for the microconstancy of the colliding evolution function, as required for independence, on the approach taken here.

We have here, then, the makings of an independence result for microconstant experiments with non-linear couplings. Independence will hold if the evolution functions for the coupled composite experiment are microconstant and have the same strike ratio as their counterparts in the uncoupled composite experiment. The previous paragraph sketched an argument that this will be so if (a) the outcome transformation is continuous, (b) it does not enlarge any part of the ic-variable space too much, and (c) it preserves strike ratios over contiguous, micro-sized regions.

Under what conditions does a non-linear transformation have property (c)? There is a very simple answer to this question. For the reasons given in the last section, a non-linear transformation will have property (c), that is, will preserve strike ratios over contiguous, micro-sized regions, if it is approximately linear (and invertibly so) over every contiguous, micro-sized region.[17] Call such a transformation a **microlinear transformation** (see also definition 3.23; note that invertibility is built into the definition of microlinearity).

This suggests the following sufficient conditions for stochastic independence: given two microconstant experiments with independent, macroperiodic ic-densities, a short-term coupling will preserve the independence of the outcomes if it effects a continuous, non-enlarging, microlinear transformation of the outcome map. Most of the rest of this section is concerned to refine this condition, in two ways. First, I deal with a complication introduced by the fact that a non-linear ic-variable transformation need not be one-to-one. Second, I rephrase the condition as a requirement on the ic-variable transformation, not its inverse, the outcome transformation.

MANY-TO-ONE IC-VARIABLE TRANSFORMATIONS

If the ic-variable transformation is many-to-one, the outcome transformation will be one-to-many over at least some micro-sized regions (remembering that I am using *transformation* in a mathematically loose sense here; see note 9). A transformation that is one-to-many over a micro-sized region cannot be approximately linear over that region; the transformation can, however, be broken up into parts that are one-to-one, and each of these may be approximately linear. If so, the transformation will preserve the strike ratios of micro-sized regions, in the sense that the image of a micro-sized region will be one or more regions each with the same strike ratio as the original, which will be adequate, in conjunction with the other conditions stated above, for independence.

Let me put this point more formally. Consider a region R of the outcome map for the non-coupled case. The outcome transformation maps R onto the inverse image of R under the ic-variable transformation. This inverse image consists of all the contiguous regions $R_1, \ldots, R_n$ that the ic-variable transformation maps onto R. (There is only one of these if the ic-variable transformation is one-to-one.) The behavior of the outcome transformation can be represented by a set of one-to-one functions $g_i(\cdot)$ such that $g_i(R) = R_i$.[18] If the g_i's are approximately linear, then the outcome transformation will preserve the strike ratios of micro-sized regions.

Incorporating the formalism of the last paragraph into the claim made at the end of the previous subsection, and making some technical, though important, additions and amendments: sufficient conditions for an outcome transformation to preserve the microconstancy and the strike ratio of the non-coupled outcome map are that, for every contiguous micro-sized region R,

1. Each $g_i(\cdot)$ for that R is approximated over R by an invertible linear, or linear plus constant, transformation,
2. The linear transformations that approximate the g_i's do not enlarge the size of their domains too much, or equivalently, have determinants not much greater than one,
3. The linear transformations do not come close to effecting absolute dependence of the variables, that is, they are not close to one of the non-invertible linear transformations that effect a relationship of the form $\zeta_1' = k\zeta_2'$, and
4. Each $g_i(\cdot)$ is continuous.

Condition (3) is new. It is necessary because, when a transformation is close to effecting absolute dependence, the horizontal and vertical lines in the checkerboard that is the uncoupled outcome map become so skewed in the transformed outcome map that they are almost parallel to one another; as a result, the slight curvature of the lines, which is due to the inexactness of the linearity, may have a great impact on the ratio of gray to white in the transformed outcome map. In what follows, I will count such a case as one where the outcome transformation is not approximated sufficiently well by an invertible linear transformation, thus rolling condition (3) into condition (1). Another simplification is also possible: a function that approximates a linear transformation is close enough, for my purposes, to being continuous, so condition (4) may also be rolled into (1).

This leaves, as the only conditions on the outcome map, conditions (1) and (2). Provided that they are met, the image, under the outcome transformation, of any micro-sized region in a constant ratio partition for the uncoupled outcome map, is one or more micro-sized regions, each with the same strike ratio, in the coupled outcome map. Thus the image of a micro-sized constant ratio partition for the uncoupled outcome map is a micro-sized constant ratio partition for the coupled outcome map, with the same strike ratio, as desired.

Reformulation in Terms of the IC-Variable Transformation

From the conclusion of the preceding subsection, sufficient conditions for independence in the case of non-linear couplings are easily derived: all that need be added is the requirement that the ic-densities of the coupled experiments are, together with their joint ic-density, macroperiodic. But the conditions formulated in the preceding subsection are conditions on the outcome transformation, whereas what I require when I put these results to work in chapter four is a set of conditions on the ic-variable transformation.

What kind of ic-variable transformation, then, will induce an outcome transformation satisfying conditions (1) and (2) above? The ic-variable transformation is in effect made up of the inverses of all the g_i's. Since the inverse of an invertible linear transformation is another invertible linear transformation, one obtains almost the same conditions on the ic-variable transformation as are required of the outcome transformation.[19] Specifically, the ic-variable transformation must have the following property: for every micro-sized region R_i in its domain, the ic-variable transformation must

1. Approximate an invertible linear (or linear plus constant) transformation, and
2. The linear transformation must not have a determinant much less than one, that is, it must not shrink R_i too much.

The principal requirement, then, is that the ic-variable transformation should be microlinear. The other requirement, that the determinant of the ic-variable transformation be well above zero, might be stated thus: the ic-variable transformation must not be deflationary, or equivalently, it must be *non-deflationary*.

Sufficient Conditions for Stochastic Independence

Now, at last, let me state the independence result (formalized, with comments on approximation, in section 3.B6). A collision between two tossed coins will not affect their stochastic independence if

1. The joint and individual ic-densities for the tosses are macroperiodic,
2. The ic-variable transformation effected by the collision is microlinear, and
3. The ic-variable transformation is non-deflationary.

The main limitation of this result is that it applies only to couplings that occur at the beginning of a trial. This will be sufficient for an understanding of the grounds of EPA. It is interesting to ask, however, in what way the limitation might be removed. The beginnings of an answer will be found in the extended discussion of colliding coins in website section 3.6B.

How difficult is it to satisfy the stated conditions for independence? The requirement that the evolution function be non-deflationary seems quite strong. There is, though, as pointed out above, an opportunity to reduce the stringency of the non-deflation requirement at the expense of increased stringency in the macroperiodicity requirement on the joint ic-density. A deflationary ic-variable transformation creates an evolution function with a rather large optimal constant ratio partition, but if the joint ic-density is sufficiently macroperiodic, it is roughly uniform over even large partitions, and so the deflation does not compromise independence. In cases where there is macroperiodicity to spare, then, a moderate amount of deflation can be allowed.

3.66 An Alternative Approach

I now want to sketch a different way of establishing the results described above. This alternative approach introduces a way of thinking about independence in systems with tightly correlated ic-values that will predominate in section 3.7; however, since that section is self-contained, what follows may be skipped on a first reading.

In the presentation above, I incorporated the effect of the short-term coupling into the evolution function, so that the probabilities attached to the composite experiment were determined by the original joint density and a transformed evolution function. Another approach is to incorporate the effect into the ic-densities, so that the probabilities are determined by a transformed joint density—the joint ic-density over the transformed ic-variables—and the original evolution function. This is only possible, note, if the coupling occurs at the very beginning of a trial, a serious limitation on the scope of the technique.

I emphasize that, in order to obtain the probability distribution over composite outcomes, it is sufficient to put the transformed density together with the evolution function for the uncoupled mechanisms. In the case of the coins, for example, the evolution function for each coin is, on this approach, just the evolution function for a coin that does not collide with any other coin. It takes only one ic-variable, its own spin speed.

What properties of the transformed joint density would guarantee independence? Because the relevant evolution functions are the same as in the uncoupled case, a sufficient condition for independence is that the transformed joint density be macroperiodic, for just the reasons given in section 3.4 (assuming that the individual ic-densities are also macroperiodic).

In section 3.B7, I prove a result (theorem 3.18) to the effect that a microlinear, non-deflationary transformation of a macroperiodic density is also macroperiodic. Thus, the transformed ic-density is macroperiodic if

1. The original ic-density is macroperiodic, and
2. The ic-variable transformation is microlinear and non-deflationary.

These, then, are sufficient conditions for independence, if the ic-densities of the individual experiments are macroperiodic. They are, of course, the same sufficient conditions for independence as were derived in the last section.

Note that, on this approach, the formal representation of the situation is identical to the representation of a situation in which the coins never interact, but in which their ic-variables are correlated because of some historical interaction. As I showed in section 3.4, in the latter case, independence obtains for microconstant experiments provided that the correlation between the ic-variables of the two experiments is only a weak correlation, that is, provided that only high-level information is correlated. Here, note, I am talking about the ic-variables of the uncoupled experiments.

It follows that, if two experiments with independent ic-variables but coupled in the short term, can be represented as equivalent to two uncoupled experiments with correlated ic-variables—and I emphasize that such a representation is straightforward only if the coupling occurs at the very beginning of a pair of trials—then a sufficient condition for independence when mechanisms are microconstant is that the ic-variable transformation correlate only high-level information about the ic-variables of the uncoupled experiments. (The question of the degree of correlation between what count as the ic-values for *uncoupled* trials is, note, distinct from the question of the degree of correlation between the ic-values for *coupled* trials. As pointed out above, the ic-values for coupled trials are identical, and so are completely correlated.)

The significance of this fact is not very clear until one has an answer to the question of what kinds of interactions correlate high-level information only, or to put it another way, what kinds of interactions preserve the infracritical independence of macroperiodic densities. The result stated above provides an answer to this question: non-deflationary microlinear interactions preserve infracritical independence. That is, if two macroperiodically distributed ic-variables start out independent, and are then correlated by a microlinear, non-deflationary transformation, their infracritical independence is preserved.

3.7 Chains of Linked IC-Values

3.71 Chains Defined

A **deterministic chain** is a series of trials, the outcome of one of which completely determines the initial conditions of the next. A chain may consist of repeated trials on the same experiment, or of trials on a number of different experiments.

Interactions at different times between enions in a deterministic complex system would seem to be links in a deterministic chain, since the outcomes of earlier interactions determine the values of the system's microvariables that will act as the initial conditions for later interactions. It is in large part because of chaining that it is hard to see, at first, how probabilistic processes in a complex system could be stochastically independent. The purpose of this section is to examine the conditions under which there can be independence in chains.

The simplest kind of chain consists of repeated trials on the same experiment. The experiment must, of course, be capable of producing an ic-value or a set of ic-values that will serve as the initial conditions for the next trial. That is, it must be capable of feeding itself.

As an example of such an experiment, consider a wheel of fortune, marked as usual with alternating red and black sections of equal size, but marked also with a scheme so that every point on the wheel corresponds to a possible spin speed. At the end of any trial on the wheel, the pointer indicates not only a color but a spin speed. This spin speed may be used as the ic-value for the next trial on the wheel. Call such a device an *autophagous* wheel of fortune.

To construct a deterministic chain using the autophagous wheel, conduct a series of trials in which the spin speed for each trial except the first is that indicated by the pointer at the end of the previous trial. The spin speed of the very first trial in such a series, then, will determine the spin speeds for all subsequent trials.

The ic-values for the trials that make up this chain cannot, of course, be stochastically independent, since they are all entirely determined by the very first ic-value. What I will investigate in this section is whether the outcomes—red or black—of the trials in the chain can be independent.

It turns out that there is a broad range of conditions that guarantees independence. When these conditions hold, the distribution over the effective ic-values for the ith trial in a chain, conditional on the outcomes of all $i - 1$ previous trials, is the same, or at least is macroperiodic, no matter what the outcomes of the previous trials. Conditioning on previous outcomes, then, does not affect the effective ic-value distribution for the ith trial, and so does not affect the probability distribution over the possible outcomes of the ith trial. The outcome of the ith trial is, in other words, stochastically independent of the outcomes of all the previous trials in the chain.

Observe that this result depends on two key facts:

1. Independence of effective ic-values is sufficient for independence of outcomes (theorem 3.2), and
2. The correlation between the full ic-values in a chain need not carry over to the effective ic-values: although the full ic-values are completely correlated, in the sense that the first determines the rest, the effective ic-values are, in the right circumstances, as uncorrelated as you like.

The overall structure of this section is much like that of section 3.6. In section 3.72 I introduce a very simple autophagous wheel, chained trials on which I show to be stochastically independent. I then generalize the properties of this experiment to obtain a sufficient condition for independence in a chain, and examine the generalization further in section 3.73. The sufficient condition is rather restrictive, however, so in sections 3.75, 3.76, and 3.77, I consider another experiment, show that chained trials on this experiment are independent, and generalize to obtain a much broader sufficient condition for independence. It is this second, broader result that is put to use in the explanation of the foundations of EPA in chapter four. The one other section in what follows is section 3.74; it provides an alternative approach to independence in chains, intended to deepen the reader's understanding of the independence results, and may be skipped on a first reading.

The reader in a hurry might read the subsection titled "The Straight Wheel," immediately below, and then go to sections 3.75 and 3.76, which present the main claims of this section.

3.72 Independence in the Straight Wheel

THE STRAIGHT WHEEL

Consider the following special kind of autophagous wheel of fortune, which I will call the *straight wheel*, and on which deterministic chains of trials produce stochastically independent outcomes. On the straight wheel, (a) all colored sectors are the same size, (b) the same range of possible initial spin speeds is marked within each sector, and (c) the speeds are spread evenly within each sector. For example, if the range of speeds is from 0 to 100, then the speeds 0 through 100 are marked out evenly in every red and every black sector of the wheel. As a consequence, the point, say, 34.5678% of the way around any sector, red or black, corresponds to the speed 34.5678, and so on. When a chain of trials is conducted on the wheel, this is the speed used for the next trial.

To see why outcomes on a straight wheel will be independent, consider the following version of a straight wheel. Say that the wheel has ten colored sections in all, five red and five black, of equal size, and that the mechanics of the wheel are such that an initial spin speed of ζ will cause the wheel to turn exactly $\zeta/10$ times. For example, an initial speed of 12 will cause the wheel to revolve 1.2 times.

The integer part of the spin speed corresponds to the total number of complete colored sections that the wheel revolves through before it comes to a halt. If the initial speed is 12, the wheel turns through 12 sections, stopping just as it enters its thirteenth section, that is, just as it reenters the third section on the wheel (since the wheel only has 10 sections). Suppose that the sections are numbered from 0 to 9, so that the first section is section 0, the third section is section 2, and so on. Then a speed of 12 causes the wheel to stop just as the pointer enters section 2, a speed of 34 causes the wheel to stop just as the pointer enters section 4, and so on.

What if the spin speed is not an integer? Then, after turning through a number of colored sections equal to the integral part of the spin speed, the wheel will turn partway through one more colored section. The proportion of this final section traversed by the pointer is equal to the fractional part of the spin speed. When given an initial spin speed of 12.34, for example, the wheel will turn through twelve complete sections, the last of these being section 1, and will then turn through exactly 34% of section 2. In effect, the integer part of the IC-value for a trial is "used up" by determining which colored section the pointer will indicate; the position of the pointer within that colored section is then entirely determined by the fractional part of the IC-value.

Now suppose that the wheel is given an initial spin speed of 12.345678, and that chained trials are conducted thereafter. On the first spin, the wheel will stop in section 2 and the spin speed indicated for the next trial will be 34.5678. Given this new spin speed, on the second spin, the wheel will stop in section 4 and the spin speed indicated for the next trial will be 56.78. Over the course of four trials, the pointer will indicate sections 2, 4, 6, and 8, corresponding to the second, fourth, sixth, and eighth digits of the initial spin speed. If the trials are continued, the section indicated by the pointer at the end of the nth trial is given by the $2n^{\text{th}}$ digit of the initial spin speed.

Since the outcome of any trial on the wheel, red or black, is determined by the section indicated by the pointer, the outcomes of a series of chained trials on the wheel will be determined by the second, fourth, sixth, and so on digits of the initial spin speed. There is, of course, normally no reason to expect the different digits of the initial spin speed to be correlated. Thus there is, normally, every reason to expect the outcomes of a series of chained trials on the wheel to be independent. The precise content of the assumption that the digits are not correlated will be examined below; until then, I simply take for granted that it is a reasonable assumption.

THE EFFECTIVE IC-VALUES OF THE STRAIGHT WHEEL

In order to generalize from the very particular straight wheel of fortune just described to all straight wheels of fortune, it is necessary to describe the properties in virtue of which my straight wheel produces independent outcomes in a suitably general way, so that it can be seen that these are properties possessed by all straight wheels.

Let me begin by characterizing independence on my straight wheel in terms of the notion of an effective ic-value of a trial (section 2.71). I remind the reader that the set of effective ic-values for an experiment is always characterized relative to a designated outcome or set of outcomes. I will not mention this fact again; the designated outcomes, throughout the treatment of chains, are just those outcomes whose independence is in question, usually red and black.

The effective ic-value of a trial on the straight wheel fixes the trial's spin speed with the minimal amount of precision necessary to determine its outcome. This means that the effective ic-value conveys the minimum amount of information necessary to determine the number of colored sections the wheel turns through before stopping. For the straight wheel described above, that information is contained in the integer part of the spin speed; it is natural, then, to take the integer part of a trial's spin speed as its effective ic-value.

In chains of trials on the straight wheel, the effective ic-value for the first trial is entirely determined by the first two digits of the initial spin speed, the effective ic-value for the second trial is entirely determined by the second two digits of the initial spin speed, the effective ic-value of the third trial by the third pair of digits, and so on. More generally, the effective ic-values of two different trials in a chain are determined by distinct parts of the decimal expansion of the initial full ic-value; see figure 3.7.

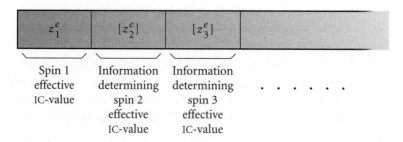

Figure 3.7 Source of effective ic-values for chained trials on a straight wheel of fortune. The pieces of information determining effective ic-values for successive spins—the ancestor ic-values—occupy successively lower levels of the initial full ic-value. The effective ic-value for the first trial is z_1^e, that for the second is z_2^e, and so on. The square brackets are there to remind the reader that what is represented is the ancestor ic-value, the information that determines the named ic-value, not the ic-value itself.

Provided that the distinct parts of the initial full ic-value are not correlated, then, the effective ic-values of chained trials on the straight wheel will be independent, and so the outcomes determined by these effective ic-values will be independent. In effect, the initial spin speed for such a chain functions as a list of random, two-digit numbers that determine in turn the effective ic-value of each next trial.

The version of the straight wheel described above exhibits this fact in a particularly perspicuous way: the effective ic-values of future trials can be seen sitting in the decimal expansion of the initial full ic-value. But what is important is not that the values themselves are, in a literal sense, present in different parts of the initial ic-value, but that something that determines them is present. This something might undergo a number of mathematical transformations on its way from its place in the initial ic-value to its role as the effective ic-value of a later trial, but all that matters is that the following conditions hold:

1. It begins life as a part of the initial ic-value distinct from the parts that determine other effective ic-values, and
2. The transformations that it undergoes are independent of the outcomes of previous trials in the chain.

If these conditions hold, and if the different parts of a chain's initial ic-value are themselves stochastically independent (an assumption examined below), outcomes in a chain of trials will be stochastically independent.[20]

I do not require, then, that the effective ic-values themselves constitute different parts of the initial ic-value, but that the ancestor ic-values do so, where the **ancestor ic-value** of an effective ic-value for a chained trial is defined to be that part of the initial ic-value that entirely determines the effective ic-value. Condition (1) of the requirement just stated can be rephrased as follows: the ancestor ic-values of the different trials in a chain must not, in a certain sense, overlap.

All straight wheels satisfy the sufficient conditions for independence stated above. The reason for this will become clearer in section 3.73. In the remainder of this section I clarify two aspects of the independence conditions: first, the question of what, exactly, it means to say that two ancestor ic-values occupy different "parts" of the same full ic-value, and second, the nature of the assumption that needs to be made about the distribution of initial full ic-values to ensure that ancestor ic-values occupying different parts of an initial

IC-value are independent. I then restate more formally the conditions for independence.

Parts of an IC-Value

So far in this discussion, I have relied on the notion that the ancestor IC-values of different trials on the straight wheel can stand in relation to a full initial IC-value as parts to a whole. This way of thinking is, of course, inspired by the observation that, for the particular straight wheel described in this section, the decimal expansions of the ancestor IC-values literally are parts of the decimal expansion of the initial IC-value. But, as I noted in the original discussion of effective and critical IC-values in section 2.7, the part/whole way of talking is, in most cases, only a metaphor.

In that same section, I introduced an alternative way to think about different parts of the same full IC-value: different parts correspond to information at different levels. In the case of the straight wheel, for example, the effective IC-value for the first trial is at a higher level than the ancestor IC-value for the second trial, which is in turn at a higher level than the ancestor IC-value for the third trial, and so on. The notion of levels of information is given a precise mathematical meaning in sections 3.B2 and 3.B4. The mathematical treatment can be extended, first, so as to define what it is for two levels to be entirely distinct, and second, so as to show that under a certain, very broad assumption about the distribution of full IC-values, information at distinct levels is independent. The extended treatment will be found in section 3.B5; the definition of distinctness is definition 3.18; the independence result is theorem 3.15.

The Initial IC-Value Assumption

My treatment of the straight wheel has made the apparently reasonable assumption that distinct levels of information about a chain's initial full IC-value—corresponding, roughly, to distinct parts of the initial IC-value's decimal expansion—will tend not to be correlated, that is, will tend to be stochastically independent. In this subsection, I ask under what conditions the assumption is likely to hold.

The answer, stated in section 3.B5 as theorem 3.15, is that the assumption will hold if the probability distribution over the initial full IC-values is macroperiodic relative to the partition corresponding to the effective IC-values of the first trial in the chain. The assumption requires, then, that the distribution be approximately uniform over any region corresponding to an effective IC-value for the first trial. (This is, as with most conditions stated in this study, a sufficient but not a necessary condition.)

The reason that the theorem holds is, roughly, that the vast majority of real numbers in any interval have uncorrelated parts (Chaitin 1990; see also corollary 3.16), in the sense that the information about these numbers at different levels is, if written out in order of decreasing level, probabilistically patterned. For the density of initial ic-values to be skewed in favor of those with correlated parts, it would have to contain very narrow spikes. Thus, a number picked from a macroperiodic distribution, which by definition has no such spikes, will with very high probability have uncorrelated parts.

Note that the distribution of interest is over only the possible initial full ic-values of a chain, and not the later full ic-values that are produced by trials in the chain. To use the terms introduced in section 2.43, the full ic-value of the first trial in a chain is *raw*, while the full ic-values for all later trials are *cooked*. The probability distribution over these cooked ic-values is determined by the distribution over the initial, raw ic-value and the mechanics of the experiments in the chain. It is useful, then, to have a way of talking about the distribution over the initial ic-value. To this end, I introduce the notion of an **initial ic-variable** for a type of chain, which is an ic-variable that takes as its values just the initial ic-values for that type of chain. It is the distribution over a chain's initial ic-variable that provides the probabilistic ingredient of all other probability distributions attached to the chain.

Sufficient Conditions for Independence

I now state more formally the sufficient condition for independence satisfied by all straight wheels. Given a designated outcome e, a series of n chained trials on an experiment (or on a series of experiments) will produce e with independent probabilities if:

1. There exists a division of the initial, full ic-value of the chain into n distinct levels of information, conveyed by variables $\omega_1, \ldots, \omega_n$, such that whether or not e occurs on the ith trial is entirely determined by the value of ω_i. More formally, there should exist n functions $f_1, \ldots, f_n$ mapping to $\{0, 1\}$, such that, on the ith trial, e occurs just in case $f_i(\omega_i) = 1$.[21]

2. The probability distribution over the chain's initial ic-variable is macroperiodic with respect to the partition formed by the effective ic-values for the first trial in the chain.

The value of ω_i, it will be seen, is just the ancestor IC-value of the ith trial in the chain.

This is a sufficient condition for the stochastic independence of the effective IC-values for each trial. It does not guarantee that the probability distribution over the effective IC-values will be the same from trial to trial, or even that it will remain macroperiodic—see note 20—but I defer this worry until section 3.75.

The chief limitation of the sufficient conditions stated above is, as will become clear in section 3.75, the requirement that the outcome of the ith trial be *entirely* determined by ω_i. In that same section I show—this is my main result concerning chains—that under a wide variety of conditions, it is sufficient for independence that the outcome of the ith trial be only partly determined by ω_i.

Before moving on to this discussion, however, I look more closely at the reason that straight wheels satisfy condition (1) of the sufficient conditions, in section 3.73, and I sketch an alternative approach to thinking about independence in chains, in section 3.74.

3.73 Outcome-Independent Inflation

I have asserted that all straight wheels satisfy condition (1) of the sufficient conditions for independence stated in the last section. Although it is clear that this is true for the particular example I used above, it is perhaps not so clear that the claim can be generalized to all straight wheels, let alone to other kinds of probabilistic experiments. This section sketches the basis for a generalization.

All straight wheels satisfy condition (1) because their *IC-evolution* is both *outcome independent* and *inflationary*. Let me define these terms and explain their relevance.

For any chained experiment, there is a function, determined by the experiment's mechanism, that takes as its argument the IC-values for one trial in the chain, and yields the IC-values for the next trial in the chain. I call this the **IC-evolution function** of the experiment, written $H(\cdot)$. In the case of an autophagous wheel, for example, if ζ_1 and ζ_2 are, respectively, the speeds of the first and second spins on the wheel, then $\zeta_2 = H(\zeta_1)$. The way in which one trial in a deterministic chain causally depends on all preceding trials in the chain is entirely captured, then, by the IC-evolution function.

When the IC-evolution function for an experiment is the same, regardless of the outcome of the experiment, I say that the experiment has **outcome-independent IC-evolution**. The straight wheel described above has outcome independent IC-evolution because, whatever the outcome of a trial on the wheel, the relation between the IC-value for that trial and the IC-value for the next trial in the chain is given by the function

$$H(\zeta) = 100\zeta \bmod 100.$$

Contrast this case with a different autophagous wheel, the zigzag wheel, which will be discussed in section 3.75, and on which the IC-evolution function for a trial depends on the outcome of that trial:

$$H(\zeta) = 100\zeta \bmod 100, \qquad \text{if } \zeta \text{ produces a red outcome;}$$
$$H(\zeta) = 100 - (100\zeta \bmod 100), \quad \text{if } \zeta \text{ produces a black outcome.}$$

The IC-evolution function for the zigzag wheel, then, is not outcome independent.[22] All straight wheels have outcome independent IC-evolution; specifically, they have IC-evolution functions of the form

$$H(\zeta) = a\zeta \bmod b + c. \text{ [23]}$$

Outcome independent IC-evolution is more or less required of the experiments in a chain, if the chain is to satisfy the sufficient conditions for independence stated above.

The second property of straight wheels, in virtue of which they satisfy condition (1) of section 3.72's sufficient conditions for independence, is that they have *inflationary* IC-evolution. Consider again the IC-evolution function for my particular version of the straight wheel:

$$H(\zeta) = 100\zeta \bmod 100.$$

This is a classic *stretch-and-fold* function. The stretching is effected by the multiplication of ζ by 100, which increases the distance between nearby values by a factor of 100. The folding is effected by the modulo operation, which lops off all but the last two digits of the integer part of the stretched value.[24] In effect, the elongated domain of the function is being folded so that it fits into the interval [0, 100), the function's range. Both the stretching and the folding are responsible for the fact that distinct parts of a chain's initial ic-value deter-mine the effective ic-values of different trials. The stretching brings ever lower level information about the initial ic-value up to the effective level; the fold-ing discards the part of the ic-value that served as the effective information for the previous trial.[25] In this and later sections I will refer to ic-evolution that stretches the set of ic-values in such a way as to bring lower and lower levels of information into play in determining effective ic-values as **inflationary**.

In general, a stretch-and-fold ic-evolution function will serve as a founda-tion for independence in chains provided that, first, the stretch-and-fold effect is the same whatever outcome is obtained on a trial—that is, that ic-evolution is outcome independent—and second, that the stretch is sufficiently great and the fold just right, that an entirely new set of information comes to occupy the effective level with each iteration of the function. These conditions could be stated more precisely, of course, but any formalization would be superseded by the results presented in section 3.76, so I will settle, at this stage, for an informal statement.

Notice that the folding of the ic-evolution function for the straight wheel is in a certain respect redundant: even if the information lost in a fold were not thrown away—even if the ic-evolution function for a wheel were, say, $H(\zeta) = 100\zeta$—the outcomes of trials would still be independent, since, in a microconstant experiment such as the straight wheel, the high-level informa-tion in the effective ic-value does not make a difference to the distribution of outcomes. Thus, in the microconstant case, inflation renders high-level infor-mation irrelevant whether or not it is literally removed by folding. The more important component of the ic-evolution function, it seems, is the stretch—and indeed, it turns out that the key to generalizing the results of section 3.72 lies mainly in an examination of the properties of the stretch.[26]

Because of the emphasis on the stretch, I say that the kind of ic-evolution function satisfying the informal conditions just stated induces *outcome-independent inflationary ic-evolution*. The conclusion of this section, then, may be stated as follows: the right sort of outcome-independent inflationary ic-evolution is sufficient, given a macroperiodic distribution over a chain's initial ic-variable, for the independence of chained trials.

I will finish with a few words clarifying one aspect of the nature of the stretch, or inflation, that provides independence in the way described above. (The clarification of other aspects, in particular the question of how much inflation is enough, must wait until section 3.76.) Consider two possible def-initions of what it is for a transformation to be inflationary relative to a given partition $\mathcal{U}$:

1. Weak inflation: a transformation f is weakly inflationary relative to $\mathcal{U}$ if it maps every member U of $\mathcal{U}$ to a larger set, that is, if for every U, $f(U)$ is larger than U.
2. Strong inflation: a transformation is strongly inflationary if it maps every subset of its domain to a larger set, that is, if for every subset V of the domain, $f(V)$ is larger than V.

A weakly inflationary transformation, then, may deflate some subsets of a given U provided that it inflates the rest sufficiently to compensate for the deflationary effect. Strong inflation does not allow this.

Must ic-evolution be strongly inflationary to achieve independence in the way described above? The answer, in general, is yes. If deflationary zones are allowed, information will be recycled in ways that can undercut independence. But I will not go into details here, because, when the independence result is later generalized, strong inflation is required for other reasons (see the remarks in section 3.76).

3.74 Multi-Mechanism Experiments

This section sketches an alternative approach to establishing the independence of chained trials. It is intended to deepen the reader's understanding of independence in chained trials; it is also used briefly in chapter four (see sections 4.54 and 4.55).

The idea is to consider the outcome of, say, the third trial in a chain on an autophagous wheel, as the outcome of a single experiment that consists of the first three trials in the chain. A trial on this experiment, then, notionally consists of spinning the wheel, taking the indicated spin speed, spinning the wheel again, taking the indicated spin speed, and spinning the wheel for a third time. The outcome of this third spin is the outcome of the experiment. More generally, an n spin experiment consists of the first n trials in a chain, with the outcome being that of the nth trial and the ic-value being the initial ic-value of the chain. In a given chain, then, all such experiments, regardless of the value of n, have the same ic-value, namely, the initial ic-value of the chain. My general name for an experiment of this sort is a **multi-mechanism experiment**.

On the multi-mechanism approach, the question of whether, say, the outcomes of the third and fourth chained trials on an autophagous wheel are independent, becomes the question of whether the outcomes of the three- and four-spin experiments in a given chain are independent. Because they share the same ic-value, the question of the experiments' independence will be resolved by an examination of their evolution functions. (This approach to the independence of chained trials is to the main approach of sections 3.72 and 3.73 as the main approach to short-term coupling is to the alternative approach of section 3.66: the first approach of the pair examines the effect that causal commonalities have on evolution functions, the second the effect that causal commonalities have on IC-densities.)

It will be useful for graphical purposes to consider a special version of the straight wheel of fortune. In this wheel, (a) there are only two colored sections: half the disk is red and half is black, (b) initial velocities are limited to those that cause the wheel to perform one revolution or less, that is, if a speed of v causes exactly one revolution, all velocities are in the range $0 \ldots v$, and therefore (c) within each colored section, markings corresponding to the range of velocities $0 \ldots v$ are evenly distributed.

The probabilities of red and black for a single spin on such a wheel are not microconstant. I will show, however, that as the number of spins in a multi-spin experiment increases, the evolution function takes on a microconstant aspect.

Suppose that a speed between 0 and $v/2$ yields a red outcome, and a speed between $v/2$ and v yields a black outcome. Then the evolution function for a single spin, taking red as the designated outcome, is

$$h(\zeta) = \begin{cases} 1, & \text{if } 0 < \zeta \leq v/2; \\ 0, & \text{if } v/2 < \zeta \leq v. \end{cases}$$

This function is graphed in figure 3.8(a).

Now consider a two-spin experiment. Assuming that the velocities are marked on the wheel in ascending order in the direction of spin, the evolution function for the two-spin experiment is:

$$h(\zeta) = \begin{cases} 1, & \text{if } 0 < \zeta \leq v/4; \\ 0, & \text{if } v/4 < \zeta \leq v/2; \\ 1, & \text{if } v/2 < \zeta \leq 3v/4; \\ 0, & \text{if } 3v/4 < \zeta \leq v. \end{cases}$$

This function is graphed in figure 3.8(b).

The pattern is clear. The evolution functions for three- and four-spin experiments are shown in figure 3.8(c) and (d). In general, for an n spin experiment, the evolution function will oscillate between zero and one $2^n - 1$ times. Thus, as the number of spins increases, the evolution function becomes more and

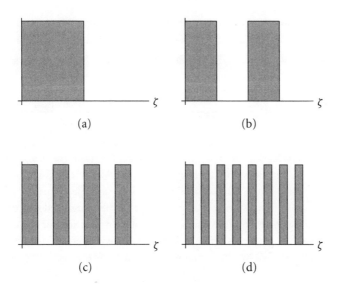

Figure 3.8 Evolution functions for multi-mechanism experiments consisting of several spins on a certain straight wheel of fortune: (a) one spin, (b) two spins, (c) three spins, (d) four spins. In each case, ζ is the speed of the first spin in the chain.

more microconstant, but maintains the same strike ratio of 1/2. (The process resembles that by which some fractals are created; for the relation between stretch-and-fold dynamics and fractals, see Stewart (1989).)

What does this have to do with independence? Consider the mth and n^{th} outcomes of a deterministic chain of trials. These can be considered as the outcomes of two multi-spin experiments, one of m spins and one of n spins, that share the same ic-value, namely, the initial ic-value of the chain. The outcomes are independent, then, if the outcomes of the multi-spin experiments are independent. Assuming that the chain's initial ic-variable has a macroperiodic distribution, a sufficient condition for independence is that the relevant composite experiment is microconstant with a strike ratio equal to the products of the strike ratios for the individual multi-spin experiments. In what follows, I show that there are systematic reasons why this will be so.

Consider, as an example, the probability of obtaining red outcomes from multi-mechanism experiments of m and $m + 1$ spins on the straight wheel described in this section. These multi-spin experiments are independent if the composite evolution function for red outcomes on both experiments is microconstant with a strike ratio of 1/4 (assuming a macroperiodic distribution over

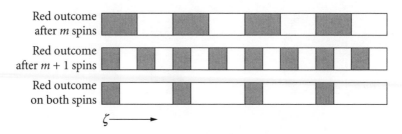

Figure 3.9 Representations of parts of the evolution functions for m and $m + 1$ spin experiments on a certain straight wheel. Each of the long rectangular boxes is, in effect, a compressed graph of an evolution function: spin speed increases from left to right; gray areas mark spin speeds giving rise to the designated outcome, red. Also shown is the corresponding part of the evolution function for the composite experiment consisting of trials on m and $m + 1$ spin experiments in the same chain. The designated outcome for the composite experiment is the event of obtaining red on both trials.

the chain's initial ic-variable). To see if this condition for independence holds, I construct a part of the composite evolution function in figure 3.9. Pictured in the figure are parts of the evolution functions for red on m and $m + 1$ spin multi-mechanism experiments, and the corresponding part of the composite evolution function for the event of obtaining red outcomes on both multi-mechanism experiments. As can be seen, the composite evolution function is microconstant with a strike ratio of 1/4, as desired for independence.[27]

This is not just a happy coincidence. The evolution functions for multi-spin experiments fulfill, as do the evolution functions for all multi-mechanism experiments satisfying the sufficient conditions to be developed and stated in section 3.76, a rather abstract condition sufficient for the strike ratio of the composite evolution function to equal the product of the strike ratios for the individual evolution functions. This is the condition that the members of a constant ratio partition for one of the experiments, namely, the experiment consisting of more trials in the chain, each fall entirely within a member of the outcome partition (that is, a maximal contiguous gray or white interval) for the other experiment. One might say that the two evolution functions are **harmonic**, since one is, in the technical sense, a harmonic of the other. Because of this harmony, the strike ratio of the composite evolution function bears the desired relation to the strike ratios of the individual multi-mechanism experiments. The harmonic relation also guarantees that, if the two multi-mechanism experiments are microconstant, the composite experiment is microconstant, since any constant ratio partition for the evolution function of the experiment consisting of fewer trials in the chain is also a constant ratio partition for the composite evolution function.

3.75 Independence in the Zigzag Wheel

The sufficient conditions for the independence of chained trials stated in section 3.72 are quite limited: they require that the way in which the IC-values for the $(i + 1)^{\text{th}}$ trial in a chain are determined is independent of the outcome of the ith and all previous trials; that is, they require that IC-evolution is outcome independent. The majority of real-life chains do not meet this requirement. An autophagous wheel, for example, will violate the requirement if there is any difference in the way that spin speeds are marked from colored section to colored section. When an experiment fails to meet the requirement, I say that its IC-evolution is **outcome dependent**. In the following sections, I will explore the conditions under which stochastic independence can exist in chains of trials with outcome-dependent IC-evolution.

THE ZIGZAG WHEEL

Consider the following example of an autophagous wheel that has outcome dependent IC-evolution, but that, when chained, produces independent outcomes. The wheel is a very slightly modified straight wheel. I assumed, recall, that the velocities on the straight wheel were marked on each colored section in the same order, relative to the direction of spin. The modified wheel has a paint scheme in which velocities are marked in ascending order on red sections but in descending order on black sections. Call this the *zigzag wheel*.

In the case of the straight wheel of fortune, to calculate the effective IC-value of the next trial, you do not need to know which colored section the pointer indicates, only which point in that section (relative to its boundaries) the pointer indicates. More abstractly, you only need to know low-level information about the IC-value of the trial. But this is not so for a zigzag wheel. Suppose I tell you that the pointer of a zigzag wheel indicates a spot 3 mm to the right of the section boundary. This information is not sufficient to determine the IC-value of the next trial. Because the order of speed markings depends on the color of the section, the spot that is 3 mm from the right in a red section will indicate a different speed from the spot that is 3 mm from the right in a black section.

To determine the IC-value for the next trial, then, you must know the outcome of the current trial. More abstractly, you must know high-level as well as low-level information about the IC-value of the current trial.

This is reflected in the fact that there are two IC-evolution functions for the zigzag wheel, one for trials that produce red outcomes and one for trials that produce black outcomes. I presented them in section 3.73; here they are again:

$$H(\zeta) = 100\zeta \bmod 100, \qquad \text{if } \zeta \text{ produces a red outcome;}$$
$$H(\zeta) = 100 - (100\zeta \bmod 100), \qquad \text{if } \zeta \text{ produces a black outcome.}$$

In effect, IC-evolution on the zigzag wheel can be conceived of as having two stages: first, depending on the outcome of a trial, one or other IC-evolution function is chosen, and second, that IC-evolution function is applied to determine the IC-value of the next trial.

As a result, the course of IC-evolution over a series of chained trials on the zigzag wheel will depend on the outcomes of those trials. Thus the IC-value of, say, the ith trial will depend on the outcomes of the $i - 1$ earlier trials, a situation that one would expect to undermine independence.[28]

Why, then, are the outcomes of chained trials on the zigzag wheel independent? The answer is that, although IC-evolution on the zigzag wheel is different for a trial that yields a red outcome than for a trial that yields a black outcome, the two forms of IC-evolution produce the *same probability distribution* over effective IC-values for the next trial in the chain, namely, a uniform distribution over all speeds from 0 to 100—just the distribution produced by the straight wheel. (This assumes that the distribution over the relevant ancestor IC-values is uniform; more on this below.)

For the purposes of determining the distribution over the effective IC-values for the ith trial, then, the outcomes of the previous $i - 1$ trials do not matter. Any sequence of outcomes will produce the same distribution of effective IC-values, and thus, the same probability distribution over the outcomes of the ith trial. It follows that the probability of red (or of black) on the ith trial is the same whatever the outcomes of previous trials in the chain or, in other words, that the outcome of the ith trial is stochastically independent of the outcomes of all previous trials.

The independence of trials on the zigzag wheel depends, then, on two important facts already encountered in this chapter:

1. Independence of effective ic-values is sufficient for independence, and
2. Matters that make a difference to particular ic-values, such as a choice of ic-evolution function, do not necessarily make a difference to the *distribution* of those same ic-values.

The reason that chained trials on a zigzag wheel are independent is also, broadly speaking, the reason that consecutive interactions between enions in a complex system are independent. In the remainder of this section, and in section 3.76, my aim is to generalize and to formalize what I have said about the zigzag wheel to obtain an independence result of broad enough scope to deal with enion interactions.

SUFFICIENT CONDITIONS FOR INDEPENDENCE

I will begin by stating some very abstract sufficient conditions for independence that are similar to, but more general than, the conditions stated at the end of section 3.72. Most of the rest of this chapter will be an attempt to show under what kinds of circumstances these abstract conditions are likely to be satisfied. Readers who prefer the specifics first may want to skip this section on a first pass.

Given a designated outcome e, a series of n chained trials on an experiment will be independent with respect to e if:

1. There exists a division of the initial, full ic-value of the chain into n distinct levels of information, conveyed by variables $\omega_1, \ldots, \omega_n$, such that the distribution over the effective ic-values for the ith trial, conditional on all previous outcomes, is entirely determined by the distribution over the information at the ith level, that is, over ω_i.[29] I mean *determined* here in a mathematical rather than a causal way: there must be some transformation of the distribution over ω_i that yields the distribution over the effective ic-values for the ith trial, no matter what outcomes come in between.
2. The probability distribution over the chain's initial ic-variable is macroperiodic with respect to the partition formed by the effective ic-values for the first trial in the chain. This condition ensures a lack of correlation between the distributions of information at different levels.

When the chained trials are microconstant, condition (1) may be relaxed considerably. It is sufficient for the independence of microconstant trials that each have an effective ic-variable distribution, whatever the outcomes of the previous trials, that is macroperiodic. Thus (1) can be loosened so as to require only that the distribution over ω_i be sufficient, together with the IC-evolution function, to guarantee that there is a macroperiodic distribution over the effective IC-values for the ith trial; no particular form for that distribution need be determined.

Because the conditions are rather more lax in the microconstant than in the non-microconstant case, I will treat the two separately. Section 3.76 considers how chained microconstant trials might satisfy the stated conditions; non-microconstant trials are not discussed in this book but in website section 3.7A. Before bifurcating, though, I want to introduce a notion that will play an important role in both parts of the discussion, the notion of a *restricted IC-evolution function*.

THE RESTRICTED IC-EVOLUTION FUNCTION

In what follows, I define the notion of a restricted IC-evolution function, and I put the definition to use to explain the independence of chained trials on setups like the zigzag wheel.

Consider one more time the functions that describe IC-evolution on the zigzag wheel:

$$H(\zeta) = 100\zeta \bmod 100, \qquad \text{if } \zeta \text{ produces a red outcome;}$$
$$H(\zeta) = 100 - (100\zeta \bmod 100), \qquad \text{if } \zeta \text{ produces a black outcome.}$$

I have been speaking as though these are two separate IC-evolution functions, but the reader might object that, in a sense, there is just one IC-evolution function with two cases. To eliminate the possibility of confusion, I introduce the notion of a **restricted IC-evolution function**, which is a function that describes the IC-evolution of a *subset* of the full range of possible IC-values. What are displayed above, then, can be understood as two restricted IC-evolution functions, one taking as its domain all IC-values that yield a red outcome on the zigzag wheel, the other taking as its domain all IC-values yielding a black outcome.

The IC-evolution of an experiment can be fully characterized by a set of restricted IC-evolution functions whose domains form a partition of the range of possible IC-values for the experiment. I will now use the term *IC-evolution function* to refer to the union of all the restricted IC-evolution functions. On the new usage, then, the IC-evolution of the zigzag wheel is determined by a single IC-evolution function made up of two restricted IC-evolution functions, one for red IC-values and one for black IC-values.

There are, of course, infinitely many different ways to divide up an IC-evolution function into several restricted IC-evolution functions. One way is to make the division according to the outcome produced by an IC-value, as I have done with the zigzag wheel above. On this scheme, there will be as many restricted IC-evolution functions as there are designated outcomes. In what follows, I will employ a finer division: I will allocate one restricted IC-evolution function for each effective IC-value. That is, each restricted IC-evolution function will have as its domain the full IC-values corresponding to a single effective IC-value for a chained experiment.

In the case of the zigzag wheel, for example, the effective IC-values correspond to the intervals $[n, n + 1]$, where n is a whole number; each restricted IC-evolution function, then, will be defined over just one such interval. Since speeds on the zigzag wheel can vary from 0 to 100, the IC-evolution function for the wheel will be divided into 100 separate restricted IC-evolution functions. Most of these functions will be identical; specifically, all functions for an even n will be identical to one another, and all functions for an odd n will likewise be identical to one another. Why divide IC-evolution into 100 parts, rather than just two, as above? Because I want the domains of the restricted IC-evolution functions to be contiguous, for reasons that should soon become clear.[30]

That concludes the definition of a restricted IC-evolution function; now I put the notion to work. In the discussion that follows, I will chiefly be concerned with the distribution over the IC-values for the $(i + 1)^{\text{th}}$ trial in a chain, conditional on the effective IC-values for the previous i trials. Provided that this conditional density has the right property—which in the microconstant case is macroperiodicity—trials in the chain will be independent. The restricted IC-evolution functions will play a key role in understanding the properties of the conditional distribution, for the following reason.

Suppose that we know the distribution over the IC-values for the ith trial in a chain. Suppose also that the effective IC-value for the ith trial was z. Then the distribution over the IC-values for the $(i + 1)^{\text{th}}$ trial, conditional on the fact that the IC-value for the ith trial was z, can be constructed as follows.

1. Take the restricted IC-evolution function corresponding to the effective IC-value z, that is, the restricted IC-evolution function that has as its domain all and only those full IC-values that fall into the interval corresponding to z.
2. Take the density over the IC-values for the ith trial, and consider just the portion of the density that spans the interval corresponding to z.

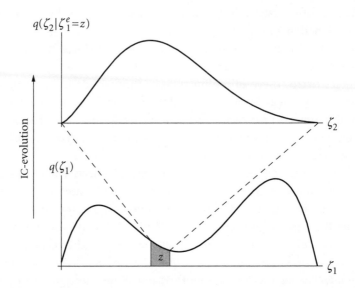

Figure 3.10 A restricted IC-evolution function acts on a small portion of the IC-density for the ith trial to determine the conditional density $q(\zeta_{i+1} \mid \zeta_i^e = z)$ for the $(i + 1)^{\text{th}}$ trial. The segment in question (shaded) spans the interval corresponding to the effective IC-value z.

3. Apply the restricted IC-evolution function to that portion of the density. The result is the desired conditional density, that is, the density over the IC-values for the $(i + 1)^{\text{th}}$ trial, conditional on the effective IC-value for the ith trial having been z.

The procedure is illustrated in figure 3.10.

The shape of the conditional density, then, is entirely determined by (a) the portion of the IC-density for the previous trial that spans z, and (b) the restricted IC-evolution function for z.

This fact can be used to explain the independence of trials in the zigzag wheel. The explanation is not materially different from that offered above; the point is just to show how the explanation goes when working with restricted IC-evolution functions divided up along the lines of the effective IC-values.

Suppose that the IC-density over the initial IC-variable for a chain of trials on the zigzag wheel is macroperiodic. Call the effective IC-value of the very first trial z_1. What is the IC-density for the second trial, conditional on the effective IC-value for the first trial having been z_1? It is determined by the portion of the initial IC-density that spans z_1 and by the restricted IC-evolution function for z_1. Because the zigzag wheel is microconstant, the interval corresponding to z_1 is micro-sized. Thus, the portion of the initial IC-density spanning that interval is uniform (since the distribution over the initial IC-variable is macroperiodic). The conditional IC-density, then, is the result of applying the relevant restricted IC-evolution function to a uniform density. The restricted IC-evolution function has one of two forms, depending on whether the outcome of the first trial is red or black. But both of these forms have exactly the same effect on a uniform density: they transform a uniform density into another uniform density over the entire range of speeds, from 0 to 100. Regardless of the effective IC-value for the first trial, then, the conditional density for the second trial is uniform over the full range of possible speeds. Consequently, regardless of the outcome of the first trial, the probability distribution over the outcomes for the second trial is the same, namely, 1/2 for red and 1/2 for black.

Call the effective IC-value of the second trial z_2. What is the IC-density for the third trial, conditional on the effective IC-value of the first trial having been z_1 and that of the second trial having been z_2? By the same reasoning as in the last paragraph, the conditional density for the third trial will be uniform. Thus the probabilities for red and black on the third trial will be the same regardless of the outcomes of the first two trials. And so on.

The independence of all trials in the chain, then, follows from the fact that every restricted IC-evolution function for the zigzag wheel has the same effect on a uniform density.[31] One might put the point thus: the zigzag wheel's different restricted IC-evolution functions have different effects on particular IC-values, for which reason the IC-evolution of particular IC-values is outcome dependent. But the different restricted IC-evolution functions have the same effect on IC-value *distributions*, for which reason the IC-evolution of distributions is outcome *independent*. Because probabilities depend on distributions, not on particular IC-values, the probabilities are also outcome independent, that is, they are stochastically independent.

In the next section I use this form of reasoning to derive some very general conditions for the independence of chained microconstant trials. Somewhat less general conditions for the independence of chained non-microconstant trials will be found in website section 3.7A.

3.76 Microconstant Outcome-Dependent IC-Evolution

Consider a deterministic chain consisting of repeated trials on a microconstant experiment. The probability of a given outcome, conditional on all previous outcomes in the chain, will be the same for all trials, and equal to the strike ratio for the outcome, provided that the distribution over the effective ic-values for any trial, conditional on the outcomes of all previous trials, is macroperiodic. (For the sense in which a discrete distribution, such as that over effective ic-values, can be said to be macroperiodic, see section 2.72.)

It follows, for reasons given in the last section, that the outcomes of trials in such a chain will be independent if, given a macroperiodic distribution over the full ic-values for the ith trial, every restricted ic-evolution function produces a distribution that is macroperiodic over the full ic-values for, and hence over the effective ic-values for, the $(i + 1)^{th}$ trial. (I am assuming a macroperiodic distribution over the chain's initial ic-variable.) In this section I investigate some fairly general conditions on the restricted ic-evolution functions in virtue of which the desideratum will be satisfied.

First Condition: Weak Inflation

The domain of a restricted ic-evolution function is a region of full ic-values that spans only a single effective ic-value, and that is therefore micro-sized. If the restricted ic-evolution function is to produce a macroperiodic distribution, its range must be macro-sized. Thus the restricted ic-evolution function must be inflationary (as shown in figure 3.10). More exactly, it must be sufficiently inflationary to blow up even the narrowest effective ic-value into a range of values wide enough to count as macroperiodic, that is, wide enough to justify the application of the results asserting the equality of probability and strike ratio (theorems 2.2 and 2.3).

Note that inflation is required only in the weak sense (see the end of section 3.73), because, while it is required that the range of the restricted ic-evolution function be greater than its domain, it is not required that every subset of the domain have its measure increase under ic-evolution. Strong inflation will, however, turn out to be important in the discussion of the second condition.

Second Condition: Microlinearity

Inflation is not sufficient to guarantee macroperiodicity, because a sufficiently inflated distribution might not be sufficiently smooth. What is needed, then, is some condition on the restricted ic-evolution functions that ensures that the inflated distribution will be approximately uniform over micro-sized regions. There is no condition that can provide such an assurance if there is no constraint on the ic-density that serves as input to the restricted ic-evolution function, but the fact that the IC-density at each point in the chain is supposed to be macroperiodic provides a strong constraint: since the regions corresponding to effective IC-values are micro-sized, macroperiodicity of the IC-density as a whole entails approximate uniformity over the domain of each restricted IC-evolution function. What is needed, then, is for restricted IC-evolution functions to transform uniform distributions into macroperiodic distributions.

In the zigzag wheel, this need is met in virtue of the fact that each restricted IC-evolution function is linear: a linear transformation of a uniform distribution is always uniform. The zigzag wheel's IC-evolution, then, produces conditional distributions that are not just macroperiodic, but uniform. If every restricted IC-evolution function is linear, then, because the domain of each of these functions is a member of the experiment's outcome partition, the full IC-evolution function is microlinear with respect to the outcome partition. The converse also holds, thus, the microlinearity of the IC-evolution function with respect to the outcome partition is a sufficient condition for adequately smooth IC-evolution.

This suggests the following sufficient conditions for the independence of chained trials on a microconstant experiment: (a) the restricted IC-evolution functions are sufficiently inflationary (in the weak sense), (b) the full IC-evolution function is microlinear with respect to the experiment's outcome partition, and (c) the probability distribution over the chain's initial IC-variable is macroperiodic. (For a similar formal result, see theorem 3.18.)

I make three comments on these sufficient conditions. First, to express the conditions more compactly, say that if the restricted IC-evolution functions of an experiment are inflationary in the weak sense, then the experiment's complete IC-evolution function is inflationary in the weak sense. (To say that the complete IC-evolution function is weakly inflationary, then, is to say that it is weakly inflationary with respect to the outcome partition; see the partition-relative definition of weak inflation at the end of section 3.73.) The first condition can now be written: (a) the IC-evolution function must be sufficiently inflationary (in the weak sense).

Second, the reader may wonder whether folding plays any role in the independence of chained trials satisfying these conditions. The answer is no; folding is unnecessary for reasons given in section 3.73: microconstancy renders high-level information irrelevant for the purposes of determining a distribution of outcomes, therefore independence holds whether or not folding removes high-level information. In the discussion of chains of non-microconstant experiments in website section 3.7A, however, folding does matter.

Finally, note that if a restricted ic-evolution function is both weakly inflationary and linear, then it is strongly inflationary (a consequence of lemma 3.7). Thus, the stated conditions for independence implicitly require strong inflation.

Let me now try to broaden the conditions for independence. If a chain of microconstant experiments satisfies the conditions stated above, the distribution over the effective ic-values for any trial after the first, conditional on all previous outcomes, will be uniform. But it would be sufficient for independence if the distribution were merely macroperiodic. This suggests that the stated conditions can be considerably weakened. The weakened conditions will require, roughly, only that the restricted ic-evolution functions for the experiment are microlinear, rather than that they are linear. But some care must be taken in spelling out the scope of this microlinearity.

What is wanted is a kind of microlinearity that is sufficient to take a uniform distribution as input and to produce, as output, a distribution over the effective ic-values for the succeeding trial with the following property: it should be macroperiodic with respect to an optimal constant ratio partition for that trial.

This desideratum will be satisfied if each restricted ic-evolution function $H(\cdot)$ is microlinear with respect to a partition constructed as follows (see figure 3.11):

1. Take an optimal constant ratio partition for the next trial.
2. For each set V in the optimal constant ratio partition, take the inverse image of V under $H(\cdot)$. Divide each inverse image into maximal contiguous parts. Take the set of all the parts for every V; it is a partition of the domain of $H(\cdot)$. It is with respect to this partition that $H(\cdot)$ should be microlinear, or in other words, $H(\cdot)$ must be approximately linear over every set in this partition.

A condition that is slightly weaker uses the formal notion of a seed set, explained in section 3.B6 (see especially definition 3.19): each restricted ic-evolution function should be approximately linear over every seed set of each member of the optimal constant ratio partition.

Whatever the details, call the union of the partitions constructed in step (2), which is itself a partition of the domain of the complete ic-evolution function,

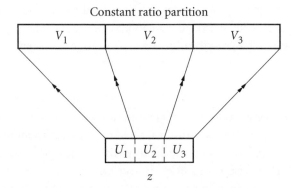

Constant ratio partition

Figure 3.11 Creating macroperiodicity from uniformity. The arrows represent the effect of the restricted IC-evolution function for the effective IC-value z, transforming a distribution over the IC-values for the ith trial (bottom box) into a distribution over the IC-values for the $(i + 1)^{th}$ trial (top box). The regions at the top—V_1, V_2, and V_3—are the members of the optimal constant ratio partition for the $(i + 1)^{th}$ trial, across each of which a macroperiodic distribution must be uniform. The region at the bottom is the effective IC-value z for the ith trial, across which the distribution is assumed to be uniform. To get there from here, the restricted IC-evolution function must be linear on each U_i, where U_i is the inverse image of V_i under the restricted IC-evolution function.

the inverse constant ratio partition. Then the new, weaker requirement on IC-evolution is that the IC-evolution function must be microlinear relative to the inverse constant ratio partition. Because any member of the inverse constant ratio partition is wholly enclosed in a member of the outcome partition, this condition is strictly weaker than the requirement that the IC-evolution function be microlinear relative to the outcome partition.

SUFFICIENT CONDITIONS FOR INDEPENDENCE

I can now state some fairly general sufficient conditions for the independence of chained trials on a microconstant experiment. Trials are independent if

1. The experiment's IC-evolution function is weakly inflationary to a sufficient degree (that is, sufficient for any restricted IC-evolution function to create a distribution that is wide enough to count as macroperiodic),

2. The experiment's IC-evolution function is sufficiently microlinear (that is, at least microlinear relative to an inverse constant ratio partition, or even better, relative to the outcome partition), and, as before,
3. The distribution over the initial IC-variable for the chain is macroperiodic.

These conditions are, I note, not difficult to satisfy.

Three remarks. First, the conditions stated here are sufficient for the maintenance of macroperiodicity essentially in virtue of the fact that a linear transformation of a uniform distribution is also uniform. In practice, no distribution will be perfectly uniform, of course, and no microlinear transformation will be everywhere perfectly linear. Because of these slight imperfections, the transformed distribution will be imperfectly uniform; see approximations 3.18.1 and 3.18.2. This in itself is no cause for worry.

In a deterministic chain, however, the same imperfectly microlinear transformation will be applied to the same imperfectly uniform distribution over and over again. This raises the possibility that the imperfections will grow more serious with iteration. There is, however, a powerful force counteracting the corruption of uniformity: strong inflation. The bumps and hollows caused by deviations from linearity and uniformity will tend to be stretched flat during each iteration by the strong inflationary process that produces the conditional distribution.[32] When the degree of inflation is high, as it must be to re-create a macroperiodic distribution from a distribution over a single member of the outcome partition, it will counteract even quite considerable imperfections in the IC-evolution function and the initial IC-density. Some comments along the same lines are made in approximation 3.18.3.

Second, I should make explicit, if only informally, the connection between the sufficient conditions just stated for the independence of chains of microconstant trials, and the sufficient conditions for the same stated in section 3.75. The conditions given in the earlier section require that the initial IC-value for a chain be divisible into levels of information such that the macroperiodicity of the distribution over the effective IC-values for the ith trial depends only on the distribution over the ith level of information in the initial IC-value.

A chain with a sufficiently inflationary, microlinear IC-evolution function satisfies this condition for the following reason. Because IC-evolution is inflationary, the effective IC-value of each next trial is determined by ever lower level information in the initial IC-value of the chain. The function that maps the information at the ith level in the initial IC-value onto the effective IC-value for the ith trial is the composition of a series of $i - 1$ restricted IC-evolution functions, the members of the series being determined by the outcomes of the first $i - 1$ trials. Because the restricted IC-evolution functions are linear, or sufficiently microlinear, no matter which $i - 1$ functions are chosen, a uniform distribution over the information at the ith level in the chain's initial IC-value is transformed into a macroperiodic distribution over the effective IC-variable for the ith trial. Thus, the macroperiodicity of the distribution over the effective IC-values for the ith trial depends only on the distribution over the information at the ith level in the initial IC-value. Provided this distribution is uniform, as it is if the chain's initial IC-variable has a macroperiodic distribution, the macroperiodicity of the distributions for all later trials, and thus independence, is assured. To be sufficiently inflationary and microlinear, then, is one way for a chain to satisfy the conditions for independence given in section 3.75.

Third, when the designated outcomes of a chained experiment are ranges of IC-values for the next trial, a case of special interest in the study of the probabilistic dynamics of complex systems, the experiment's evolution function will be determined by the same underlying mechanics as its IC-evolution function. It is rather intriguing that the properties required for the independence of chained microconstant trials—inflationary and microlinear IC-evolution—are also conducive to a microconstant evolution function. This topic is discussed in website section 3.7B.

The Multi-Mechanism Approach

I next discuss independence in chained microconstant trials from a different perspective, built around the notion, introduced in section 3.74, of a multi-mechanism experiment. As in section 3.74, the intention is not to obtain any new results, but to deepen the reader's understanding of the foundation of the results already obtained.

Let me begin with a simple visual demonstration, in which I juxtapose the evolution function for a one-spin experiment on the zigzag wheel with that for a two-spin experiment. (The two-spin experiment, recall, consists of spinning the wheel once, noting the speed z indicated by the pointer, spinning the wheel with speed z, and then noting the outcome, red or black.) The evolution

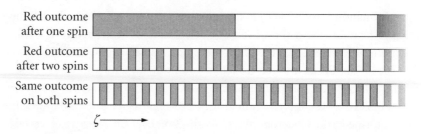

Red outcome after one spin

Red outcome after two spins

Same outcome on both spins

$\zeta \longrightarrow$

Figure 3.12 Representations of the evolution functions for one- and two-spin experiments on a zigzag wheel, together with the evolution function for the composite experiment consisting of trials on one- and two-spin experiments in the same chain. The designated outcome for the composite experiment is the event of obtaining the same outcome on both trials. For the interpretation of the figure, see the caption to figure 3.9.

functions are shown in figure 3.12. Also pictured is the evolution function for the composite experiment consisting of one- and two-spin experiments in the same chain. The designated outcome for the composite experiment is the event of obtaining the same outcome on both experiments.

It will be seen that the evolution functions for the one- and two-spin experiments are harmonic in the same sense as are the evolution functions for multi-mechanism trials on the straight wheel (see section 3.74, especially figure 3.9). As a result, the strike ratio for obtaining the same outcome on both experiments is equal to 1/2, just what, if the outcomes of the experiments are to be independent, the probability of the composite outcome must be. Thus, given a macroperiodic distribution over the initial conditions for the experiments—that is, a macroperiodic distribution over the chain's initial ic-variable—independence will obtain.

This observation generalizes. Multi-mechanism experiments involving microconstant mechanisms with sufficiently microlinear and inflationary ic-evolution functions, outcome independent or not, will have harmonic, microconstant evolution functions with the same strike ratios. The reasons for this are:

1. The structure of deterministic chains, together with inflationary ic-evolution, creates a recursive nesting effect, where the evolution function for a multi-mechanism experiment of i trials is created by embedding a compressed version of the entire evolution function for the single

mechanism experiment inside each member of the optimal constant ratio partition of the evolution function for the $i - 1$ trial multi-mechanism experiment. This nesting guarantees that the evolution functions are harmonic.

2. Microlinearity guarantees that the compression of the evolution function preserves the strike ratio, for exactly the reasons given in section 3.6, namely, that the part of a microlinear ic-evolution function that is responsible for the compression of a particular member of the optimal constant ratio partition is linear, and thus preserves ratios of areas.

The reason that an inflationary ic-evolution function is said to have a compressing or deflationary effect is that we are dealing with an outcome map. For reasons given in note 9, an ic-variable transformation—in this case, the transformation effected by an episode of ic-evolution—is manifested as the inverse transformation of an outcome map.

3.77 Generalization

CHAINS OF MIXED MECHANISMS

What are the prospects for generalizing the conditions for independence stated in section 3.76?

The most important generalization is to the case where a chain contains more than one kind of experiment. In such a chain, the outcome of one experiment must provide the ic-values for the next experiment even if the next experiment is of a wholly different kind. The easiest way to arrange this is, of course, to choose experiments all of which have the same kinds of ic-variables. Chains of events in complex systems, it will become apparent in chapter four, satisfy this description.

There are two cases to consider. In the first, the experiment for the ith trial in the chain is chosen according to some rule that does not take into account previous outcomes. The rule might, for example, take into account only the value of i, choosing one apparatus for, say, odd values of i and another apparatus for even values.

It should be clear that the sufficient conditions stated above are easily extended to such a chain. When all experiments in the chain are microconstant, each need simply maintain a macroperiodic distribution over the ic-variables. Assuming that the standard for macroperiodicity is roughly the same for all experiments (that is, that they all have optimal constant ratio partitions with similar CRIs), for an experiment to maintain macroperiodicity for the sake of a different experiment is no harder or easier than maintaining macroperiodicity for the sake of another iteration of itself. When there are different standards of macroperiodicity, the requirements for macroperiodicity maintenance will depend on details such as the ordering of experiments within the chain, but despite the complications, there is nothing particularly new or interesting to say about these cases.

If some experiments in the chain are non-microconstant, things are less straightforward. But everything that can usefully be said about such cases follows straightforwardly from the discussion of non-microconstant chains in website section 3.7A.

The second case to consider is a chain in which the experiment chosen for the ith trial depends in part on the outcomes of one or more of the previous $i - 1$ trials. This is the sort of chain found in complex systems, where the outcome of one interaction may well determine what kind of interaction next occurs.

The case of complex systems suggests an examination of chains in which the experiment chosen for the ith trial depends on the outcome of the $(i - 1)^{\text{th}}$ trial. Consider, for example, the following chain constructed from two wheels of fortune, called R and B. Let the first trial of the chain be conducted on R, and thereafter, let the ith trial be conducted on R if the outcome of the $(i - 1)^{\text{th}}$ trial is red, B if black.

If the wheels are not identical, the ic-evolution function that transforms the full ic-value will change from trial to trial. Thus the identity of the restricted ic-evolution function that determines the conditional distribution over effective ic-values for the $(i + 1)^{\text{th}}$ trial will depend not only on high-level information in the full ic-value for the ith trial, but on information that is normally no longer in that ic-value: the $(i - 1)^{\text{th}}$ trial's outcome.

This chain in fact offers no new complications. Provided that all the restricted ic-evolution functions for both R and B are sufficiently inflationary and sufficiently microlinear, a macroperiodic conditional distribution will be maintained throughout the existence of the chain, and the probabilities of outcomes in the chain will be equal to the corresponding strike ratios, and therefore independent of one another.

Of course, to say that the probabilities are independent is not to say that the probability of obtaining, for example, red on the second trial is independent of the outcome of the first trial. Rather, the probability of red *conditional on the identity of the experiment on which the trial is taking place* is independent of the outcome of the first trial. (This conditional aspect is built into the definition of complex probability: all complex probabilities are conditional on an experiment type.) For example, if the probability of red on *B* is 1/2 and that of red on *R* is 5/6, then the probability of red on *B* followed by red on *R* is 1/2 multiplied by 5/6, or 5/12. Independence of this sort is all that can reasonably be expected in such cases, and more importantly, it is all that is necessary for the explanation of the foundations of EPA in chapter four.

The same sorts of comments apply to more complicated chains, in which non-microconstant experiments may be chosen or in which choices depend on outcomes other than the immediately preceding outcome. But as before, I omit the details.

MULTIVARIABLE MECHANISMS

All the examples in this section have involved experiments with one IC-variable. How does generalization to experiments with two or more IC-variables affect the results? The sufficient conditions for independence are the same in the multivariable case as in the single-variable case. In the micro-constant case, these conditions will require that IC-evolution inflate every IC-variable in an inflationary and microlinear way, so that the joint IC-density, conditional on the previous outcomes in a chain, remains macroperiodic.

This may appear to be an unrealistically strong requirement, since experiments in complex systems have many IC-variables, and some of these appear not to be inflated (see especially sections 4.92 and 5.4). The requirement can be relaxed considerably, however, by appealing to the notion of a strike set (see section 2.25). The probability of an outcome is equal to its strike ratio, recall, if the distribution over the IC-variables in a strike set is, in the relevant sense, macroperiodic.[33] Independence in a chain of multivariable experiments, then, can be achieved by maintaining a macroperiodic distribution over just those IC-variables in a strike set for the experiments. Thus, only the IC-evolution of the variables in the strike set need be inflationary and microlinear. The practical significance of this result will depend, of course, on the number of IC-variables in the strike set. Worked examples will be found in sections 4.8 and 4.9 (though the approach in section 4.8 is more sophisticated than is indicated by this brief subsection).

BEYOND MICROLINEARITY

The remaining two subsections discuss the prospects for extending the independence results derived above to deterministic chains with IC-evolution that is not microlinear or not inflationary.

I begin with microlinearity. The role of microlinearity in the sufficient conditions for independence is to ensure that ic-evolution preserves macroperiodicity. Can the requirement of microlinearity be relaxed in any way? A microlinear transformation preserves macroperiodicity because it is linear over any microregion, and linear transformations preserve the uniformity of a distribution. One approach to the question is to ask how a transformation that is not linear over a microregion might nevertheless preserve uniformity. The concept of *patchwork linearity* (definition 3.6) provides a modification of linearity that preserves uniformity, but I doubt that it has any physical interest. Otherwise there is not much prospect for a straightforward generalization.

I think that there is more to say, however. One interesting line of thought is that the effects of different non-microlinear ic-evolution functions operating in succession may tend to cancel one another out, so that uniformity may be, over the course of several trials, preserved under ic-evolution (see website section 3.6A for some further comments in this vein). An attempt to make good formally on this comment would require, perhaps, an assumption that non-microlinear ic-evolution functions are in some sense randomly selected; I am not sure how such an assumption would be justified.

Another, less dramatic possibility is exemplified later in this study, in the discussion of statistical physics: in particular cases particular techniques may allow the application of the sufficient conditions developed above even where microlinearity does not quite hold true (section 4.85).

Beyond Inflation

The question whether inflation is necessary in order to ground independence in deterministic chains is far more interesting. All the results about chains derived above depend on the ability of ic-evolution to mine an endless supply of new, randomly distributed material from ever lower levels of the initial, full ic-value of a trial. This information is brought to the surface by means of inflationary ic-evolution, which, as it goes on, elevates lower and lower level pieces of information to the effective level.

Can non-inflationary ic-evolution provide a basis for independence in chains? More specifically, can non-inflationary ic-evolution produce something that simulates the properties of the genuinely random low-level information delivered by inflationary ic-evolution?

There is a kind of ic-evolution that is not inflationary, but which may generate sequences of outcomes that are independent with respect to many, but not all, selection rules. I have in mind functions that *shuffle* the information in the initial ic-value that lies above a certain level. Consider a function that, rather than deleting the two most significant digits of its argument, moves those digits to the seventh decimal place and moves all the digits before and including those in the seventh place up two places to make room. For example, $H(12.34567890) = 34.56789012$. The first five trials on a wheel of fortune with such an ic-evolution function will be independent (assuming that the effective ic-value for a trial on the wheel is the integer part of the full ic-value). In a longer sequence of trials, however, there will be an evident pattern. For any i, the ith and $(i + 5)^{th}$ trials will have the same outcome. If an ic-evolution function shuffles the full ic-value in more sophisticated ways, the pattern— unavoidable with any shuffling function—will be far less evident.

From this line of inquiry, several questions immediately arise:

1. What kinds of functions have desirable shuffling properties?
2. Are some of these functions also microlinear?
3. Is there any reason to think that any of these functions describe ic-evolution in real, physical chains?

I will say something about each of these questions in turn.

The mathematics of shuffling functions is investigated in the study of pseudorandom number generators used by computers. (Knuth 1998 provides a good overview.) This is an active, but not a particularly extensive area of research, and given the practical concerns of most of the work, not many different kinds of functions have been surveyed. By far the most well studied shuffling functions, and the most popular pseudorandom number generators, use what are called *linear congruential* functions. They have the form

$$f(x) = (ax + c) \bmod m$$

where x, a, c, and m are integers. A sequence of pseudorandom numbers is generated by taking an arbitrary initial number and applying f repeatedly. Many other pseudorandom number generators are variations on this technique. For judicious choices of a and m, the linear congruential method and its relatives provide quite good pseudorandom sequences, that is, sequences that are by most measures very close to being probabilistically patterned.

A linear congruential function, it will be observed, is simply a stretch-and-fold function, where a determines the linear stretch and m determines the folding scheme. It is very similar in mathematical structure, then, to the inflationary ic-evolution functions discussed above. But, because it involves only integer arithmetic, it is not inflationary: the least significant digit in the $(i + 1)^{\text{th}}$ member of a linear congruential sequence is always determined by information at that level and above in the ith member of the sequence. A similar effect could be achieved by operating with reals but rounding off after each iteration. I will return to this point shortly (see especially note 34).

Because the stretch in a linear congruential function is linear, the function itself is microlinear in the sense required for the preservation of macroperiodicity. This suggests that a more complicated random number generator might use a function of the form

$$f(x) = g(x) \bmod m$$

where $g(x)$ is a stretch that is microlinear. (The linear congruential function is just the special case in which $g(x)$ is linear or linear plus constant.) A small amount of research has been done on polynomial choices of $g(x)$, and they seem often to be good generators of random numbers (Knuth 1998, 26, 552–553), but for the most part, workers in the field prefer the faster generators based on linear functions.

There are encouraging answers, then, to questions (1) and (2): some non-inflationary functions are microlinear and have powerful shuffling abilities. Is there any reason to think that such functions might be encountered in nature? I will discuss just one way in which they could turn up. In section 4.8, I argue that ic-evolution in some systems of statistical physics is inflationary and microlinear, and so, using the arguments developed above, that these systems have various independence properties. The application of the argument to any particular kind of system depends, therefore, on the availability of infinite amounts of low-level information in any ic-value of the system, for example, in the values of the microvariables giving the position of a particular molecule of a gas. But what if, say, space is discrete? Then, after a point, there is simply no new low-level information to draw on. In such a case, microlinear, inflationary ic-evolution would become the kind of microlinear, non-inflationary process represented by the non-linear generalization of the linear congruential function.[34] The same observation would be true of a complex system in which access to very low level information is limited for other, metaphysically less dramatic, reasons.

Can shuffling ic-evolution generate more or less the same kinds of probabilistic patterns as inflationary ic-evolution? I will not undertake a theoretical investigation of this question here, but there are reasons to think that the answer is yes. Any computer simulation of a physical system is, in effect, a simulation of a system in a world where microvariables can have values of only limited precision. Yet such simulations generate the usual statistics, for example, the Maxwell-Boltzmann distribution over the positions and velocities of the molecules in a confined gas. In computer simulations that I have conducted, the positions and velocities of gas molecules were represented by *integers;* the Maxwell-Boltzmann distribution was nevertheless observed.[35] These probabilistic patterns must be pseudorandom, rather than truly random. That is, there must necessarily be some predictable relation between the outcomes of periodic measurements of the simulated molecules. But it is far too complicated to discern in the data. For all we know, the same is true of the actual world.

APPENDIX

3.A Conditional Probability

On several occasions I make use of conditional complex probability distributions. This appendix explains how conditional distributions are to be understood.

In order to satisfy the axioms of the probability calculus, the definition of conditional complex probability must entail the following identity:

$$(\text{CP}) \quad \text{cprob}(e|f) = \frac{\text{cprob}(ef)}{\text{cprob}(f)}$$

in those cases where $\text{cprob}(f)$ and $\text{cprob}(ef)$ are well defined and $\text{cprob}(f)$ is not equal to zero.

The most straightforward way to satisfy CP is to take the identity itself as the definition of conditional probability. This will not do for my purposes, however, because for various reasons, my use of the notion of conditional probability several times assumes an alternative definition:

$$\text{cprob}(e|f) = \int h_e(\xi)\, q(\xi \mid \eta \in I_f)\, d\xi$$

where ξ is the IC-variable responsible for e outcomes, η is the IC-variable responsible for f outcomes, and I_f is the set of all f-values, that is, the set of all values of η that produce an f outcome. Note that the conditional probabilities I use in this study are often relativized to a selection; in such cases, the IC-density should, of course, be relativized to the same selection.

I ought to show that my definition satisfies CP. A sufficient condition that it does so is that the conditional IC-density $q(\xi \mid \eta \in I_f)$ satisfies CP, so that (with I_e defined analogously to I_f)

$$Q(\xi \in I_e \mid \eta \in I_f) = \frac{Q(\xi \in I_e, \eta \in I_f)}{Q(\eta \in I_f)}$$

from which it follows immediately that cprob($e|f$) satisfies CP. Ultimately, then, CP is satisfied if the IC-densities of the relevant basic IC-variables satisfy CP. The content of this requirement depends on how the basic IC-densities are interpreted. I assume that whatever interpretation is given to the basic densities, they satisfy CP, and hence that complex probabilities satisfy CP.

3.B Proofs

The following definitions, proofs, and so on formalize some of the claims made about independence in the main text. Section 3.B1 covers results stated in sections 3.2, 3.3, and 3.4 that do not invoke the notion of low-level information. Low-level information is characterized in sections 3.B2 and 3.B4, and various results about low-level information are proved in those and other sections, covering results stated in the main text in sections 3.4, 3.6, and 3.7.

3.B1 Stochastic Independence from Causal Independence

The three goals of this section are to provide a formal counterpart to the definition of a composite experiment in section 3.2, to show that independent IC-densities entail independent outcomes, the result that underlies most of section 3.3, and to prove that composite evolution functions inherit the microconstancy of their constituent experiments, as asserted in section 3.4.

In all the following theorems, X and Y are probabilistic experiments with IC-variables ξ and η and designated outcomes e and f respectively, and s is a selection. For most theorems, I assume one-variable experiments; generalization is straightforward.

Definition 3.1 Given probabilistic experiments $X = \langle q(\xi), S_X, H_X \rangle$ and $Y = \langle q(\eta), S_Y, H_Y \rangle$, and a selection rule s, the *composite experiment* XY_s is the probabilistic experiment $\langle q^s(\xi, \eta), S_X \times S_Y, H\rangle$, where the evolution functions in H are defined so that for any e in S_X and f in S_Y,

$$h_{ef}(\xi, \eta) = h_e(\xi)\, h_f(\eta).$$

The probability of a designated outcome type ef on the composite experiment XY_s, written cprob(ef_s), is defined to be

$$\mathrm{cprob}(ef_s) = \iint\limits_V h_{ef}(\xi, \eta)\, q^s(\xi, \eta)\, d\xi\, d\eta$$

where V is the domain of $q^s(\xi, \eta)$.

As with any complex probabilistic experiment, I regard the domain of a composite experiment's ic-density and evolution functions as restricted to possible values of the ic-variables, in this case, pairs of values of ξ and η that might possibly satisfy s.

The reasoning behind the definition of the composite experiment's evolution function is given in section 3.2. Note that both the notion of a selection rule, and the only part of the composite experiment that makes reference to a selection rule, its ic-density $q^s(\xi, \eta)$, are left undefined. What matters for the purposes of the definition is that for any selection rule, there is a corresponding ic-density, so that for any selection rule, there is a composite experiment.

Definition 3.2 Outcomes e and f, produced by probabilistic experiments X and Y respectively, are *stochastically independent* relative to a selection s just in case

$$\mathrm{cprob}(ef_s) = \mathrm{cprob}(e)\, \mathrm{cprob}(f)$$

where cprob(ef_s) is the probability of the event ef occurring on the relevant composite experiment XY_s.

Theorem 3.1 *Outcomes e and f, produced by probabilistic experiments X and Y respectively, are stochastically independent relative to a selection s if*

$$q^s(\xi, \eta) = q(\xi)\, q(\eta)$$

where $q^s(\xi, \eta)$ is the ic-density of the composite experiment XY_s.

Proof. Taking all integrals over the domains of the relevant ic-densities,

$$\mathrm{cprob}(ef_s) = \iint h_{ef}(\xi, \eta)\, q^s(\xi, \eta)\, d\xi\, d\eta$$

$$= \iint h_e(\xi)\, h_f(\eta)\, q(\xi)\, q(\eta)\, d\xi\, d\eta$$

$$= \int h_e(\xi)\, q(\xi)\, d\xi \int h_f(\eta)\, q(\eta)\, d\eta$$

$$= \mathrm{cprob}(e)\, \mathrm{cprob}(f)$$

as desired. ∎

Approximation 3.1.1 If the joint ic-density is everywhere slightly different from the product of the individual densities, the probability of *ef* will be different to an equally slight degree from the product of the individual probabilities.

Approximation 3.1.2 If the joint ic-density is in certain *bad regions* quite different from the product of the individual densities, the probability of *ef* will tend to deviate from the product of the individual probabilities in proportion to the probability of the bad regions. It is possible that the effects of different bad regions will to some degree cancel one another out, but cancelling out cannot, of course, be relied on.

The next theorem states that experiments with independent effective ic-variables have independent outcomes. The proof uses no particular formal definition of effective ic-values; all that is assumed is that the information in an effective ic-value is sufficient to determine the outcome of an experiment. The questions as to how this information is conveyed, and whether any surplus information is conveyed, need not be answered to take advantage of the result. The theorem is equally true on any of the various schemes for assigning effective ic-values discussed in section 2.7. Also assumed is the relation between the probability of an outcome *e* and the probability distribution over its effective ic-variables stated in section 2.72, namely, in the one variable case that

$$\mathrm{cprob}(e) = \sum_V h(\zeta^e) Q(\zeta^e)$$

where V is the set of possible effective ic-values for the relevant experiment.

Theorem 3.2 *Outcomes e and f, produced by probabilistic experiments X and Y respectively, are stochastically independent relative to a selection s if*

$$Q^s(\xi^e, \eta^f) = Q(\xi^e)\,Q(\eta^f)$$

where ξ^e and η^e are the effective IC-variables for X and Y, and $Q^s(\xi^e, \eta^f)$ is the effective IC-variable distribution for the composite experiment XY_s.

Proof. Let U be the domain of $Q(\xi^e)$, that is, the set of possible effective IC-values for X, and V the corresponding set for Y. The proof follows the reasoning of theorem 3.1:

$$\begin{aligned}
\operatorname{cprob}(ef_s) &= \sum_U \sum_V h_{ef}(\xi^e, \eta^f)\,Q^s(\xi^e, \eta^f) \\
&= \sum_U \sum_V h_e(\xi^e)\,h_f(\eta^f)\,Q(\xi^e)\,Q(\eta^f) \\
&= \sum_U h_e(\xi^e)\,Q(\xi^e) \sum_V h_f(\eta^f)\,Q(\eta^f) \\
&= \operatorname{cprob}(e)\,\operatorname{cprob}(f)
\end{aligned}$$

as desired. ∎

The next two results, theorems 3.3 and 3.4, establish the claim that composite experiments inherit the microconstancy of their constituent experiments.

Definition 3.3 The *partition product* of two partitions $\mathcal{U}$ and $\mathcal{V}$ is the set

$$\{\, U \times V : U \in \mathcal{U},\, V \in \mathcal{V} \,\}.$$

Obviously, this definition does not take advantage of the fact that $\mathcal{U}$ and $\mathcal{V}$ are partitions, but where I use the notion of a partition product in what follows, the sets of sets involved are, in fact, always partitions. Note that the partition product of partitions of sets X and Y is itself a partition of $X \times Y$.

Theorem 3.3 *Let X and Y be probabilistic experiments with constant ratio partitions $\mathcal{U}$ and $\mathcal{V}$ respectively for outcomes e and f, such that all members of $\mathcal{U}$ and $\mathcal{V}$ have measure greater than zero. Then the partition product of $\mathcal{U}$ and $\mathcal{V}$ is a* *constant ratio partition for the corresponding composite evolution function, and*

$$\text{srat}(ef) = \text{srat}(e)\,\text{srat}(f).$$

Proof. It suffices to show that the strike ratio for ef over any member of the partition product of $\mathcal{U}$ and $\mathcal{V}$ is equal to $\text{srat}(e)\,\text{srat}(f)$. Consider a member U of $\mathcal{U}$ and a member V of $\mathcal{V}$. Writing $M(R)$ for the measure of a region R, and ξ and η for the sets of ic-variables of X and Y respectively, the strike ratio for ef over the partition product member $U \times V$ is

$$\text{srat}(ef) = \iint\limits_{U \times V} h_{ef}(\xi, \eta)\, d\xi\, d\eta \Big/ M(U \times V)$$

$$= \iint\limits_{U \times V} h_e(\xi)\, h_f(\eta)\, d\xi\, d\eta \Big/ M(U \times V)$$

$$= \int\limits_{U} h_e(\xi)\, d\xi \int\limits_{V} h_f(\eta)\, d\eta \Big/ M(U)\, M(V)$$

$$= \text{srat}(e)\,\text{srat}(f)$$

as desired. ∎

Approximation 3.3.1 Departures from a constant strike ratio in individual experiments lead to commensurate departures from a constant strike ratio in the composite experiment.

The theorem, although it allows that X and Y may have many variables, is not as strong as the result for multivariable experiments stated in the main text. The theorem is useful only for cases where there are constant ratio partitions all of whose members have a measure greater than zero. But a constant ratio partition relative to a strike set that excludes some ic-variables has members all of which have measure zero (see definition 2.12). Another result is needed, then, to deal with multivariable experiments that have microconstancy only relative to a subset of their ic-variables, such as the tossed coin discussed in sections 2.25 and 2.B.

Theorem 3.4 *Let X and Y be complex probabilistic experiments, and ξ and η strike sets for X and Y respectively, relative to outcomes e and f. Then*

1. *The union of ξ and η is a strike set for any composite experiment XY relative to the composite outcome ef.*
2. *The composite experiment's constant strike ratio for ef over $\xi \cup \eta$ is the product of X's strike ratio for e and Y's strike ratio for f, that is, is equal to $\mathrm{srat}(e)\,\mathrm{srat}(f)$.*
3. *The partition product of constant ratio partitions for X and Y is a constant ratio partition for the composite experiment.*

Proof. To prove the theorem, it is sufficient to show that for any pair of restrictions a of X and b of Y, the partition product of a constant ratio partition for $h_e[a](\xi)$ and a constant ratio partition for $h_f[b](\eta)$ is a constant ratio partition for $h_{ef}[a, b](\xi, \eta)$, with a strike ratio of $\mathrm{srat}(e)\,\mathrm{srat}(f)$.

For any a and b,

$$h_{ef}[a, b](\xi, \eta) = h_e[a](\xi)\, h_f[b](\eta).$$

Since $h_e[a](\xi)$ and $h_f[b](\eta)$ have constant ratio partitions with strike ratios of $\mathrm{srat}(e)$ and $\mathrm{srat}(f)$, and with no members of measure zero (see definition 2.10), by theorem 3.3, the partition product of those partitions is a constant ratio partition of $h_{ef}[a, b](\xi, \eta)$ with strike ratio $\mathrm{srat}(ef) = \mathrm{srat}(e)\,\mathrm{srat}(f)$, as desired. ∎

The next proposition is used in the main text only incidentally (section 3.42).

Proposition 3.5 *Let ξ and η be random variables with a joint density $q(\xi, \eta)$, and let $\mathcal{U}$ and $\mathcal{V}$ be partitions of the possible values of ξ and η, respectively. The following are equivalent:*

1. *$q(\xi, \eta)$ is macroperiodic over the partition product $\mathcal{W}$ of $\mathcal{U}$ and $\mathcal{V}$.*
2. *For all values y of η, $q(\xi \mid \eta = y)$ is macroperiodic over $\mathcal{U}$, and for all values x of ξ, $q(\eta \mid \xi = x)$ is macroperiodic over $\mathcal{V}$.*

Proof. (1) entails (2): By (1), over any member W of $\mathcal{W}$, $q(\xi, \eta)$ is uniform, equal to, say, k. Thus within W,

$$q(\xi \mid \eta = y) = \frac{q(\xi, y)}{q(y)}$$

$$= \frac{k}{q(y)}.$$

It follows that $q(\xi \mid \eta = y)$ is constant for values of ξ within any W in $\mathcal{W}$, and so is macroperiodic over $\mathcal{U}$. A parallel argument shows that $q(\eta \mid \xi = x)$ is macroperiodic over $\mathcal{V}$.

(2) entails (1): For any two points (a, b) and (c, d) within a member W of $\mathcal{W}$,

$$
\begin{aligned}
q(a, b) &= q(\xi = a \mid \eta = b)\, q(\eta = b) \\
        &= q(\xi = c \mid \eta = b)\, q(\eta = b) \quad \text{(macroperiodicity of } q(\xi \mid \eta = y)) \\
        &= q(c, b) \\
        &= q(\eta = b \mid \xi = c)\, q(\xi = c) \\
        &= q(\eta = d \mid \xi = c)\, q(\xi = c) \quad \text{(macroperiodicity of } q(\eta \mid \xi = x)) \\
        &= q(c, d).
\end{aligned}
$$

Thus any two points in W have the same density, so $q(\xi, \eta)$ is uniform over any W in $\mathcal{W}$, and so is macroperiodic over $\mathcal{W}$. ∎

3.B2 Complete Low-Level Information

In this section I define the notion of *complete low-level information* concerning a set X. I assume that X is a real space $\mathbf{R}^n$, although the definition could easily be made more general. The main result of this section, theorem 3.10, states, roughly, that a macroperiodic distribution over a variable entails a uniform distribution over low-level information about that variable.

Definition 3.4 Let X be a real space $\mathbf{R}^n$. A *domain of complete low-level information concerning X* is a triple $\Omega = \langle Z, \mathcal{U}, \{\lambda_U\}\rangle$ consisting of:

1. A set Z, called the *index set*,
2. A partition $\mathcal{U}$ of X into contiguous, measurable regions, called the *level partition*, and
3. For each set U in the level partition $\mathcal{U}$, a bijection $\lambda_U : U \to Z$, called the *index function* for U. The third member of Ω is the set of these index functions.

The index functions may be put together to form what I call the *full index function*:

$$
\lambda(x) = \lambda_U(x)
$$

where U is the member of $\mathcal{U}$ into which x falls. For any x in X, $\lambda(x)$ gives the *complete low-level information* about x.

On the conception of low-level information inherent in this definition, high-level information about a member x of X determines into which member of the level partition x falls, while low-level information determines, given the high-level information, exactly where in the member of the level partition x falls. For further informal discussion, see section 2.7.

Example 3.1 The fractional part of a real number gives complete low-level information about that number. The index function maps a real number onto its fractional part, the index set is $[0, 1)$, and the level partition is $\{ [n, n+1) : n \in \mathbf{I} \}$, where $\mathbf{I}$ is the set of integers. The low-level information about x does not specify into which interval $[n, n+1)$ the value x falls, but it does specify x's position within that interval, whichever interval it is.

The requirement that the sets in the level partition be contiguous plays no role in the results proved below. It is included so that the definition better captures the informal notion of high-level information deployed in the main text, according to which high-level information about a point in a space determines that the point belongs to a particular locale in that space. For certain purposes, it may be useful to drop the contiguity requirement. Measurability is required here for later convenience.

The size of the sets in the level partition determines the "lowness" of the level of information. The smaller the sets, the lower the level of the information.

Whereas complete low-level information determines an exact position within a member of the level partition, partial low-level information, to be characterized in definition 3.13, determines an approximate position within a level partition member. Critical information about an IC-value (see section 2.73) is a kind of partial low-level information, as is, for example, the information about a real number conveyed by the eighth digit after the point in the number's decimal expansion. Infracritical information (see section 3.43) is, by contrast, complete low-level information, and it is in part to prove results about infracritical information in section 3.B3 that the notion of complete low-level information is developed here.

Definition 3.5 Let X be a real space $\mathbf{R}^n$. A domain of complete low-level information concerning X is *linear* just in case its index set is a subset of X and its index functions are all invertible linear or linear plus constant transformations.

A linear plus constant transformation is a linear transformation with a constant added. That is, a transformation $f : X \rightarrow X$ is linear plus constant just in case there exists a linear transformation $l(x)$ of X and a member a of X such that for all x, $f(x) = l(x) + a$.

Example 3.2 Construct a domain of complete low-level information about the positive real numbers as follows. Let the level partition $\mathcal{U}$ be $\{ [10^n, 10^{n+1}) : n \in \mathbf{I} \}$, let the index function for a member of the partition $U = [10^n, 10^{n+1})$ be

$$\lambda_U(x) = \frac{x}{10^n}$$

and let the index set therefore be $Z = [1, 10)$. This kind of low-level information does not cut off any significant digits of x, but it is low level because it does not convey into which of the members of the level partition x falls.

Linear low-level information about a macroperiodically distributed quantity is, I will later show (theorem 3.10), uniformly distributed. This is a consequence of the following property of linear domains: in a linear domain, a uniform distribution over any member of the level partition imposes a uniform distribution over the index set.

Although the notion of a linear domain of low-level information is quite adequate for my purposes when dealing with information about single-valued variables, it is not sufficiently flexible to deal with information about variables that take on vector values, that is, information about members of $\mathbf{R}^n$ with $n \geq 2$. The problem is that the various members of the level partition may have shapes sufficiently different from the index set that no linear function can bijectively map one to the other.

The solution to this problem is to extend the notion of a linear domain, as follows. Any given member U of a level partition is cut up into small regions of non-zero measure, say, cubes. The index set is cut up into an equal number of cubes, and these cubes are paired up. An invertible linear or linear plus

The constant a is equal to the magnitude of the determinant of the transformation.

Lemma 3.7 can be extended to patchwork linear transformations, as follows.

Lemma 3.8 *For any patchwork linear transformation $f : Y \to Z$ on X, there exists a constant a such that, for any measurable set R in Y,*

$$M(f(R)) = aM(R).$$

The constant a is equal to the determinant of the transformation.

Proof. Let $\mathcal{V}$ be the partition of Y over each member of which f is linear (guaranteed to exist by the patchwork linearity of f). Define $\mathcal{N}$ as the set of those parts of the members of $\mathcal{V}$ that belong to a given subset R of Y. That is,

$$\mathcal{N} = \{ V \cap R : V \in \mathcal{V} \}.$$

$\mathcal{N}$ is a partition of R. From this fact and the definition of a patchwork linear transformation, it follows that the images under f of the sets in $\mathcal{N}$ make up a partition of $f(R)$. Thus,

$$
\begin{aligned}
M(f(R)) &= \sum_{N \in \mathcal{N}} M(f(N)) \\
&= \sum_{N \in \mathcal{N}} M(l_V(N)) \quad \text{for } V \text{ such that } N = V \cap R \\
&= \sum_{N \in \mathcal{N}} aM(N) \quad \text{(lemma 3.7)} \\
&= a \sum_{N \in \mathcal{N}} M(N) \\
&= aM(R)
\end{aligned}
$$

as desired. ∎

Because a linear transformation is trivially a patchwork linear transformation, lemma 3.8 supersedes lemma 3.7.

I am now almost ready to state the main result of this section, the theorem stating that a macroperiodic probability distribution over a variable induces a uniform distribution over patchwork linear complete low-level information concerning that variable. I first prove a lemma that will be invoked in the proof of the theorem.

Lemma 3.9 *Let $\langle Z, \mathcal{U}, \{\lambda_U\}\rangle$ be a linear or patchwork linear domain of complete low-level information concerning a set X. Then, for any U in $\mathcal{U}$, there exists a constant a_U such that, for any subset H of Z,*

$$M(U \cap H^{-1}) = a_U M(H)$$

where H^{-1} is the inverse image of H under the full index function.

Proof. Observe that, because λ_U is surjective on Z, hence on H,

$$\lambda_U(U \cap H^{-1}) = H.$$

Now

$$M(U \cap H^{-1}) = \frac{1}{a}M(\lambda_U(U \cap H^{-1})) \qquad \text{(lemma 3.8)}$$

$$= \frac{1}{a}M(H).$$

Setting $a_U = 1/a$, the result is proved. ∎

Theorem 3.10 *Let $\langle Z, \mathcal{U}, \{\lambda_U\}\rangle$ be a linear or patchwork linear domain of complete low-level information concerning a set X, and q(x) a probability density over X. If q(x) is macroperiodic relative to $\mathcal{U}$, then the probability density $q_\lambda(z)$ imposed over the complete low-level information by q(x) is uniform.*

Proof. Let H be some subset of Z, and let H^{-1} be the inverse image of H under the full index function, that is, the set of all x such that $\lambda(x)$ is in H. The probability of H is

$$Q_\lambda(H) = Q(H^{-1})$$

$$= \sum_{U \in \mathcal{U}} Q(U \cap H^{-1})$$

$$= \sum_{U \in \mathcal{U}} k_U M(U \cap H^{-1}) \qquad \text{(lemma 3.6 and macroperiodicity)}$$

$$= \sum_{U \in \mathcal{U}} k_U a_U M(H) \qquad \text{(lemma 3.9)}$$

$$= M(H) \sum_{U \in \mathcal{U}} k_U a_U$$

$$= cM(H)$$

where c (equal to $\sum_{U \in \mathcal{U}} k_U a_U$) takes the same value for any H, and so is a constant. By lemma 3.6, then, the distribution over Z is uniform. ■

Approximation 3.10.1 As in the case of other results that take macroperiodicity as a prerequisite, the conditions for the application of this theorem could fail in two ways: the density might not be exactly uniform over each set in the partition, or there might be some members of the partition over which the density is nowhere near uniform. In either case, because the density is linearly transformed, the effect of any deviation from macroperiodicity will be, at worst, directly proportional to the deviation itself. I say *at worst* because, as with theorems 2.2 and 2.3, there exists considerable potential for the effects of different deviations to cancel one another out.

3.B3 Infracritical Independence

In what follows I define and put to use the notion of *infracritical independence*, which is the independence of a certain kind of low-level information that I call, of course, infracritical information. The aim is to show, in theorem 3.12 and proposition 3.13, that infracritical independence is equivalent, in the cases of interest, to weak independence (defined in section 3.4; see also proposition 3.5).

What counts as infracritical information about an ic-variable depends on what counts as the critical level, which in turn depends on a mechanism and a designated outcome (section 2.73). In the formalization that follows I represent this relativization as a dependence on what I will call a critical-level partition for an ic-variable. Information about a value above the critical level is information that determines into which member of the critical-level partition the value falls; information at the critical level or below is information that determines exactly where in that critical-level partition member the value falls.

For the purpose of applying the results stated in this section to problems concerning independence, the critical-level partition is just an optimal constant ratio partition for the relevant mechanism and outcome. However, by relativizing to a partition rather than explicitly to a mechanism and outcome, I make it clear that (a) the only role of the mechanism and outcome in these results is to determine a partition, and thus that (b) if any other partition into micro-sized regions were determined, the results would hold just as true. Consequently, in what follows, the only constraint on the critical-level partition is that its members are all micro-sized. As in section 2.C, even this constraint is only implied: what is actually required is that the densities in question are macroperiodic over the critical-level partition.

Definition 3.10   Given a real-valued ic-variable ζ and a critical-level partition $\mathcal{U}$ of ζ, the *infracritical information* about ζ relative to $\mathcal{U}$ is provided by a linear or patchwork linear domain of complete low-level information constructed as follows:

1. The index set is the unit interval $[0, 1]$.
2. The level partition is the critical-level partition.
3. The index functions are any linear or patchwork linear transformations mapping the level partition members to the index set.

The definition can be extended to multivariable experiments by choosing an index set of the appropriate dimensionality. Because there is more than one choice for the index functions that satisfies the definition, there is more than one domain of low-level information that can convey what I refer to as *the* infracritical information about a variable. When I later define the independence of critical-level information, it will be seen that it is sufficient that for some choice of low-level domains, the independence relation holds. In any case, if the information is independent for one choice, it is independent for all choices. For this reason, it is unnecessary to relativize infracritical information to a choice of domain.

Definition 3.11 Given two ic-variables ξ and η and critical-level partitions $\mathcal{U}$ and $\mathcal{V}$ for ξ and η respectively, the *joint infracritical information* about the two variables relative to the two partitions is provided by a linear or patchwork linear domain of complete low-level information constructed as follows:

1. The index set is the Cartesian product of the index sets for the infracritical information concerning ξ and η, that is, the unit square $[0, 1]^2$.
2. The level partition is the partition product of $\mathcal{U}$ and $\mathcal{V}$. That is, the level partition will consist of all sets $W = U \times V$, where U and V are members of $\mathcal{U}$ and $\mathcal{V}$ respectively.
3. The index functions are the appropriate combinations of the index functions for the infracritical information concerning ξ and η. That is, if the index functions of the infracritical information for ξ and η are $\mu(\cdot)$ and $\nu(\cdot)$, then the index function for the joint infracritical information is

$$\lambda(\xi, \eta) = \big(\mu(\xi), \nu(\eta)\big).$$

This definition can be generalized in obvious ways to take care of multivariable experiments, as can the definitions and results stated in the rest of this section.

The usefulness of the notion of the joint infracritical information depends on the level partition for the joint infracritical information being a constant ratio partition for the corresponding composite experiment. Assuming that $\mathcal{U}$ and $\mathcal{V}$ are constant ratio partitions for the individual experiments in question, this is guaranteed by theorems 3.3 and 3.4.

I now define infracritical independence, beginning with an informal characterization. Two ic-variables ξ and η have infracritical independence relative to a selection s just in case (a) infracritical information about ξ is independent relative to s of infracritical information about η, and (b) infracritical information about ξ and η is independent relative to s of higher-level information about ξ and η, that is, of information about which regions in their respective critical-level partitions ξ and η fall into. (All infracritical information is relative to a critical-level partition, hence infracritical independence is relative to two critical-level partitions as well as to s.)

To define infracritical independence more formally, I use the following notation. I will write p to represent infracritical information about ξ and r to represent infracritical information about η. The density imposed over p by

$q(\xi)$ is written $q_\mu(p)$ and that imposed over r by $q(\eta)$ is written $q_\nu(r)$. Consider the joint infracritical information for ξ and η. Define

$$q_\lambda^s(p, r \mid U, V)$$

as the density imposed over the joint infracritical information by

$$q^s(\xi, \eta \mid \xi \in U, \eta \in V)$$

where U and V are members of the critical-level partitions for ξ and η respectively.

Definition 3.12 Given two IC-variables ξ and η, critical-level partitions $\mathcal{U}$ and $\mathcal{V}$ of ξ and η respectively, and a selection s, ξ and η have *infracritical independence* relative to s, $\mathcal{U}$, and $\mathcal{V}$ just in case for all U in $\mathcal{U}$ and V in $\mathcal{V}$,

$$q_\lambda^s(p, r \mid U, V) = q_\mu(p)\, q_\nu(r).$$

The main result in this section states that if two macroperiodically distributed IC-variables have infracritical independence relative to two critical-level partitions, then their joint density is macroperiodic relative to the joint critical-level partition (that is, relative to the partition product of the two individual critical-level partitions). Thus, macroperiodic IC-densities with infracritical independence are weakly independent.

Lemma 3.11 Let $\langle Z, \mathcal{U}, \{\lambda_U\}\rangle$ be a domain of low-level information over a set X, and $q(x)$ a probability density over X. Then for any member U of the level partition $\mathcal{U}$ and any subset R of U,

$$Q(R) = Q(U) \int_{\lambda(R)} q_\lambda(z \mid U)\, dz$$

where $q_\lambda(z \mid U)$ is the density imposed over the index set by the conditional density $q(x \mid x \in U)$.

Proof. Writing $Q(R \mid U)$ for $Q(x \in R \mid x \in U)$,

$$Q(R) = Q(U)\, Q(R \mid U).$$

Now

$$Q(R \mid U) = Q(x \in R \mid x \in U)$$
$$= Q_\lambda(\lambda(R) \mid U)$$
$$= \int_{\lambda(R)} q_\lambda(z \mid U)\, dz.$$

From the two equalities, the desired result follows.　■

Theorem 3.12 *If two IC-variables ξ and η have densities that are macroperiodic relative to critical-level partitions $\mathcal{U}$ and $\mathcal{V}$, and are infracritically independent relative to $\mathcal{U}$ and $\mathcal{V}$ and to a selection s, then for any U in $\mathcal{U}$ and V in $\mathcal{V}$, $q^s(\xi, \eta)$ is uniform over $W = U \times V$.*

Proof. The proof proceeds by showing that, for any subset R of W, $Q^s(R)$ is proportional to $M(R)$. The result then follows by lemma 3.6.

W is a member of the level partition for the joint infracritical information concerning X and Y. Thus, writing $\lambda(\cdot)$ for the index function of the joint information, by lemma 3.11,

$$Q^s(R) = Q^s(W) \int_{\lambda(R)} q^s_\lambda(z \mid W)\, dz$$
$$= Q^s(W) \int_{\lambda(R)} q^s_\lambda(p, r \mid U, V)\, dp\, dr.$$

Now, because of the macroperiodicity of $q(\xi)$, the density $q_\mu(p)$ over the infracritical information concerning ξ is uniform (theorem 3.10), and thus everywhere equal to 1 (since the index set is $[0, 1]$). By parallel reasoning, the density $q_\nu(r)$ over the infracritical information in η is also uniform and equal to 1. By infracritical independence, the joint density $q^s_\lambda(p, r \mid U, V)$ is therefore also equal to 1 everywhere. Thus

$$\int_{\lambda(R)} q^s_\lambda(p, r \mid U, V)\, dp\, dr = \int_{\lambda(R)} dp\, dr$$
$$= M(\lambda(R))$$
$$= aM(R) \qquad \text{(lemma 3.8)}$$

for some a. It follows that

$$Q^s(R) = a\, Q^s(W)\, M(R).$$

Thus the probability of any R in W is proportional to the measure of R, which implies uniformity, as desired. ■

Approximation 3.12.1 The conditions of application for this theorem hold only approximately if the infracritical information concerning the variables is only approximately independent. Small deviations in independence amount to small deviations in the uniformity of the joint infracritical density. Because this density is related to the joint density by linear transformations, deviations in the one are directly proportional to deviations in the other.

I now prove what is, on the assumption that the ic-variables in question are macroperiodically distributed, the converse of theorem 3.12: weak independence entails infracritical independence.

Proposition 3.13 *If two ic-variables ξ and η have distributions $q(\xi)$ and $q(\eta)$ that are macroperiodic relative to partitions $\mathcal{U}$ and $\mathcal{V}$ respectively, and if $q^s(\xi, \eta)$ is macroperiodic relative to the partition product of $\mathcal{U}$ and $\mathcal{V}$ (in other words, ξ and η are weakly independent relative to s and the partition product), then ξ and η have infracritical independence relative to s, $\mathcal{U}$, and $\mathcal{V}$.*

Proof. Let U and V be members of $\mathcal{U}$ and $\mathcal{V}$ respectively. Because of its macroperiodicity, $q^s(\xi, \eta)$ is uniform over $U \times V$. Hence, the density $q^s(\xi, \eta \mid \xi \in U, \eta \in V)$ is uniform. By the linearity or patchwork linearity of the index function over $U \times V$, the density $q^s_\lambda(p, r \mid U, V)$ imposed on the joint infracritical information is also uniform (lemmas 3.6 and 3.8). Thus $q^s_\lambda(p, r \mid U, V)$ is everywhere equal to 1 (since the index set is the unit square $[0, 1]^2$).

Now, by the macroperiodicity of $q(\xi)$ and theorem 3.10, the density imposed over the infracritical information about ξ is uniform, and so is equal everywhere to one (since the index set is the unit interval $[0, 1]$). The same is true for η. Thus

$$q^s_\lambda(p, r \mid U, V) \;=\; 1 \;=\; q_\mu(p)\, q_\nu(r)$$

as desired. ■

Theorem 3.12 and proposition 3.13 together entail the result stated in the main text: if two IC-variables have macroperiodic distributions then they have infracritical independence relative to a pair of partitions just in case they are weakly independent relative to (the partition product of) those partitions.

3.B4 Partial Low-Level Information

In this section I define the notion of *partial low-level information* concerning a set X. Partial low-level information gives some, but not all, of the complete low-level information about a member of X. The definition of a domain of partial low-level information differs from the definition of a domain of complete low-level information in only one respect: the index functions are required to be surjections, but not bijections, on the index set.

Definition 3.13 Let X be a real space $\mathbf{R}^n$. A *domain of partial low-level information concerning X* is a triple $\Omega = \langle Z, \mathcal{U}, \{\lambda_U\}\rangle$ consisting of:

1. A set Z, called the *index set*,
2. A partition $\mathcal{U}$ of X into contiguous, measurable regions, called the *level partition*, and
3. For each set U in $\mathcal{U}$, a surjection $\lambda_U : U \to Z$, called the *index function* for U. The third member of the triple is the set of these index functions.

The index functions may be put together to form the *full index function*:

$$\lambda(x) = \lambda_U(x)$$

where U is the member of $\mathcal{U}$ into which x falls. For any x in X, the full index function gives *partial low-level information* about x.

When the index functions λ_U are bijections, the lower-level information is complete. Thus complete lower-level information is a special case of partial low-level information. However, in what follows I generally use the term *partial low-level information* to refer to information that is strictly partial, that is, not complete.

In what follows, I normally assume that the index set of a domain of partial (but not complete) low-level information is finite. This assumption is not absolutely necessary for the purposes I have in mind, but when dealing with probabilities over partial low-level information, it is not easy to relax elegantly.

Example 3.3 Construct a domain of partial low-level information about the positive reals as follows. Let the level partition $\mathcal{U}$ be

$$\{ [n, n+1) : n \in \mathbf{I} \}.$$

Let the index function for a member of the level partition $U = [n, n+1)$ be

$$\lambda_U(x) = \lfloor 10(x - n) \rfloor$$

where $\lfloor x \rfloor$ is the floor of x, that is, the greatest integer less than x. Then $\lambda(x)$ gives the first digit in x's decimal expansion after the decimal point. Finally, let the index set Z be the integers between 0 and 9. The domain so constructed is, of course, a formalization of the information conveyed about a number by the first digit after the decimal point.

Definition 3.14 Let $\Omega = \langle Z, \mathcal{U}, \{\lambda_U\} \rangle$ be a domain of partial low-level information concerning a set X, with Z finite, and let $q(x)$ be a probability density over X. Then the *probability distribution imposed by $q(x)$ over the partial low-level information* supplied by Ω is defined to be the function:

$$Q_\lambda(z) = \int_{z^{-1}} q(x) \, dx$$

where $z \in Z$ and $z^{-1} = \{ x \in X : \lambda(x) = z \}$. The definition assumes that the sets z^{-1} are measurable.

The imposed probability distribution can also be written:

$$Q_\lambda(z) = \sum_{U \in \mathcal{U}} Q(U) \, Q(x \in U_z \mid x \in U)$$

where U_z is the set of members of U with index z, that is $U \cap z^{-1}$. I use this formulation in the proof of this section's main theorem.

I now define the notion of a well-tempered domain of partial low-level information, the counterpart of a linear or patchwork linear domain of complete low-level information. The definition is meant to apply only to a domain of partial low-level information with a finite index set; it does not apply, then, to a domain of complete low-level information. It also assumes, as will be seen, that every member of the level partition has a non-zero measure.

Definition 3.15 A domain $\Omega = \langle Z, \mathcal{U}, \{\lambda_U\}\rangle$ of partial low-level information concerning a set X, with Z finite, is *well tempered* just in case for every U in $\mathcal{U}$, the proportion of members of U that have a given index is the same. That is, Ω is well tempered just in case for every z in Z there exists a constant a_z such that, for every U in $\mathcal{U}$,

$$\frac{M(U_z)}{M(U)} = a_z$$

where U_z is, as elsewhere, the set of members of U that have index z.

Example 3.4 The partial low-level information about a real number conveyed by the first digit after the decimal point (see example 3.3) is well tempered. The ratio a_z for any digit is, of course, $1/10$.

Theorem 3.14 *Let* $\langle Z, \mathcal{U}, \{\lambda_U\}\rangle$ *be a well-tempered domain of partial low-level information concerning a set X. Then there exists a probability distribution $Q_\lambda(z)$ over the partial low-level information such that any probability density over X that is macroperiodic relative to $\mathcal{U}$ imposes the distribution $Q_\lambda(z)$ over the low-level information.*

Proof. Let $q(x)$ be a probability density over X that is macroperiodic relative to $\mathcal{U}$. Then by lemma 3.6, for any U in $\mathcal{U}$ there exists a constant k_U such that, for any $V \subseteq U$,

$$Q(V) = k_U M(V).$$

It follows that, for any z in Z,

$$\begin{aligned}
Q(x \in U_z \mid x \in U) &= \frac{k_U M(U_z)}{k_U M(U)} \\
&= \frac{M(U_z)}{M(U)} \\
&= a_z
\end{aligned}$$

for the a_z guaranteed by the definition of well-tempered information. Then

$$Q_\lambda(z) = \sum_{U \in \mathcal{U}} Q(U) \, Q(x \in U_z \mid x \in U)$$

$$= \sum_{U \in \mathcal{U}} Q(U) \, a_z$$

$$= a_z.$$

Thus all probability densities over X that are macroperiodic relative to $\mathcal{U}$ induce the same probability distribution over Z:

$$Q_\lambda(z) = a_z$$

as desired. ∎

The following example shows that critical-level information about the IC-value of a trial on a microconstant experiment, assigned according to the teleological scheme (section 2.73), is a kind of well-tempered partial low-level information about that IC-value.

Example 3.5 Given a one-variable, deterministic microconstant experiment $X = \langle q(\zeta), S, H \rangle$ and a designated outcome e, construct a domain of partial low-level information about X's IC-variable ζ as follows. Let the level partition $\mathcal{U}$ be an optimal constant ratio partition for e, let the index function be the evolution function for e, that is

$$\lambda(\zeta) = h_e(\zeta)$$

and let the index set therefore be $Z = \{0, 1\}$. Then $\langle Z, \mathcal{U}, \{\lambda_U\} \rangle$ is a well-tempered domain of partial low-level information about X's IC-variable ζ. The ratio a_1 (the proportion of each member of the level partition mapping to 1) is, obviously, the strike ratio for e; the ratio a_0 is $1 - \text{srat}(e)$. The information conveyed by this domain of low level information is just the critical information about the IC-value of X, assigned according to the teleological scheme.

The probability of e is, of course, equal to the probability that the teleologically assigned critical IC-value ζ_e for a trial is equal to 1. If $q(\zeta)$ is macroperiodic relative to the optimal constant ratio partition, then

$$\text{cprob}(e) = Q_\lambda(\zeta_e = 1)$$

$$= a_1 \qquad \qquad \text{(theorem 3.14)}$$

$$= \text{srat}(e).$$

This constitutes an alternative proof of theorem 2.2. The same result can be obtained assuming only that the discrete density (see section 2.72) over the critical IC-values is uniform, a strictly weaker condition than macroperiodicity over the constant ratio partition.[36]

The alternative proof makes explicit a rather important fact foreshadowed in section 2.73: the value of a microconstant probability is determined by a probability distribution over the relevant experiment's critical IC-values, that is, by a probability distribution over partial low-level information about the experiment's IC-variable. It is this simple truth that underlies the major independence results of sections 3.4, 3.6, and 3.7: the results of sections 3.4 and 3.6 concerning microconstant experiments and causal coupling depend on the fact that information above the critical level makes no difference to the probability, and the results of section 3.7 concerning chains depend on the fact that information below the critical level makes no difference to the probability.

3.B5 Nested Information

The aim of this section is to provide, for the purposes of section 3.72, a precise definition of what it is for two levels of information to be distinct, and to prove an independence theorem about distinct levels of information. When a lower level of information about a variable is distinct from a higher level, I say that the lower level is *nested* in the higher level, for reasons that will soon become clear.

I begin by defining the *lower-level partition* of a domain of partial low-level information. (My use of this definition will assume that the domain is strictly partial, that is, not complete.) A lower-level partition is created by dividing each member of the level partition into maximal parts that each map entirely onto a single member of the index set.

Definition 3.16 Let $\Omega = \langle Z, \mathcal{U}, \{\lambda_U\}\rangle$ be a partial domain of low-level information concerning a set X. The *lower-level partition* of Ω is the set of maximal sets V that satisfy the following conditions:

1. Each V is entirely contained in some member of the level partition U, and
2. For each V, there is a member z of the index set such that, for all x in V, $\lambda(x) = z$.

That is, the lower-level partition of Ω is the set

$$V = \{ \lambda^{-1}(z) \cap U : z \in Z, U \in \mathcal{U} \}.$$

The notion of a lower-level partition is used in the following gloss of partial low-level information: partial low-level information about x specifies into which member of the lower-level partition x falls, given the high-level information specifying into which member of the level partition x falls. The level partition and the lower-level partition, then, mark the upper and lower limits of the information provided by a domain of partial low-level information.

For example, the critical information about an IC-value of a microconstant experiment has as its level partition an optimal constant ratio partition for the experiment, and as its lower-level partition the outcome partition for the experiment.[37]

A slightly different definition of the lower-level partition would require that every partition member be contiguous. Contiguity has its advantages, but one major conceptual disadvantage: if contiguity is required, it is no longer quite correct to say that the lower-level partition corresponds to the lower limit on the partial low-level information.

Two domains of partial low-level information are nested if the lower-level partition for one contains, in a certain sense, the level partition for the other. The appropriate sense of containment is specified in the following definition.

Definition 3.17   A partition V of a set X is *nested* in a partition $\mathcal{U}$ of X just in case every member of V is entirely contained within some member of $\mathcal{U}$.

Definition 3.18   A domain of low-level information Ω (partial or complete) is *nested* in a domain of partial low-level information Φ just in case Ω's level partition is nested in Φ's lower-level partition.

For example, the information provided by the third digit after the decimal point in a number's decimal expansion is nested in the information provided by the first or second digits after the decimal point.

If the lower-level partition for one domain is identical to the level partition for another, then the two domains are *exactly nested*. The sequence of digits in the decimal expansion of a real number conveys a series of pieces of ever lower, exactly nested information about the number, in the sense that the information conveyed by the $(i + 1)^{\text{th}}$ digit is exactly nested in the information conveyed by the ith digit.

I now state the main theorem of this section, concerning the independence of the information provided by nested domains of well-tempered low-level information.

Theorem 3.15 *Let* $\Omega = \langle Z, \mathcal{V}, \{\mu_V\} \rangle$ *and* $\Phi = \langle W, \mathcal{U}, \{\nu_U\} \rangle$ *be well-tempered domains of partial low-level information about members of a set X, with* Ω *nested in* Φ. *Let* $q(x)$ *be a probability distribution over X. If* $q(x)$ *is macroperiodic with respect to* $\mathcal{U}$, *the level partition of* Φ, *then the information provided by* Ω *is independent of that provided by* Φ, *in the sense that, for any z in Z and w in W, and for* $R = \{ x \in X : \mu(x) = z, \nu(x) = w \}$,

$$Q(R) = Q_\mu(z) \, Q_\nu(w).$$

Proof. Because $q(x)$ is macroperiodic with respect to $\mathcal{V}$ and Ω is well-tempered,

$$Q_\mu(z) = M(V_z)/M(V) \tag{3.1}$$

where V is any member of $\mathcal{V}$ and V_z is the set of members of V that have index z.

Let $\mathcal{V}_w$ be the set of all members of $\mathcal{V}$ that fall into regions of X that ν maps to w, that is, $\{ V \in \mathcal{V} : \text{for all } x \text{ in } V, \nu(x) = w \}$. Because Ω is nested in Φ, the combined probability of these sets is equal to the probability of w. That is,

$$Q_\nu(w) = \sum_{V \in \mathcal{V}_w} Q(V). \tag{3.2}$$

Now for any V in $\mathcal{V}$, let V_z be, as above, the set of members of V with index z. Then the probability of the set R (defined in the statement of the theorem) is

$$Q(R) = \sum_{V \in \mathcal{V}_w} Q(V_z)$$

$$= \sum_{V \in \mathcal{V}_w} \frac{M(V_z)}{M(V)} Q(V) \qquad \text{(uniformity of } q(x) \text{ over } V)$$

$$= \sum_{V \in \mathcal{V}_w} Q_\mu(z) Q(V) \qquad \text{(by 3.1)}$$

$$= Q_\mu(z) \sum_{V \in \mathcal{V}_w} Q(V)$$

$$= Q_\mu(z) Q_\nu(w) \qquad \text{(by 3.2)}$$

as desired. ■

In the application of this theorem to independence in chains, in section 3.72, the level partition $\mathcal{U}$ for Φ, the higher of the two levels of information, is the outcome partition for the first trial in a chain. Each member of the outcome partition corresponds to a possible effective ic-value for the first trial. Thus the requirement that $q(x)$ is macroperiodic relative to $\mathcal{U}$ is satisfied just in case the ic-density over the chain's initial ic-variable is macroperiodic relative to the effective ic-values for the first trial.

There is an approximation result for this theorem, although I do not use it in this study:

Approximation 3.15.1 The information given by Ω and Φ will be approximately independent even if some sets of the level partition of Ω are not entirely contained in members of the lower-level partition of Φ, as long as the overlap is proportionally small. The overlap is proportionally small if either (a) the non-contained sets are each *almost* entirely contained in a member of Φ's lower-level partition, or (b) there are sufficiently many sets of Ω's level partition, weighted by measure, that are entirely contained in Φ's lower-level partition.

Corollary 3.16 *Given a uniform distribution over the interval* $[0, 1]$, *the probability of a number's having a certain digit in the* i^{th} *place after the decimal point is independent of the probability of the number's having a certain digit in the* j^{th} *place after the decimal point, provided that* $i \neq j$.

It is a consequence of corollary 3.16 that almost all real numbers in the interval $[0, 1]$ have probabilistically patterned digits. The argument is as follows. Given a uniform distribution over $[0, 1]$, the probability of finding any given digit at any given place after the decimal point is, of course, $1/10$. Since the uniform distribution over $[0, 1]$ yields probabilities equal to the standard measure, and independence together with the law of large numbers entails that the probability of a number with probabilistically patterned digits is approximately 1, the proportion of numbers having probabilistically patterned digits is approximately 1.

3.B6 Short-Term Coupling

This section formalizes the main claims of section 3.6. I begin by defining the notion of a *seed set* of a set V relative to a transformation. The seed sets of V are, roughly, the sets obtained by breaking the inverse image of V into maximal contiguous parts each of which maps more or less one-to-one to V. The *more or less* allows parts of the domain where the transformation is constant to be included in a single seed set. The role of the notion of a seed set is specified in the main result of this section, theorem 3.17.

Definition 3.19 Let f be a piecewise continuous mapping from X to Y, and V a contiguous subset of Y contained in the range of f. A set of *seed sets* of V under f is a partition of the set $f^{-1}(V)$ into maximal subsets satisfying the following criteria:

1. Each subset is contiguous, and
2. The function f is injective (that is, one-to-one) over each subset, excluding neighborhoods where f is constant. That is, there do not exist distinct members x_1 and x_2 of the subset such that $f(x_1) = f(x_2)$, unless both x_1 and x_2 belong to a single neighborhood over which f is constant.

I assume, note, that f is piecewise continuous; the idea of dividing the inverse image into seed sets does not make much sense otherwise (and continuity is anyway assumed in theorem 3.17). The definition allows that there may be more than one way to partition the inverse image into seed sets; in what follows, however, I will for reasons of convenience talk as though there is just one way.

In what follows, I will be interested only in seed sets that have the following property.

Definition 3.20 The *seed sets* of a set V under a transformation f have the *surjective property* just in case for any seed set U, $f(U) = V$.

For arbitrary f and V, the seed sets are not particularly likely to have the surjective property. If, however, as is assumed in this and the next section, f is roughly linear over the regions containing each of the seed sets, the sets will tend to have the surjective property. It is possible to put this more formally, but to save space, I will instead explicitly assume the surjective property in the theorems that follow.

I now define the notion of a transformed experiment to represent the effect of a short-term coupling imposed at the beginning of a trial (see section 3.64).

Definition 3.21 Let $X = \langle q(\zeta_1, \ldots, \zeta_k), S, H \rangle$ be a k-variable, deterministic probabilistic experiment, and let $f_1, \ldots, f_k$ be a set of transformations with domain and range

$$f_i : \zeta_1 \times \cdots \times \zeta_k \to \zeta_i.$$

The *transformation of X by $f_1, \ldots, f_k$* is the experiment

$$X' = \langle q(\zeta_1, \ldots, \zeta_k), S, H' \rangle$$

where the evolution function for any designated outcome e is defined as follows:

$$h'_e(\zeta_1, \ldots, \zeta_k) = h_e(f_1(\zeta_1, \ldots, \zeta_k), \ldots, f_k(\zeta_1, \ldots, \zeta_k))$$

where $h_e(\cdot)$ is X's evolution function for e.

The transformations $f_1, \ldots, f_k$ can be thought of as making up a single transformation f that maps the IC-variable space, the space $\zeta_1 \times \cdots \times \zeta_k$, onto itself. It is then possible to talk about the transformation of X by f.

As in the main text, I sometimes write ζ_i' for $f_i(\zeta_1, \ldots, \zeta_k)$. Thus

$$h'(\zeta_1, \ldots, \zeta_k) = h(\zeta_1', \ldots, \zeta_k').$$

I next define the notions of microlinearity and of a non-deflationary transformation in a suitably technical way.

Definition 3.22 An *r-sized cube* in R^n is a hypercube in R^n of n dimensions with edge length r.

In the remainder of this section and the next, I assume that the set X is a subset of R^n.

Definition 3.23 A transformation f of X is *r-microlinear* just in case, for any r-sized cube R in X, there is some invertible linear or linear plus constant transformation $l(x)$ such that, for any x in R, $f(x) = l(x)$.

Note that invertibility is built into the formal notion of microlinearity.

A transformation is r-microlinear in the strict sense just defined only if it is linear. The strict notion of r-microlinearity, then, is not of much use in itself. However, the theorem later proved involving r-microlinearity is approximately true for approximately r-microlinear functions (see approximations 3.17.1 and 3.17.2). It is here that the value of the notion lies, for transformations that are far from linear may nevertheless be approximately r-microlinear. In some cases, it may be useful to liberalize the definition of microlinearity by allowing that the micro-sized regions over which a microlinear function must be linear need not be cubes.

Definition 3.24 A transformation f of X is *r-nondeflationary* just in case the seed sets under f of any r-sized cube in X are themselves contained in an r-sized cube.

I now state the main result of this section.

Theorem 3.17 *Let* $X = \langle q(\zeta_1, \ldots, \zeta_k), S, H \rangle$ *be a deterministic probabilistic experiment, and* X' *a transformation of* X *by a function* f. *For some designated outcome* e, *let* V *be a constant ratio partition for* e *on* X *with strike ratio* s *and size* r *(that is, every member of* V *is* r-*sized or less). Let* f *be an* r-*microlinear, r-nondeflationary transformation. Define* U *as the set containing the seed sets under* f *of every member of* V. *Suppose that the seed sets all have the surjective property. Then* U *is a constant ratio partition for* e *on* X' *with strike ratio* s *and size no greater than* r.

Proof. Let U be a seed set of a member V of $\mathcal{V}$. It must be shown that U is r-sized and has a strike ratio (for e on X') of s.

That U is r-sized follows immediately from the fact that V is r-sized and f is r-nondeflationary.

More work is required to show that the strike ratio is s. The strike ratio for e over U on X' is, writing ζ for $\zeta_1, \ldots, \zeta_k$,

$$\text{srat}(e) = \int_U h'_e(\zeta) \, d\zeta \Big/ M(U). \tag{3.3}$$

Now,

$$\int_U h'_e(\zeta) \, d\zeta = M(I) \tag{3.4}$$

where I is the set $\{ \zeta \in U : h_e(f(\zeta)) = 1 \}$. Define J as the set $\{ \zeta \in V : h_e(\zeta) = 1 \}$; then $I = f^{-1}(J) \cap U$. Because U is r-sized and f is r-microlinear, the mapping between U and V is equivalent to a linear function l. Given that, by the surjective property, $f(U) = V$, it follows that $I = l^{-1}(J)$, and so, from lemma 3.7, that

$$M(I) = M(J)/a \tag{3.5}$$

where a is the magnitude of the determinant of l.

Now observe that, from the surjective property and lemma 3.7,

$$M(V) = aM(U) \tag{3.6}$$

and finally, that

$$M(J) = \int_V h_e(\zeta) \, d\zeta. \tag{3.7}$$

Putting everything together,

$$\text{srat}(e) = \int_U h'_e(\zeta)\, d\zeta \Big/ M(U) \qquad \text{(by 3.3)}$$

$$= M(I)/M(U) \qquad \text{(by 3.4)}$$

$$= M(J)/aM(U) \qquad \text{(by 3.5)}$$

$$= M(J)/M(V) \qquad \text{(by 3.6)}$$

$$= \int_V h_e(\zeta)\, d\zeta \Big/ M(V) \qquad \text{(by 3.7)}$$

$$= s$$

as desired. ∎

Three remarks on this theorem. First, it assumes explicitly that the relevant seed sets have the surjective property; this follows under fairly weak conditions from the microlinearity of the transformation.

Second, note that, if the transformation is not merely non-deflationary, but is positively inflationary, then the transformation need be microlinear only for a commensurably smaller r. That is, the more inflationary the transformation, the less microlinear it need be.

Third, the usefulness of the theorem is due to the fact that it continues to hold, roughly, even for functions that are not exactly microlinear, as stated in the following approximations.

Approximation 3.17.1 The definition of r-microlinearity can be relaxed so as to allow transformations that are approximately, rather than exactly, linear over every r-sized cube. The correspondingly relaxed theorem asserts only the approximate equality of the strike ratios for the members of the constant ratio partitions of the original and the transformed experiment. The greater the deviation from linearity, the greater the potential difference between the strike ratios.

Approximation 3.17.2 The definition of r-microlinearity can be relaxed so as to allow transformations that are linear over almost all, but not all, r-sized cubes. This is important because no function is linear in the neighborhood of any of its turning points, and many transformations have turning points. The correspondingly relaxed theorem does not assert that every set in $\mathcal{U}$ has the same strike ratio for e; it allows that at least some of the seed sets of the bad regions—the regions over which the transformation is non-linear—have a strike ratio that differs from the norm.

As ever in the case of bad regions, provided that the probability that a trial's ic-values fall into a bad region is very low, the effect of the bad region on the desired property will be small. In the case of the colliding coins, for example, the probability of obtaining the same outcomes on both coins may not exactly equal $1/2$ if there are bad regions in the coupling transformation, but provided that the bad regions are small, the probability will not differ from $1/2$ by much.

3.B7 Microlinearity and Macroperiodicity

The final theorem asserts the ability of a non-deflationary, microlinear transformation to preserve a probability distribution's macroperiodicity. I appeal to this macroperiodicity-preserving property of certain transformations both in the alternative approach to short-term coupling (section 3.66) and in the results concerning chained trials (sections 3.75 and 3.76). Thus the transformation f in the statement of the theorem may be understood either as representing the effect of a short-term coupling between experiments or as an ic-evolution function for a deterministic chain.

As in the last section, I assume throughout this section that the set X is a subset of $\mathbf{R}^n$.

Definition 3.25 A probability distribution over X is *r-macroperiodic* just in case it is macroperiodic relative to any partition of X into r-sized cubes.

Theorem 3.18 *If a transformation f of a set X is r-microlinear and r-nondeflationary, and if the seed sets under f of any r-sized cube in X have the surjective property, then the transformation relative to f of any r-macroperiodic probability distribution is r-macroperiodic.*

Proof. Let $q(x)$ be an r-macroperiodic distribution over X. To show that the transformation of $q(x)$ by f is r-macroperiodic over any partition of X into r-sized cubes, it must be shown that the transformed probability distribution is uniform over any particular r-sized cube.

Let R be an r-sized cube in X. Let $U_1, \ldots, U_m$ be the seed sets of R under f. Because f is r-nondeflationary, any U_i is contained in an r-sized cube, thus for any U_i, there is some linear transformation $l_i(x)$ that approximates $f(x)$ over U_i. Because of the surjective property, $l_i(x)$ is surjective on R.

The proof now follows roughly the same lines as the proof of theorem 3.10. The proofs are so similar, in fact, that theorem 3.10 can be used to prove the current theorem, as follows.

The aim is to show that the transformed probability distribution over R is uniform. Call the set of R's seed sets $\mathcal{U}$. Observe that $\langle R, \mathcal{U}, \{l_i\}\rangle$ fulfills the conditions for being a linear domain of complete low-level information concerning R^{-1}. The index set is R. The level partition is $\mathcal{U}$. The index function $\lambda_{U_i}(x)$ for each U_i is the linear bijection $l_i(x)$. The full index function is $f(x)$. The probability distribution over the low-level information is the transformation of $q(x)$ by f. Since $q(x)$ is macroperiodic relative to $\mathcal{U}$, by theorem 3.10, the transformed probability distribution is uniform over R, as desired. ∎

The theorem guarantees macroperiodicity over the range of $f(x)$, but it does not imply anything about the size of the range. This is taken care of where necessary in the main text by, for example, the requirement that $f(x)$ be inflationary.

Approximation 3.18.1 When $q(x)$ is almost, but not quite, uniform over every r-sized cube, or f is almost, but not quite, linear over every r-sized cube, the transformed probability distribution will be not quite macroperiodic. As with other results of this sort, the transformed distribution's deviation from macroperiodicity will be proportional to the imperfections in the macroperiodicity of the original distribution (see approximation 3.10.1) and in the microlinearity of the transformation.

Approximation 3.18.2 Few transformations are strictly microlinear; microlinearity tends to fail in the vicinity of a transformation's turning points. These are the transformation's *bad regions*. Provided that there are few turning points, and few other bad regions of any sort, there will be few cubes over which the probability distribution is not uniform, and so the transformed probability distribution will be approximately macroperiodic. The deviation from macroperiodicity increases with the probability of the bad regions.

See also approximations 3.10.1, 3.17.1, and 3.17.2.

Approximation 3.18.3 One might well worry that, when a transformation that is only approximately microlinear is applied *repeatedly* to a macroperiodic distribution, the non-microlinear aspect of the transformation, however small, will bring about a gradual degradation of macroperiodicity. Where theorem 3.18 and similar results are used to establish the properties of repeated microlinear transformations in the main text, however, it is assumed that the transformation not only is non-deflationary, but is extremely inflationary. This inflation will tend to stretch out any areas of non-macroperiodicity that appear in the distribution, bringing it closer to macroperiodicity. No entirely general result is possible—it all depends on the degree of stretching versus the degree of non-microlinearity—but it will be seen that, especially in cases of chained trials (section 3.7), stretching will win out.

4

The Simple Behavior of
Complex Systems Explained

Why do complex systems behave in simple ways? I argued in chapter one that this question can be answered by analyzing the foundations of a certain scientific method, which I call enion probability analysis. An understanding of EPA's foundations requires an understanding of the basis of the very strong assumptions that EPA makes about enion probabilities, assumptions which I have combined into what I call the *probabilistic supercondition*.

The supercondition requires the independence of enion probabilities, hence enion statistics, from all microlevel facts. Because macrovariables are enion statistics, an explanation of the supercondition will also explain how macrodynamics can float free of the microlevel, and thus how there can exist laws that involve only macrovariables—the laws that assert the simple behavior whose explanation is my goal.

I developed a series of formal results concerning the properties of complex probability in chapters two and three; I now apply the results to the probabilities that govern the dynamics of complex systems, in order to show that they satisfy the supercondition. The chapter pivots on section 4.4. There I develop, in outline, a method to show that a complex system's enion probabilities are microconstant and independent. Microconstancy and independence together underwrite the probabilistic supercondition.

To make a general argument that the supercondition holds for all complex systems is impossible. Thus this chapter has two parts. The first part presents a schematic argument that the supercondition holds for any complex system. At one crucial point, the argument makes a major assumption about the dynamics of the system to which it is applied. It is this assumption that cannot be given any general justification, but which must rather be vindicated system by system. The second part of the chapter presents the vindication for two kinds of system in particular, a gas in a box and a predator/prey ecosystem. Other systems, too, will have to be treated individually to establish this step of the argument. Social systems are discussed briefly in chapter five (section 5.4).

Section by section, the structure of the chapter is as follows. Section 4.1 introduces a more formal way of representing the kinds of complex systems discussed in chapter one, making use of the language of dynamic systems theory. In section 4.2, I show that it is easy to assign a complex probability distribution over the behavior of an enion in a complex system. Section 4.3 discusses some important properties of the microdynamics of complex systems. Section 4.4, as I have said, presents the form of the argument that enion probabilities are microconstant and independent. This argument depends on the independence of what I call microdynamic probabilities, to be established using the independence results of chapter three. Section 4.5 provides the schematic argument for microdynamic independence. In section 4.6, I discuss the aggregation of enion probabilities, and in so doing, I solve a serious problem encountered in section 4.4. Section 4.7 gathers together the most important assumptions made in the previous sections. The second part of the chapter applies the schematic argument to the case of a gas in a box, in section 4.8, and then to a predator/prey ecosystem, in section 4.9.

4.1 Representing Complex Systems

The broad outlines of a complex system can be described with a single phrase: "one-hundredth of a mole of hydrogen enclosed in a one-liter cubical box" or "the ecosystem in my back garden." Such a description determines, either explicitly or implicitly, what kind of stuff makes up the system, where the boundary of the system is drawn, and what is fixed and what is variable in the system.

The more conspicuously variable parts of a system are its enions. The state of each enion in a system is represented by a set of variables, quantifying such things as the position or satiety of a particular organism, the current rate of consumption of an economic actor, or the momentum of a gas molecule.

Changes in enion state take place against a backdrop, the *background environment* (the garden, the nation, the box). Parts of the background environment may themselves vary, for example, the weather in the garden. These variations can be represented by what I will call *background variables*. Changes in the value of background variables may be the result either of the activity of enions within the system, as perhaps in the case of vegetation cover, or of processes that go on outside the system, as in the case of weather, or both.

In posing the problem of the simple behavior of complex systems, I have distinguished two levels at which the changes in a system can be described: the microlevel and the macrolevel. Microlevel changes are changes in the states of individual enions and changes in the background that affect individual enions. The microlevel variables can be divided into two classes: first, the microvariables representing the state of each enion, and second, the background variables. The complete microlevel state of a system is called its **microstate**. I call the laws that govern changes in microvariables, and thus in a system's microstate, the laws of **microdynamics**. In a garden, laws of microdynamics concern the behavior of individual organisms; in an economy, individual economic actors; in a gas, individual molecules.

Macrolevel changes are changes in certain statistical facts about enions. Thus the macrolevel variables of a system are enion statistics.[1] Typical macrovariables are the population of a species in an ecosystem, the GDP of an economy, or the entropy of a gas; a system's **macrostate** is given by specifying the values of all its macrovariables. I call the laws that govern changes in macrovariables, and so govern changes in a system's macrostate, the laws of **macrodynamics**. Throughout this study I assume that a system's macrodynamic laws are determined by its microdynamic laws.

It is useful to have a way—necessarily very abstract—of talking about changes in the state of a complex system. To this end I introduce the idea of a *state space,* a generalization of the idea of a phase space in physics. A single point in a system's state space represents the values of all of its microvariables, completely fixing the microstate of the entire system. In other words, a point in state space represents the state of every enion in the system, as well as the values of any background variables. For a single point to contain all this information, a state space must have a separate set of dimensions for each one of the system's enions, as well as a set of dimensions to represent the background variables.[2] The microdynamic laws will determine the trajectory of the system through state space, from microstate to microstate.

Any given macrostate of a system corresponds to a particular region of state space, namely, the region spanning all points representing microstates that determine the same values for the macrovariables. This region is in most cases, though it need not be, connected. A change in the value of a macrovariable occurs just when the trajectory of a system crosses from one such region into another.

4.2 Enion Probabilities and Their Experiments

A probabilistic experiment is characterized in terms of mechanisms, ic-variables, and designated outcome types. In section 4.1, I characterized the behavior of enions in terms of systems, microvariables, and macrovariables. There is a rough correspondence between these two triads: for a given enion probability, the experiment mechanism will be the complex system, the ic-variables will be the microvariables, and the designated outcomes will often be the events in an enion's evolution that have an effect on the macrovariable under consideration. The rest of the section establishes this correspondence in greater detail.

Consider a typical enion probability: a rabbit's chance of being eaten over the course of a month. The designated outcome, as remarked, is an event that has an impact on a relevant macrovariable, namely, the population of rabbits. The "trial" that stands to produce this outcome consists in the rabbit's wandering the ecosystem for a month. The mechanism of the corresponding experiment, then, must include everything—foxes, food, fellow rabbits—directly or indirectly relevant to the survival of the rabbit. It is, in other words, the entire ecosystem. The ic-variables of the experiment, then, are all those variables that play a part in determining the microdynamics of the ecosystem over the course of a month, that is, all the microvariables of the system, including background variables.

Figure 4.1 shows the ecosystem considered as a part of a probabilistic network. The dashed line represents the probabilistic experiment itself. The box inside the dashed line represents the mechanism of the experiment, that is,

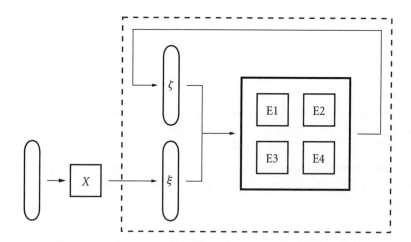

Figure 4.1 A representation of a complex system as a probabilistic network. The dashed line encloses the experiment to which the enion probabilities are attached. The ic-variables of the experiment are the microvariables of the system, namely, ζ and ξ.

the ecosystem. The boxes inside the mechanism represent enions. They are there to remind the reader that enions—even the rabbit whose life is in the balance—form only a part of the mechanism for an enion probability in a complex system. For simplicity's sake, I have included only two microvariables. One of these, the variable ζ, has its value determined by mechanisms within the system (perhaps it represents rabbit position or vegetation cover); the other, the variable ξ, has its value determined by some mechanism X outside the system (perhaps it represents weather or pollution levels). The first sort of microvariable is called **endogenous**, the second sort **exogenous**. Background variables may be endogenous or exogenous; microvariables representing enions are, in any system worthy of the name, endogenous.

One more remark: enion probabilities are, recall, functions of macrovariables, or to put it another way, they are probabilities conditional on the system's being in a particular macrostate m (section 1.23). As a consequence, the experiment for a given enion probability is limited to certain ranges of ic-values, namely, those ranges within which the system is in m. The relevant experiment type simply rules out, by stipulation, certain combinations of initial conditions. For this reason I will, when necessary, write enion probabilities $cprob(e_m)$, where m is the macrostate relative to which the probability of the designated outcome e is specified.

4.3 The Structure of Microdynamics

The argument for the microconstancy and independence of enion probabilities turns on certain properties of complex system microdynamics that are spelled out in this section. Most importantly, I will claim that

1. The microdynamics of a complex system is in large part determined by a series of probabilistic interactions (section 4.32), and
2. The interactions are causally isolated (section 4.33).

These observations are later put to use to establish a key premise in the argument for enion probability microconstancy and independence, that a complex system's microdynamics can be seen as a kind of random walk, that is, a process in which the system's microvariables undergo a series of randomly determined, stochastically independent changes as time goes on (section 4.4).

The truth of the observations I will make depends on the microlevel chosen for the system. Thus the observations are relative to the way the system is described. But what is really important is something observer-independent: that it is possible to describe a given complex system in such a way that the observations are true. Given the existence of such a description, one can proceed with the analysis of the foundations of EPA along the lines suggested in sections 4.4 and 4.5.

It is at this point in the chapter, then, that I begin to make substantive assumptions about the nature of complex systems. If there are some complex systems for which these assumptions do not hold, at least approximately, then there are some systems to which the understanding of EPA offered in this chapter simply does not apply. There is no general argument that complex systems do tend to satisfy the assumptions; these matters can only be resolved by considering complex systems domain by domain, as I do for statistical physics and population ecology in sections 4.8 and 4.9.

4.31 The Microlevel and the Fundamental Level

Call the macrolevel of the rabbit/fox ecosystem—the level at which the system is described in terms of changes in the population of rabbits, foxes, and so on—the **population level**. Call the microlevel of the ecosystem—the level at which the system is described in terms of the activities of individual organisms—the **behavioral level**.

In the ecosystem and in most other complex systems, the microlevel is not the level of fundamental physics (which I call the **fundamental level**). It follows that an ecological microstate, that is, a state of the ecosystem described at the behavioral level, does not correspond to a single fundamental-level microstate. Rather, a given ecological microstate may be realized at the fundamental level in a large number of ways. Thus, a description of the state of an ecosystem at the behavioral level leaves out a large amount of information. There are two ways in which information is left unspecified:

1. Some facts are left out altogether, such as the neurobiological facts about a rabbit's brain that help to determine its foraging decisions.
2. Some physical quantities are described only at a certain level of precision, coarser than the level of precision possible at the fundamental level. For example, the position of a rabbit may be described to the nearest square foot, but will not be described to the nearest square nanometer. Animals may be described as hungry, but the exact details as to what the animal requires in the way of nutrients will usually be omitted.

I call the information specified in an ecological microstate the *behavioral information*, and the information that is left out of such a specification the *sub-behavioral information*. Sub-behavioral information, then, includes every physical detail not included in the behavioral information.

Note that, as a consequence of the second kind of information loss, the ecosystem's microvariables are discrete, not continuous. Microdynamics consists of a series of jumps, not a continuous motion. Although it is often computationally convenient to treat behavioral variables as continuous, their discreteness is very important for the arguments that are to follow.

In summary, I have divided the information about the goings-on in an ecosystem into three levels: the population-level information, the behavioral information, and the sub-behavioral information, corresponding respectively to macrolevel information about the system, microlevel information about the system, and information about the system below the microlevel. This division is the basis for everything that follows.

4.32 Microdynamics Is Probabilistic

Suppose that sub-behavioral information affects the course of behavioral evolution, as is indeed true of the dynamics of most, perhaps all, ecosystems (section 4.9). Then the behavioral information alone cannot wholly determine the course of an ecosystem's time evolution. Ecological microdynamics is not deterministic.

In any system for which this is true, microdynamics can be characterized in terms of laws containing complex probabilities. These laws, and thus these probabilities, govern change in the system at the microlevel. I call the probabilities appearing in microdynamic laws **microdynamic probabilities**.

The purpose of this section is to make some comments about the nature of microdynamic probabilities, using the rabbit/fox ecosystem as an example. Let me give two toy examples of microdynamic laws about rabbits and foxes; the first will be developed at some length for illustrative purposes. The lack of realism is not important, as the point of the laws is to stand proxy for the real probabilistic laws of microdynamics in whatever complex system is to be analyzed.

The first law determines where a rabbit will forage. It states that, when in a certain region of the ecosystem, the rabbit is 70% likely to forage in the meadow and 30% likely to forage in the scrub. The second law governs interactions between foxes and rabbits. It states that, in certain circumstances, a

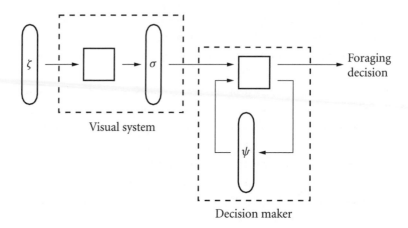

Visual system

Decision maker

Figure 4.2 The parts of a rabbit's neurobiological circuitry relevant to foraging decisions. On the left is the visual system; on the right is the decision maker itself. The IC-variable σ is an internal representation of the rabbit's surroundings; ψ represents an internal neurobiological state.

fox encountering a rabbit has a 20% chance of catching the rabbit. The probabilities mentioned in the laws, then—70%, 30%, 20%—are microdynamic probabilities.

Consider the foraging law in more detail. The 70% probability is complex (section 4.93). Like any complex probability, it is attached to a probabilistic experiment. It has as its mechanism the rabbit and (in a passive capacity) the surrounding terrain, and as its IC-variables, I will suppose, facts about the rabbit's position and neurobiological facts about its brain. A very simple version of the experiment is pictured in figure 4.2. In the figure, the experiment in the left dashed box represents the visual processing of sensory information about the rabbit's environment. The mechanism takes information about the environment, which depends on the rabbit's position ζ and the surrounding terrain, and represents it in the brain. This mental representation is encoded by the IC-variable σ. The experiment on the right represents the making of the foraging decision, which depends on the mental representation σ of the neighborhood, and on some endogenous variable, represented by ψ, which is, one may suppose, responsible for the randomization of foraging decisions (sections 4.93 and 4.94).

The most important lesson to learn from this example is that the level of the mechanics of microdynamic experiments is *lower* than the microlevel. This is inherent in the fact that the ic-variables for microdynamic experiments represent a finer level of detail than the ic-variables for enion probabilities. For example, the neurobiological ic-variable is not a microvariable of the ecosystem; the information in this ic-variable is entirely sub-behavioral. And while position is a behavioral variable, the position variable of the foraging decision maker may, if the decision maker is sensitive to differences in position too subtle to appear in a behavioral description, contain some sub-behavioral information.

All three levels of description characterized in section 4.31 are now in play. From highest to lowest they are:

1. The population level, at which change in populations, but not the behavior of individual enions, is described. This is the macrolevel of the rabbit/fox ecosystem.
2. The behavioral level, at which the behavior of individual enions is described. This is the microlevel of the rabbit/fox ecosystem, the level at which the ic-variables of enion probabilities are specified, and the level at which the outcomes of microdynamic experiments have their effect.
3. The level at which the ic-variables and the mechanics of the microdynamic experiments are described, which I am calling the sub-behavioral level.

These three levels of description do not in themselves determine the level of description of the outcomes of microdynamic experiments. There is an obvious choice, however, for this level: the outcomes should be specified just precisely enough that, in conjunction with the current microstate of the enions involved in an interaction, they determine the next microstate. In everything that follows, I will assume that microdynamic outcomes are described at this level. It is worth noting, however, that in principle, they could be described at a coarser level, so that an outcome together with an enion's current microstate determines, not the next microstate, but a probability distribution over possible next microstates.

Let me now introduce some sophistication into the picture of microdynamic evolution. The one microdynamic experiment I have considered in any detail so far, the foraging-decision experiment, directly determines the value of the microvariable on which it has an effect, namely, the position of the rabbit. In other words, the outcome of the experiment is simply a new value for the position. But this is not the only way to conceive of the microdynamic experiments' influence on a system's microvariables. There are a number of other ways in which microdynamic experiments can play a role in determining the course of a system's time evolution. I will consider three.

First, and simplest, the outcomes of some microdynamic experiments may best be thought of not as new values for microvariables, but rather as changes in the values of microvariables. For example, rather than a trial's outcome being the event of a rabbit's coming to occupy a given position in the ecosystem, the outcome might be the event of the rabbit's position changing by a particular amount—say, the rabbit's moving ten feet to the left. When a rabbit is fleeing from a predator, to take just one example, it may be more natural to describe the rabbit's actions in this way. Representing microdynamic outcomes as relative rather than absolute in this case would allow rabbit/fox interactions in different parts of the system to be represented as trials on the same kind of experiment.

Second, some microvariables may be affected only indirectly by the outcomes of microdynamic experiments. For example, an experiment may have a direct probabilistic effect on one microvariable, which then has a direct deterministic effect on a second microvariable. The net effect is a probabilistic distribution over both microvariables. Suppose, for example, that there is a microvariable in the representation of a rabbit/fox ecosystem that tracks a rabbit's exposure to a certain environmental pollutant. The exposure increases in direct proportion to the time that the rabbit spends in the polluted part of the ecosystem. There may be no microdynamic probabilistic experiment that has a direct effect on a rabbit's exposure. But if the microdynamic probabilities affect the rabbit's position, they thereby indirectly affect its exposure, as a result of which there will be some sort of probability distribution over the exposure microvariable.

The third possibility is a variation of the second. It may be that for certain periods in an enion's time evolution, the enion's microvariables evolve deterministically rather than probabilistically. For example, a rabbit may return home at an appointed time at the end of each day by turning in the direction of its warren and traveling in a straight line. During this interval of time, the rabbit's position will change in a way that is fixed deterministically by its position at the moment of its turn homeward. Even this sort of time evolution will be affected by microdynamic probabilities, however, as it is microdynamic probabilities that determine the rabbit's position at the time of the homeward turn. There will therefore be a probability distribution over the direction and duration of the rabbit's journey home, even though both are, proximally, deterministically fixed.

The difference between this and the last possibility is that, in the last, the deterministic evolution of one microvariable, exposure to the pollutant, proceeds concurrently with the probabilistic evolution of the other microvariables, such as position, while in the sort of case exemplified by the rabbit's return home, the deterministic evolution may be the only change that is occurring in the rabbit's state; it occurs, as it were, in a probabilistic lull.

It is not absolutely necessary that the changes that take place in a rabbit's state due to periods of deterministic evolution be even indirectly affected by microdynamic probabilities. Some changes may take place in exactly the same way or with exactly the same result irrespective of the outcomes of the preceding microdynamic experiments. For example, at the end of its homeward journey, the rabbit always arrives at its warren; it seems that there is no nontrivial probability distribution, then, over the place in which the rabbit spends the night. This introduces an apparent problem in the treatment of population ecology, which is described and discussed in section 4.9.

All of this suggests that microdynamics may be conceived of—in the most schematic way—as having two aspects: a probabilistic aspect, in which microdynamic experiments probabilistically bring about certain outcomes, and a deterministic aspect, in which these outcomes in turn affect a system's microvariables.

4.33 Microdynamics Consists of Small, Causally Isolated Steps

A microdynamic probability, such as the probability that a rabbit decides to forage in the meadow, differs from an enion probability, such as the probability that a rabbit dies over the course of a month, in a number of ways. First, as noted in the last section, microdynamic probabilistic experiments sit at a finer level of description than enion probabilistic experiments; in particular, microdynamic experiments' ic-variables contain more information than the ic-variables of the experiments to which enion probabilities are attached. Second, the experiments to which enion probabilities are attached take place over a relatively long period (in the example, one month), whereas the experiments to which microdynamic probabilities are attached take place over a comparatively short period (foraging decisions and fox encounters may be over in a matter of seconds). Third, the mechanism to which an enion probability is attached is the whole system with all its enions, but the mechanism for a microdynamic probability is only a small part of the system, containing perhaps one or two enions (one enion for the foraging decision, two for the fox en-counter, and so on).

It is the third observation I wish to emphasize in this section, but the third clearly depends on the second. In the long-period experiments to which enion probabilities are attached, the enion in question is able, in principle, to inter-act with every other enion in the system. (In section 4.42 I show that enion probabilistic experiments *must* be long-period experiments.) In short-period microdynamic experiments, however, only local elements have a chance to play a role in an enion's fate.

As a consequence of their short period, the mechanisms of distinct micro-dynamic trials are causally isolated from one another. It follows that, provided that the IC-variables of the trials are stochastically independent, the outcomes of these trials are also independent. This provides one half—by far the easier half!—of the argument for the independence of microdynamic probabilities that will later play an important role in this chapter. The difficult half is, of course, showing that the IC-variables of distinct microdynamic trials are in-dependent. (Note that when I say that an experiment is causally isolated, I mean that its mechanism is causally isolated from the mechanisms of other experiments. Causal isolation, then, is necessary but not sufficient for causal independence, as defined in section 3.33; microdynamic trials in a complex system normally share IC-values and so are not causally independent.)

I now want to say something about the way in which microdynamic prob-abilistic experiments work together to determine the course of a complex system's microlevel time evolution. The microlevel trajectory of a particular system is determined by three things:

1. The system's initial microstate,
2. The operation of the deterministic aspect of microdynamics, and
3. The outcomes of the microdynamic probabilistic experiments that occur along the way.

Only (3) is probabilistic. The contribution of the probabilistic element of mi-crodynamics, then, to the microlevel trajectory taken by a particular system over time, is contained in the outcomes of the microdynamic experiments. Given an initial microstate, the outcomes of these experiments, and the laws governing the deterministic aspect of microdynamics, one can predict or re-construct the microlevel trajectory of the system.

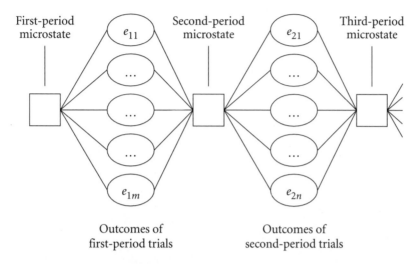

First-period microstate · Second-period microstate · Third-period microstate

e_{11} · ... · ... · ... · e_{1m}

e_{21} · ... · ... · ... · e_{2n}

Outcomes of first-period trials

Outcomes of second-period trials

Figure 4.3 Microlevel time evolution over a long period is determined by successive sets of simultaneous, causally isolated, trials on microdynamic probabilistic experiments, with outcomes represented as e_{ij}.

Putting together the two observations I have made in this section:

1. Trials on microdynamic experiments are governed by causally isolated mechanisms involving only a few enions, and
2. The trajectory of a system is determined by its initial microstate, the deterministic microdynamic laws, and the outcomes of the microdynamic trials,

one obtains a picture—somewhat idealized, of course—of microlevel time evolution as the cumulative result of a series of steps, each involving many simultaneous, causally isolated microdynamic trials, as shown in figure 4.3. Each set of outcomes determines, in concert with the previous microstate and the deterministic microdynamic laws, the next microstate. Microlevel time evolution consists, notionally, of a sequence of such steps.

This picture puts the outcomes of microdynamic probabilistic experiments at the center of attention. I should point out, however, that there is nothing in what I have said that requires that all, or even most, microvariables should be affected by the outcomes of microdynamic experiments. Any microvariable

may evolve entirely deterministically, and thus independently of the probabilistic aspect of microdynamics. At a later stage of the discussion, however,
I will require that all microvariables be affected in a certain way by microdynamic probabilities (see section 4.42).

4.34 Single Microstate Enion Probabilities

If a system's microdynamics is at all probabilistic, then even given a complete
set of microlevel information—an exact initial microstate—the future of an
enion is not entirely determined. In the rabbit/fox ecosystem, for example,
given an initial microstate z, there is a certain probability, neither zero nor one,
that a particular rabbit will die in the course of a month. I call this probability
a **single microstate enion probability**.

This is not the sort of enion probability required for EPA, as it depends
explicitly on microlevel information, that is, on z. But what I will do in section 4.4 is show that, under certain conditions satisfied by the complex systems
in which I am interested, the single microstate enion probabilities for every
microstate in any given macrostate are the same. I will conclude that the corresponding enion probabilities do not depend on microlevel information. The
notion of a single microstate enion probability thus plays an important part in
my argument.

What is the relationship between a single microstate enion probability and a
system's microdynamic probabilities? Let z be, as above, a microstate of a given
ecosystem. Let $t_1, \ldots, t_k$ be all the trajectories of the ecosystem, described at
the microlevel, that start out at z and lead to the rabbit's death in the course
of a month. Then the single microstate enion probability, relative to z, of the
rabbit's dying at some point during the month is

$$P_z(death) = P(t_1) + \cdots + P(t_k),$$

since the trajectories are mutually exclusive.

As observed in the last section, the particular trajectory taken by a system is
determined by its initial microstate, the deterministic aspect of microdynamics, and the outcomes of the relevant microdynamic probabilistic experiments.
Because the initial microstate z and the deterministic dynamics are fixed, the
probability of a given trajectory t_i is equal to the probability of the sequence
of outcomes that would, given the initial microstate, guide the system along t_i.

For example, if these outcomes are, as shown in figure 4.3, the events e_{11}, e_{12}, and so on, then the probability of t_i is just the combined probability of the e_{ij}'s.

Now assume that, as I will argue in section 4.5, the outcomes of microdynamic experiments are independent. Then the probability of an individual microlevel trajectory t_i can be expressed as follows:

$$P(t_i) = \underbrace{P(e_{11}) \ldots P(e_{1m})}_{\text{First step}} \underbrace{P(e_{21}) \ldots P(e_{2n})}_{\text{Second step}} \ldots$$

On the assumption of microdynamic independence, then, there is a simple relationship between the single microstate enion probabilities and the microdynamic probabilities.

4.4 Microconstancy and Independence of Enion Probabilities

It is easy to find a complex probability to play the role of an enion probability in a given complex system. What is not so easy is to find a complex probability that is microconstant and independent of other enion probabilities, properties essential for EPA. In this section and the next I will provide a schematic argument for demonstrating that the enion probabilities of ecology, statistical physics, economics, sociology, meteorology, and other sciences are microconstant and stochastically independent, in virtue of some of the very properties that are responsible for the complexity of the systems that they describe.

The argument is divided into two parts. First, I show that both microconstancy and independence depend on the stochastic independence of the microdynamic probabilities introduced in section 4.3. Then, in section 4.5, I provide a schematic argument for microdynamic independence itself.

4.41 Varieties of Microconstancy

Before laying out the argument for the microconstancy of enion probabilities, I must pause to consider which evolution function is to be shown to be microconstant, and which of two possible varieties of microconstancy is to be pursued.

Consider an enion probability $\text{cprob}(e_m)$, the chance of an enion's experiencing a type e event, given that the system starts out in macrostate m. To show that the probability is microconstant, I must show that the evolution function for e is microconstant over m.

I will take the rabbit/fox ecosystem as my paradigmatic complex system, and the probability of a rabbit's death over the course of a month as my paradigmatic enion probability. My goal, then, is to show that the evolution function for the death is microconstant over any region of the ecosystem's microvariables corresponding to a single ecological macrostate. Exactly what counts as an ecological macrostate is something that I do not want to fix in advance, but I assume that, at the very least, the rabbit and fox population levels are macrovariables, so that states of an ecosystem that differ with respect to either the number of rabbits or the number of foxes, or both, are different macrostates.

The first preliminary choice to make is that of the microlevel. Earlier—in section 4.3, in particular—I assumed that the microlevel of the ecosystem was what I called the behavioral level. This is indeed the choice I will make. But I want to point out that there are other possible choices. One particularly interesting choice is to take as the microlevel the level of fundamental physics. If this level were chosen, the microvariables of the ecosystem would concern the movements of individual fundamental particles.

Let me cast the choice between microlevels as a decision between two different evolution functions for the designated outcome of rabbit death. One evolution function takes as its arguments behavioral level variables such as the position and the hunger level of the rabbit. The other takes as its arguments fundamental-level variables. I call the first the *biological evolution function*, and the second the **complete evolution function**. I want to show that the evolution function for death is microconstant; the choice, then, is between trying to show that the biological evolution function is microconstant and trying to show that the complete evolution function is microconstant.

I will do both in this chapter, discussing the microconstancy of the biological function in this section and the microconstancy of the complete function in section 4.55. The reason to begin with the biological function is obvious: tractability. The reason to consider the microconstancy of the complete function is that the microconstancy of the biological function will turn out to be, in a certain important sense to be explained in section 4.55, less explanatory than that of the complete function.

I begin with the microconstancy of the biological evolution function. Given that the microlevel is the behavioral level, the microdynamics of a rabbit/fox ecosystem is, for reasons stated in section 4.32, probabilistic. Thus, the biological evolution function is probabilistic. I will focus on the version of the evolution function that takes an explicitly probabilistic form, that is, that maps not only onto 0 and 1 but onto all real numbers in between, the number for a particular set of IC-values being the probability of obtaining the designated outcome on a trial that has those IC-values. (The notion of an explicitly probabilistic evolution function was introduced in section 2.61.)

I have identified two ways in which an explicitly probabilistic evolution function $h_e(\zeta)$ can be microconstant over a macrostate m:

1. Oscillating microconstancy: It may be that $h_e(\zeta)$ oscillates between zero and one with the same strike ratio over all small regions of m.
2. Steady-state microconstancy: It may be that $h_e(\zeta)$ is reasonably constant—that is, is approximately equal to some probability p—over all of m.

In what follows, I construct an argument that the biological evolution function has steady-state microconstancy.[3] In section 4.55 I will then be able to show, as noted above, that the complete evolution function for an ecosystem has either oscillating or steady-state microconstancy, depending on the role played by simple probability at the fundamental level.

Why have I written so much about oscillating microconstancy if it is steady-state microconstancy that drives my argument? There are two reasons. First, oscillating microconstancy turns up elsewhere in what follows, for example, in the discussion of statistical physics in section 4.8. Second, in a deterministic system, all microconstancy is based, at the fundamental level, in oscillating microconstancy. In a sense that will become clear in section 4.55, steady-state microconstancy in a deterministic system is only a convenient higher-level representation of the effects of this fundamental-level oscillating microconstancy.

4.42 Microconstancy of the Biological Evolution Function

My paradigmatic biological evolution function, the evolution function for rabbit death, has, I argue in this section, steady-state microconstancy. The argument makes a number of assumptions, some of which are not justified in this section and some of which probably do not hold for a real rabbit/fox ecosystem. The most important assumption not argued for in this section is the assumption of the independence of microdynamic probabilities. The argument is deferred to later parts of the chapter: it is discussed in section 4.5 generally, and in sections 4.8 and 4.9 for two particular kinds of systems, including the rabbit/fox system. The discussions all draw heavily on chapter three. Two assumptions that may not hold for a real rabbit/fox ecosystem are discussed in section 4.43; both problems are resolved, one in that section and one in section 4.6.

Let $h(\cdot)$ be the biological evolution function for rabbit death over the course of a month in a rabbit/fox ecosystem. As section 4.2 shows, a set of IC-values for an enion probability specifies a microstate, and vice versa. Thus the biological evolution function can be written as a function that ranges over microstates of an ecosystem, rather than over IC-variables. For any such microstate z, $h(z)$ is the probability that a system starting in the microstate z will realize the designated outcome. It follows that $h(z)$ is the single microstate enion probability defined in section 4.34.

To demonstrate steady-state microconstancy, I must show that for any microstate z that realizes a given macrostate m, the single microstate enion probability of rabbit survival $h(z)$ is the same.

Let the argument begin. By a *single microstate experiment*, I mean an experiment to which a single microstate enion probability is attached. A single microstate trial, then, consists in taking a system in a given microstate z and allowing it to evolve for the requisite time; in my example of rabbit death, for a month.

As observed in section 4.33, the outcome of a single microstate trial depends on three things: the initial microstate z, the laws governing the deterministic aspect of microdynamics, and the outcomes of the microdynamic probabilistic experiments—foraging decisions, predator/prey interactions, and so on—that occur along the way. Taking the deterministic laws of microdynamics as given, the value of the corresponding single microstate probability depends on z and on the relevant microdynamic probabilities, stated in the probabilistic laws of microdynamics.

If the value of $h(z)$ is to be the same for any z in a given macrostate, it must be that the role of z in determining the single microstate enion probability is negligible.[4] Thus, $h(z)$ must be entirely determined by the microdynamic laws. This is the case when the following four conditions are satisfied:

A. The values of the microdynamic probabilities for any microdynamic experiment depend only on the current microstate,
B. There are no *fencing* effects within the initial macrostate m (fencing effects are defined below),
C. The microdynamic probabilities are independent, and
D. The trial time (for the enion probability, and thus for the single microstate probability) is sufficiently long.

If the conditions hold, the cumulative randomizing effect of the microdynamic probabilities will "wash out," or render irrelevant, the starting point of the system within the macrostate. Satisfaction of the conditions, then, is sufficient for the steady state microconstancy of enion probabilities. In the rest of this section, I explain the individual significance of conditions A–D.

Condition A

The purpose of condition A is to ensure that the starting position z of a trial does not fix the microdynamic laws in a certain way for the duration of the trial. The import of the condition is as follows: if two rabbits start out in systems with the same background environment but with different microstates, and the systems arrive, at some time, at the same microstate, the same microdynamic laws must apply to both rabbits at that time.[5] The condition is satisfied due to the definition of a microvariable. Anything, aside from the fixed background, that is relevant at a given time to the operation of the microdynamic laws counts as a microvariable, so there can be nothing relevant to the operation of the laws that is not specified by the current microstate.

Condition B

The initial state of a system may never lose its significance if there are *fences* within the system that are straddled by the macrostate m. A system contains a fence if microstates accessible from some points in state space are not accessible from others. A macrostate straddles a fence if microstates accessible from some points in the macrostate are not accessible from others.

Let me put this more exactly: condition B requires that, if a system starting out in a microstate z in m has a non-negligible probability of ending up in another microstate z' by the end of the time allowed for the experiment (in the case of rabbit death, one month), then a system starting out in any other microstate in m must also have a non-negligible probability of ending up in z' by the end of the experiment.

An important example of fence-straddling will be discussed in section 4.43. At this stage, I will give only a toy example. Suppose that the rabbit/fox ecosystem were divided into two areas, east and west, by an impassable wall. Consider two microstates: a microstate in which all foxes are on one side of the wall and all rabbits on the other, and a microstate in which both rabbits and foxes are equally distributed on both sides of the wall. Neither microstate is accessible from the other: if the system starts out in one, it cannot, in the course of a month, get to the other.

A macrostate m straddles the wall if there are microstates in m which differ as to which side of the wall a particular enion is on. This will be the case if the only macrovariables are the sizes of the rabbit and fox populations, since populations of the same sizes could be distributed in either of the ways just described. When there is fence-straddling, a system's future possibilities are strongly constrained by its initial microstate, and so the influence of the microstate cannot be washed out in the time available. This will likely make enion probabilities sensitive to microlevel information. For example, in a system where there are more foxes on one side of the wall than on the other, the probability of a rabbit's dying will depend on which side of the wall it lives on, a piece of microlevel information.

One way for condition B to be satisfied is for there to be no fences in a system at all, that is, for every microstate in the system to be accessible from every other. This is true for a system such as a gas in a box.

The other way to satisfy condition B is to carefully individuate macrostates so that none straddles a fence. In some cases, this is straightforward. In other cases, it will involve reassigning far too much information from the microlevel to the macrolevel. In the example above, for example, to avoid fence-straddling, a macrostate would have to specify for each fox and each rabbit whether it was on the east or the west side of the wall, since microstates in which a certain enion is on one side of the wall are inaccessible from microstates in which the same enion is on the other side of the wall.

It may occur to the reader that, in the wall case, it would be sufficient, for the purposes of stating a macrolevel law about the system, that a macrolevel description of the system distinguish just four populations: foxes on the west side, foxes on the east side, rabbits on the west side, and rabbits on the east side. Facts about the disposition of particular enions need not be specified. It is true that a macrolevel law can be stated that involves only these macrovariables, but the reason for this is not microconstancy. In the walled ecosystem I have described, the enion probability of rabbit death will necessarily be affected by the microlevel information concerning the rabbit's position with respect to the wall, but this microlevel dependence can be removed at the stage where enion probabilities are aggregated, as explained in section 4.6.

One property required of microdynamic evolution that is implicit in condition B but perhaps not immediately obvious, is that every microvariable in the system must be randomized, in some sense, by microdynamic probabilities.[6] The randomization need not be direct; it may be the result of the outcomes of microdynamic experiments operating through one or more microlevel intermediaries, as in the cases of a rabbit's exposure to pollution or its path home at the end of the day, discussed at the end of section 4.32. But if a microvariable is not randomized in some way or other, there will be values of the variable that are not probabilistically accessible from all initial microstates. An example is given in section 4.43.

CONDITIONS C AND D

Given A and B, conditions C and D work together to "wash out" the influence of the initial microstate. The information about the initial state is slowly swamped by the randomizing effects of the microdynamic probabilities. Thanks to these probabilities and their independence, the system is in effect taking a random walk through state space, or at least, through those parts of state space accessible from the initial macrostate; because the space is bounded, the position at which the walk starts gradually loses importance. Once the walk has gone on for long enough, there will be no significant correlation between the system's current state and its initial state z. It will be the patterns of outcomes of microdynamic experiments that determine the destiny of any enion; consequently, it will be the probability distributions attached to those experiments—the probabilities stated by the microdynamic laws—that determine the single microstate probability of any given destiny. In short, for a sufficiently long trial time, $h(z)$ will not depend on the initial microstate z, but only on the initial macrostate m and the microdynamic laws. Thus $h(z)$ will have approximately the same value for any z in m.[7, 8]

Suppose, to take an extremely simple example, that nothing affects the position of a rabbit but its decision as to where to forage, and that there is a 30% chance of a rabbit's deciding to forage in the scrub and a 70% chance of its deciding to forage in the meadow. Take a group of different rabbits starting out at different places in different systems. Then, after a month had elapsed, one would expect about 30% of rabbits in each system to be in the scrub and 70% to be in the meadow, regardless of the starting configuration.

Condition D requires that the trial time be long enough. How long is long enough? That depends on the microdynamic laws. Roughly speaking, however, the trial time must be at least long enough for each enion to complete a circuit of the territory, that is, long enough for its microvariables first to take on any given possible set of values, and then to revert to their original values.

SATISFYING CONDITIONS A–D

As I have remarked already, condition A is automatically satisfied in the formalism I have adopted. Conditions B and D can be satisfied by a judicious choice of experiment type for enion probabilities. B is satisfied by individuating macrostates sufficiently finely that they do not straddle fences; D is satisfied by choosing a sufficiently long trial time.[9]

The satisfaction of condition C, however, depends on the nature of the microdynamics. Showing that microdynamic probabilities are independent is by far the most difficult part of the argument, and I have left it to a separate section, namely, section 4.5 (which consists for the most part of invocations of the results in chapter three). For the remainder of this section, I simply assume microdynamic independence. (Some readers may wish to move immediately to section 4.5 at this point, returning when they are satisfied that there are reasons to regard microdynamic experiments as independent.)

4.43 Failures of Microconstancy

There are some microstates in an ecosystem that are far more dangerous to a rabbit than others. Two examples:

1. A microstate in which the rabbit is very close to a hungry fox, and
2. A microstate in which the rabbit is in very poor health.

In either case, the rabbit starts out at a significant disadvantage relative to the other rabbits in the system. This disadvantage must surely be manifest in the single microstate probabilities: a rabbit starting out in a particularly dangerous microstate will have a lower probability of surviving the month than if it had started out in some other microstate. The probability of death is therefore not the same for every microstate in a macrostate, and so the evolution function for death cannot have steady state microconstancy. (Exercise for the reader: which of conditions A to D fails in the above cases?)[10] This reflects the enion probabilities' sensitivity to microlevel information about fox proximity and health.

In this section, I present a solution to the problem raised by fox proximity, and discuss further, without solving, the problem raised by health. The solution to the health problem will be given in section 4.6.

The case of fox proximity, and similar problems, can be approached as suggested in approximation 2.2.2. Because the especially dangerous areas of state space are very small, it is reasonable to assume that the proportion of the ic-density that spans the dangerous areas is equally small. Thus the deviation from microconstancy will have a negligible effect on the overall enion probabilities, and can for all practical purposes be ignored.

The case of health is more interesting, and more troubling. Let me begin by describing the problem in more detail. Assume that the behavioral level of description includes a microvariable representing a rabbit's state of health. The health variable has two noteworthy properties. First, the initial state of the health microvariable will have a lingering effect over the trial time of a month, because health changes fairly slowly. Second, health will affect such important microdynamic probabilities as the probability of surviving an encounter with a fox.

As a consequence of these facts about health, a rabbit which begins a trial in a microstate where it has low health will have a lower probability of surviving than it would if it had begun in a different microstate, in the same macrostate, where it had high health. It follows that the probability of survival is not microconstant. Put very abstractly, the reason for this is that condition B does not hold: the ecosystem's state space contains "health barriers" that often take longer than a month for a rabbit (and thus the system containing the rabbit) to cross.

The problem cannot be dealt with in the same way as fox proximity, unless almost all rabbits are almost always equally healthy, which I assume is not the case. It could be dealt with by individuating macrostates more finely, as suggested in the discussion of condition B in section 4.42. But in order to remove any fence-straddling, a macrostate would have to specify the health level of every rabbit in the system. While it is certainly possible to open up the macrolevel to this kind of information, the result will be a very unwieldy macrolevel law, which is a bad thing for two reasons. First, there are the obvious practical disadvantages of unwieldiness. Second, and more important, we know that rabbit/fox ecosystems obey macrolevel laws that do not specify the health level of every rabbit. We want to understand why health does not appear in such laws. So far, EPA does not appear to offer the resources for doing so.

It turns out, however, that the disappearance of health from the macrolevel can be explained within the framework of EPA. The removal of health occurs not at the stage of EPA where microconstant enion probabilities are assigned, the current topic of discussion, but at the stage where enion probabilities are aggregated. I therefore postpone the treatment of the health problem to the discussion of aggregation in section 4.6.

4.44 Independence of Enion Probabilities

I now argue that, given the stochastic independence of a system's microdynamic probabilities, its enion probabilities are approximately independent. Much of the work is done by the assumption of microdynamic independence, which will be established in section 4.5 using the results of chapter three.

My strategy is to show that any two single microstate enion probabilities are independent; because the enion probabilities are equal to the single microstate probabilities (as argued in section 4.42), the independence of the enion probabilities follows.

Consider, then, the enion probability $\mathrm{cprob}(e_m)$ that a particular rabbit is eaten by a fox in the course of a month, given that the system is in macrostate m. This probability is equal to the single microstate enion probability of e for any microstate z in m (section 4.42). Choose such a z in m. The single microstate probability of death for z is the sum of the probabilities of all trajectories starting from z and leading to the rabbit's being eaten. The enion probability is equal to the single microstate probability for z and so is also equal to the sum of the probabilities of the trajectories:

$$\mathrm{cprob}(e_m) = P(t_1) + \cdots + P(t_k)$$

where the t_i's are the fatal trajectories originating from z. The particular trajectories will be different for different choices of z, but this does not matter, since the probability is the same for any choice of z.

I next want to express $\mathrm{cprob}(e_m)$ in terms of the relevant microdynamic probabilities. Given the stochastic independence of the outcomes of microdynamic experiments (section 4.5), the probability of each trajectory $P(t_i)$ can be expressed as a product of microdynamic probabilities:

$$P(t_i) = \underbrace{P(e_{11}) \ldots P(e_{1m})}_{\text{First step}} \underbrace{P(e_{21}) \ldots P(e_{2n})}_{\text{Second step}} \cdots$$

where the e_{ij}'s are the outcomes of trials on microdynamic experiments (see section 4.34, especially figure 4.3). Re-indexing the microdynamic outcomes to simplify the expression, I will write:

$$\mathrm{cprob}(e_m) = \underbrace{P(e_{11}) \ldots P(e_{1a})}_{P(t_1)} + \underbrace{P(e_{21}) \ldots P(e_{2b})}_{P(t_2)} + \cdots$$

There is a similar expression for any other enion probability, for example, the probability of a particular fox's starving to death in the course of a month. (Choose the same arbitrary microstate to fix the trajectories, since we are interested in the independence of outcomes occurring over the same period of time in the same system, hence originating in the same microstate.) The two probabilities will be independent if the probabilities of any two trajectories are independent,[11] which will be the case if the microdynamic probabilities of the events e_{ij} that direct the trajectories are independent.

From microdynamic independence, then, it follows that enion probabilities are independent—almost. Almost, because no trial is stochastically independent of itself, and the same microdynamic trial may direct trajectories contributing to two different enion probabilities. For example, the outcome of an encounter between a rabbit and a fox will turn up in trajectories that contribute to both the rabbit and the fox enion probabilities. This may undermine the independence of the enion probabilities: if the rabbit lives, the fox goes hungry, while if the fox eats, the rabbit dies.

As a rule, however, the probabilities attached to a particular microdynamic encounter between rabbit and fox will make only a very small contribution to the enion probabilities of rabbit and fox death, because the encounter will figure in only a small proportion of the possible trajectories leading to the rabbit's or the fox's death. (The other fatal trajectories involve the rabbit's being eaten by other foxes or the fox's failing to eat other rabbits.) Thus the enion probabilities will be approximately independent, as required by EPA.

To generalize: consider enion probabilities involving two different enions in a system. In some possible trajectories, these enions will meet. The probabilities will be (approximately) independent only if the net effect of these meetings on the probabilities is small. This will be the case if either (a) the number of trajectories in which the enions meet is a small proportion (probabilistically weighted) of the total number of trajectories, or (b) for each trajectory, there is a low probability of the meetings on that trajectory significantly affecting both of the relevant outcomes. One or both of (a) and (b) will likely be satisfied when the number of enions is large.

The fact that independence is only approximate will manifest itself in this way: macrodynamic laws based on the assumption that probabilities are completely independent will allow the violation of certain conservation laws. In a (rather idealized) ecosystem, for example, the following law must hold true: the number of rabbits eaten is equal to the number of times that foxes feed. The complete independence of enion probabilities entails a non-zero probability that the law is violated. For example, there will be a non-zero probability that all rabbits are eaten but no fox feeds.[12] This is not a conceptual problem with EPA, however, provided that the assumption of complete independence is regarded as an idealization adopted for computational convenience.

The reader may have noticed that, granted the independence of the microdynamic probabilities, the enion probabilities' independence need not be directly involved in the explanation of simple behavior. The reason is as follows. The purpose of the enion probability independence assumption is, as stated in section 1.23, to ensure the existence of a joint probability distribution over macrovariables that is independent of microlevel information. From the existence of such a joint distribution, the existence of laws containing only macrovariables can be inferred. But given microdynamic independence and the other three conditions stated in section 4.42, the existence of an appropriate joint distribution is assured regardless of whether or not enion probabilities are independent.

The joint distribution is constructed as follows. To calculate the probability of a macrostate m evolving into a macrostate m' after a given period of time, choose a microstate in m. Consider the trajectories leading from the microstate to the macrostate m' in the allotted time. Sum the probabilities of all these trajectories. By the arguments in section 4.42, this probability is the same for any choice of microstate in m. Thus there is a probability for the transition from m to m' that is independent of the microlevel. From such probabilities, the joint macrolevel probability distribution may be derived.[13]

I am not suggesting that enion probabilities are not, in fact, normally stochastically independent. Provided that the conditions stated earlier in this section obtain, independence holds, just as I have claimed. It is rather that the fact of this independence need not be invoked in the explanation of simple behavior (though the facts underlying independence certainly are invoked).

Does it follow, then, that the question of the independence of enion probabilities is a side issue? If the aim is only to explain the simple behavior of complex systems, then the answer might be yes. But I am aiming to do more than this. I am aiming to explain two additional facts: first, that complex systems' macrolevel behavior is related to statistics concerning enions in a certain way, namely, in the way made explicit in the enion probability independence assumption, and second that, as a consequence, a method for understanding the behavior of complex systems—enion probability analysis—that makes extensive use of the independence assumption, has been, and will quite likely continue to be, a success.

4.5 Independence of Microdynamic Probabilities

Everything turns on explaining the stochastic independence of microdynamic probabilities. The independence results of chapter three were developed to that end; I now put them to use to provide sufficient conditions for microdynamic independence in complex systems. I also establish the microconstancy of complete evolution functions, as promised in section 4.41.

The case for microdynamic independence is intended to be quite generic; it uses no technical tricks and takes no shortcuts. When applying the argument I will develop to particular kinds of systems, however, system-specific tricks and shortcuts will likely prove extremely useful in streamlining the procedure. In the study of gases and ecosystems in sections 4.8 and 4.9, for example, I make use of an especially useful technique: the identification of what I later call a *randomizing variable* (section 4.91). Use of a randomizing variable obviates the need to apply the argument outlined in this section to every one of a system's microvariables; rather, the argument is applied to a particular variable, the randomizing variable, and the independence of different instantiations of this variable so derived is then leveraged, in some system-specific way, to obtain a more general independence result.

The following argument is concerned exclusively with the independence of *complex* microdynamic probabilities. If microdynamic probabilities are simple, then I assume that their independence is a matter to be decided by the fundamental laws of physics.

4.51 The Problem of Correlated IC-Values

The difficulty of demonstrating the independence of complex microdynamic probabilities is starkest in a system in which simple probabilities play no role at all, that is, in which all relevant fundamental level laws are deterministic or quasi-deterministic, and all basic ic-densities are frequency-based or are given some other interpretation compatible with determinism. I will call such a system a *deterministic system;* the argument that follows assumes determinism in this sense.

There are two kinds of cases to consider in the argument for microdynamic independence: the case of the independence of simultaneous microdynamic trials, and the case of the independence of sequential microdynamic trials. (By sequential trials I mean trials involving the same organism at different times. Sequential trials need not be consecutive.) Some trials are neither sequential nor simultaneous, but their independence will follow from the arguments for the independence of sequential and simultaneous trials. In both the sequential and the simultaneous cases, the relevant mechanisms are causally isolated from one another (section 4.33), but there are good reasons to think that the ic-values are correlated.

In the case of sequential trials, ic-values are correlated because earlier values determine later values. Take a toy example, an ecosystem containing just one rabbit whose movements are entirely due to its foraging decisions. Consider two consecutive decisions by the rabbit. The decisions will be independent if their fundamental level ic-values, say z_1 and z_2, are independent. But the value of z_1 determines the value of z_2. Thus the two ic-values are as dependent, as correlated, as two ic-values can be.

In the case of simultaneous trials, ic-values may be correlated if some of the organisms taking part in one trial share a piece of causal history with some of the organisms taking part in the other. The simplest case is where organisms from both trials have interacted directly in the past. If they have, the ic-values for the later trials are very likely to be correlated, because they are both determined in part by the ic-values of the organisms' earlier interaction. The functional relation between the values is more complicated than in the case of consecutive trials, but it exists all the same.

4.52 Internal and External Sources of Randomness

Because the ic-variables of microdynamic experiments in a complex system have a tendency to get themselves correlated, there is a prima facie reason to think that the outcomes of microdynamic trials are not independent. There are, broadly speaking, two ways to show that, contrary to this expectation, independence holds. First, it may be that some external force interferes with the microdynamics of the system, destroying the correlation of the ic-values; this I call the *external account* of microdynamic randomness. Second, it may be that the correlation between full ic-values is not of the right sort to destroy the independence of the microdynamic probabilities. This I call the *internal account* of microdynamic randomness.

Consider the external account. It is certainly possible for exogenous variables to shake up a system, destroying to some extent the dependence of endogenous ic-values on their immediate predecessors. The more vigorous the shaking, the weaker the connection between one ic-value and the next. The existence of external "noise" is often used in this way to justify the so-called rerandomization posit, or the assumption of molecular chaos, in statistical physics (Sklar 1993, 250–254).

For such an effect to achieve the desired result, the exogenous variables would have to be what I call in section 4.91 *randomizing variables* for the system. This means, roughly, that they would have to supply all the randomness needed to wash out the influence of a system's initial microstate over the course of the kind of experiment to which enion probabilities are attached. In many systems, there are good reasons to doubt that the exogenous variables have what it takes to be randomizing variables. For example, foraging patterns in an ecosystem are far more variable, over time and between organisms, than the local weather, which suggests that the weather alone is not enough of a randomizer to account for probabilistic patterns in foraging behavior (for further comments, see section 4.9).

Another reason not to lean too hard on the external account is that it provides a less powerful explanation of randomness than an internal account. For external randomness to explain independence, the correlation-destroying exogenous ic-values must themselves be independent of one another, or they will bring new correlations with them. On the external account, then, the independence of outcomes within the system is explained by unexplained independence from outside the system. The internal account, by contrast, shows how certain mechanisms within the system can create independence from almost nothing, the basis for a superior explanation of simple behavior. (This does not mean, however, that the external account is explanatorily empty: it may perhaps show how independence can be contagious, a legitimate form of explanation discussed in section 2.53.)

The next two sections present the framework for an internal account of microdynamic independence based on the results of chapter three. The independence of sequential trials is examined in section 4.53 and that of simultaneous trials in section 4.54. Independence is shown to follow from the same assumptions in both cases. Since I am pursuing an internal account of randomness, I will assume for clarity's sake that the systems I consider have no exogenous variables.

The aim of these sections is to derive sufficient conditions for microdynamic independence that are fairly general, but not to show that any particular systems satisfy the conditions (that is the purpose of sections 4.8 and 4.9), or to derive anything like necessary conditions for independence. By *fairly general* I mean that large classes of possible dynamics satisfy the conditions, but certainly not that a majority of possible dynamics do so. The focus on a single set of sufficient conditions allows for a reasonably simple exposition at the expense of greater generality. There is much work that could be done on making the conditions more general. Whether that work would be worthwhile, that is, whether many actual systems that fail to meet the sufficient conditions for microdynamic independence stated below would turn out to satisfy a more general set of conditions, I do not know.

4.53 Independence of Sequential Trials

Microdynamic trials in a deterministic system with no exogenous variables in effect form a *deterministic chain*, in which the outcomes of one set of trials determine the ic-values of the next. I showed in section 3.7 that the outcomes of deterministically chained experiments can be independent if the effective ic-variable distributions of successively later trials, conditional on the outcomes of all previous trials, depend on distributions over successively finer information in the full initial ic-value.

Because I have treated independence in chains only for the case where the chained experiments are microconstant, the discussion here and in the following sections will assume that the relevant microdynamic experiments are microconstant. The extension of the argument to the non-microconstant case, which is not too complicated, is discussed in website section 4.5A.

The sufficient conditions for the independence of chained microconstant experiments stated in section 3.76, applied to the case of sequential microdynamic trials, are as follows:

1. The ic-evolution functions for all microdynamic mechanisms should be weakly inflationary to a sufficient degree,
2. The ic-evolution functions for all microdynamic mechanisms should be sufficiently microlinear,
3. The distribution over possible initial ic-values for the system should be macroperiodic.

The stated conditions not only guarantee the independence of any two sequential trials involving the same enion; they also guarantee the independence of two sequential trials involving different enions. This is because, when the conditions hold, a macroperiodic conditional distribution is generated for any trial.

Of the sufficient conditions for microdynamic independence, the most stringent—the hardest to satisfy—is condition (1), that ic-evolution functions are sufficiently inflationary. Let me explain. An ic-evolution function for a chained experiment is sufficiently inflationary if it maps the region of ic-variable space corresponding to an effective ic-value for the ith trial onto a full range of effective ic-values for the $(i + 1)^{\text{th}}$ trial. The region corresponding to an effective ic-value is a contiguous set of full ic-values all leading to the same designated outcome. The designated outcome of a microdynamic experiment is a new value for, or a change in, a microvariable, for example, a change in a rabbit's position. These changes are individuated rather finely—in rabbit position, say, to the nearest foot—so the designated outcomes are individuated finely. Consequently, the region corresponding to a given effective ic-value will be very small. The ic-evolution function must be sufficiently inflationary to map a region this small onto a region corresponding to all possible effective ic-values, or at least, sufficiently many effective ic-values that the resulting density is wide enough to count as macroperiodic.

For the same reasons that condition (1) is hard to satisfy, condition (2) is easy to satisfy. The smaller the regions corresponding to effective ic-values, the smaller the regions over which condition (2) requires that ic-evolution be roughly linear. The most pressing question, then, at least in a system in which microdynamic experiments are microconstant, is whether the system satisfies condition (1).

Inflation of the degree required by condition (1) is in fact not such a rare thing. In section 4.8, for example, I show that collisions between gas molecules are sufficiently inflationary, provided that the gas is not too dense. The reason is, as one might suppose, that the microdynamics of a gas is very sensitive to initial conditions. This sensitivity creates the inflationary pressure that results in microdynamic independence and so in simple behavior at the macrolevel—order from chaos, again.

In systems with enions that behave in more sophisticated ways than gas molecules, such as rabbits and foxes, there is also reason to expect a high degree of inflation, due to a connection between inflation and the flexibility of enion behavior. The connection is as follows. The sooner the distribution of the microvariables describing an enion's state, conditional on whatever just happened, comes to span once more the range that allows all possible outcomes in the next microdynamic trial, the more control the enion will have over what happens next. (Think of the outcomes of the next trial as different possible evasive maneuvers, for example.) The period of time required to inflate a micro-sized conditional distribution to a macro-sized range corresponds roughly, then, to what might be called the enion's refractory period, the period of time between the occurrence of a microdynamic event and the moment where the enion has recovered sufficiently that it is once more able to take advantage of the full range of physically available options. As a rule, enions have an interest in making the refractory period as short as possible. In systems where enions' interests play a role in determining the structure of enion behavior—biological and social systems—there is therefore some reason to expect a high degree of inflation. An example is discussed in section 4.9.

4.54 Independence of Simultaneous Trials

The obstacle to establishing the stochastic independence of simultaneous trials is the possibility that the IC-variables for two such trials might be correlated in virtue of a past interaction. There are two quite different ways in which this obstacle can be overcome.

SHIELDING

The first potential source of stochastic independence is causal independence. Of course, it cannot reasonably be supposed that all the IC-variables for one trial are causally independent of all the IC-variables for the other: if there has been any past interaction, this is necessarily false, on my definition of causal independence. But it may be that

1. Some IC-variables of one trial are causally independent, and thus stochastically independent, of some IC-variables of the other trial, and
2. The stochastic independence of these subsets of IC-variables is sufficient for the stochastic independence of the outcomes of the trials.

The first condition will tend to hold in practice only if the two sets of IC-variables are causally independent not only of each other, but of all other microvariables in the system, since there are otherwise too many ways that an indirect causal link might be established. When an IC-variable (or set of IC-variables) is causally independent of all other microvariables in a system, I say that it is **shielded**.

An example of a shielded IC-variable might be a variable representing certain neurobiological properties of a rabbit's brain that contribute to its foraging decisions, such as the IC-variable ψ in figure 4.2. The variable would be shielded if neither the interaction of two rabbits, nor of a rabbit and a fox, had any effect on the neurobiological properties. This may or may not be true, given the neurobiological effects of stress, excitement, and so on.

The second condition will hold if the two sets of causally independent ic-variables are strike sets for their respective experiments (section 3.45). This is true for a given experiment only if the strike ratio for all outcomes of interest is the same for any assignment of values to the variables not in the strike set that is consistent with the current microstate. In the case of the rabbit's foraging decision, this will be true if the rabbit's decisions depend on just (a) the value of the neurobiological variable, and (b) other information about the rabbit's position only at the behavioral level and above, so that small differences in position make no difference to foraging decisions. Then, a given value of the neurobiological variable will bring about the same decision in any fundamental level realization of a given ecological microstate.

MICROLINEARITY

The second potential source of independence is the result about short-term coupling derived in section 3.6. Let me first sketch the argument for independence in a single paragraph, drawing on the approach to short-term coupling described in section 3.66, and then give a longer and more careful version of the argument.

Suppose that there has been a past interaction W of two enions now involved in separate, simultaneous trials X and Y. Assume that the conditions stated for the independence of sequential trials hold. Then the probability distributions over the outcomes of X and Y depend on the probability distributions over information that resided at a low level of the ic-values for W. To destroy the independence of X and Y, then, W would have had to correlate information at this level. Given the assumption of macroperiodicity, the relevant low-level information was uniformly distributed immediately before W (theorem 3.10). Given microlinearity and inflation, the same information was uniformly distributed immediately after W (theorem 3.18). Uniformly distributed variables with a uniform joint distribution are always independent. It follows that W did not introduce a correlation, and so that it did not compromise the independence of X and Y.

Now I give the longer and more careful version of this argument. Consider all the trials involving the first enion that occur after W but before X as a single long experiment, which I will call A. The experiment A has as its ic-variables the ic-variables generated by W (as well as ic-variables involving other enions

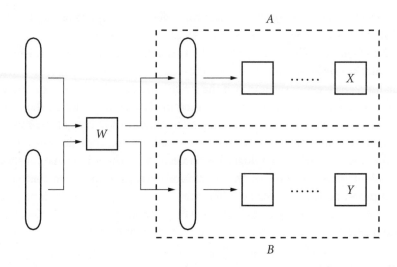

Figure 4.4 Experiments in the argument for the independence of simultaneous trials. The part of the network surrounded by the upper dashed box is the experiment A; the lower dashed box picks out B.

that participate in A), and as its outcomes the outcomes of X. Do the same for the other enion, resulting in a single long experiment B that culminates in Y, as shown in figure 4.4.

Now amend the picture a little: think of W not as an experiment in its own right, but as a mechanism effecting a causal coupling between the experiments A and B. Think, that is, of W's ic-variables as being the ic-variables of A and B, and of W's ic-evolution function as representing a process that introduces a coupling at the very beginning of A and B.

I will show that, if the conditions stated in section 4.53 for the independence of sequential trials hold, then the coupling effected by W does not destroy the independence of A and B. I conclude that, for the same reason, W does not destroy the independence of X and Y.

To keep the argument simple, I will assume that there is no other correlation that threatens the stochastic independence of A and B. In particular, I assume that A and B involve distinct sets of enions, and that any ic-variables of A and B other than the ic-variables of W are stochastically independent. If A and B are not independent, then, it is because of the coupling effected by W. I take this as a license to ignore any ic-variables of A and B other than the ic-variables coupled by W. Thus, I will treat A and B as though they are experiments that have as ic-variables only the ic-variables of W, as suggested by figure 4.4.

I showed in section 3.6 that a sufficiently inflationary, sufficiently microlinear coupling of variables with a macroperiodic joint density will not affect the independence of trials on microconstant experiments, provided that the coupling occurs at the very beginning of the trials. To use this result to derive the independence of A and B, I must establish the following:

1. The joint density over the ic-variables of W is macroperiodic,
2. The ic-variable transformation effected by W is sufficiently inflationary and sufficiently microlinear, and
3. A and B are microconstant.

The first condition I simply assume to be true for most complex systems. The second condition follows from the conditions for the independence of sequential trials, since the ic-variable transformation effected by W is identical to W's ic-evolution function, and the conditions for the independence of sequential trials require ic-evolution that is sufficiently inflationary and microlinear. The third condition follows from the microconstancy of X and Y (as explained in the discussion of multi-mechanism experiments in sections 3.74 and 3.76).[14]

I conclude that W does not destroy the independence of X and Y. When there are more complex causal commonalities in the histories of two trials—for example, when enions in both trials have previously interacted not with each other but with some third enion—the same style of reasoning can be employed. Similarly, when there have been many interactions at different times, the reasoning can be applied, first, to show that the most recent interaction does not compromise independence, then, that the second most recent interaction does not compromise independence, and so on.

Together, the arguments for the independence of sequential and simultaneous microdynamic trials amount to an argument for the independence of all trials on microdynamic experiments, as desired. Thus the conditions stated at the beginning of section 4.53—inflationary and microlinear ic-evolution of an initially macroperiodic distribution—are sufficient for the general microdynamic independence used to establish the microconstancy and independence of enion probabilities.

An important remark: if the microdynamics of a variable has both a probabilistic and a deterministic part, as described in section 4.32, then the effect of the deterministic part must be counted as a part of ic-evolution for the purpose of satisfying the conditions stated here. This may create problems if, for example, the deterministic aspect of microdynamics is deflationary. An example and further discussion of this point may be found in section 4.92.

4.55 Microconstancy of the Complete Evolution Function

I am now in a position to show, with the help of a few additional assumptions, that the complete evolution function for an enion probability—the evolution function taking as its ic-variables the quantities of fundamental physics—is microconstant. The argument is not essential to the vindication of EPA, but it is, I think, important for a deep understanding of the simple behavior of complex systems. In particular, the microconstancy of the complete evolution function provides a fuller explanation of the probabilistic patterns generated by complex systems that, at a low level, have either a deterministic or a quasi-deterministic dynamics, for the following reason.

In sections 2.24 and 3.5 I showed that oscillating microconstancy explains the probabilistic patterns. But steady-state microconstancy does not explain the patterns in the same way. To show that an experiment has steady-state microconstancy establishes that, for every possible set of ic-values, the nomic probability that the designated outcome occurs is the same, but it leaves open the question of why this constant nomic probability generates probabilistic patterns (section 2.61). In the case of complex systems, it is shown that the single microstate enion probability is the same for every microstate in a given macrostate, but it is not shown why single microstate enion probabilities generate probabilistic patterns. In order to put right this explanatory omission, I propose to show that, in a deterministic system, the complete evolution functions for single microstate enion probabilities have oscillating microconstancy. In so doing, I will have explained the probabilistic patterns of outcomes observed in complex systems, and—incidentally—I will have shown that the complete evolution functions for enion probabilities also have oscillating microconstancy.

Consider a macrostate m of a deterministic rabbit/fox ecosystem. Describe the state space at the fundamental level, then divide that part of the space corresponding to m into microregions, each corresponding to one of the ecosystem's microstates. Consider any such microregion R. Because the system is deterministic, the complete evolution function over R oscillates between zero and one. The strike ratio over R is equal to the proportion of system trajectories starting out in R that lead to the designated outcome (for example, rabbit survival) after a given period of time has expired (for example, one month).

Now, suppose that all the assumptions made concerning complex systems in previous sections hold. Suppose in addition that an evolution function that is inflationary and microlinear in the behavioral ic-variables is also inflationary and microlinear in the fundamental-level ic-variables, and that if the behavioral ic-variables are macroperiodically distributed then the fundamental-level ic-variables are also macroperiodically distributed. Given these assumptions, one can reason as follows:

1. The complete evolution function for the single microstate enion probability attached to R will be microconstant, because it is attached to a multi-mechanism experiment in a chain with inflationary, microlinear ic-evolution (see section 3.74 and the end of section 3.76).
2. The strike ratio of the complete evolution function over R will be equal to the single microstate probability for R, since the distribution of the fundamental ic-variables is macroperiodic and therefore uniform over R. (Rather than deducing the probability from the strike ratio, I am here deducing the strike ratio from the probability.) Consequently,
3. Because the single microstate enion probabilities are the same for every microregion in m, the strike ratio of the complete evolution function is the same for every microregion in m.

Thus, the complete evolution function for any enion probability has oscillating microconstancy. Note that nothing in the argument assumes that microdynamic probabilities generate probabilistic patterns.

This result may be explained as follows. When the conditions listed in the previous sections are satisfied, the point that a system reaches after a longish period of evolution, and so the realization of the designated outcome, does not significantly depend on the behavioral-level information in the fundamental states that make up R. Rather, it depends on very low level information. Thus the proportion of trajectories that realize the designated outcome and that begin in a given R depends on the distribution of very low level information in R. But this low-level information is the same in every microregion. Consequently, the same pattern of outward trajectories repeats itself from microregion to microregion.

This is, perhaps, a surprising result. From some very abstract properties of a complex system, such as the microlinearity of microdynamic mechanisms, it is possible to divine the fine structure of its complete evolution function, a function that incorporates every significant detail recognized by fundamental physics. The argument provides a striking example, then, of the power latent in EPA.

4.6 Aggregation of Enion Probabilities

Everything so far said of complex systems has concerned enion probabilities' satisfaction of the probabilistic supercondition. What remains is to explain how enion probabilities satisfying the supercondition, or almost satisfying the supercondition, are put together so as to produce macrolevel laws.

If enion probabilities satisfy the supercondition exactly—if they are entirely independent of microlevel information—then extracting the macrolevel laws is a very simple operation. But in many cases, enion probabilities are independent only of *almost* all microlevel information. For example, the probability of a rabbit's survival, I suggested in section 4.43, is sensitive to microlevel information about the rabbit's health. Enion probabilities that do not satisfy the supercondition fully can often be aggregated, I will show, so as to remove the remaining dependence on microlevel information.

The simple case of aggregation, where the supercondition is exactly satisfied, is discussed in section 4.61; more complex cases, where microlevel dependence must be removed in the process of aggregation, are discussed in sections 4.62, 4.63, and 4.64.

4.61 The Simple Case: Total Independence

The existence of microconstant, independent enion probabilities entails the existence of microconstant macrodynamic transition probabilities. A macrodynamic transition probability is the probability that a system in a macrostate m evolves into another macrostate m' after a certain fixed period of time. (Note that it is not precluded that the system passes through other macrostates on its way from m to m'.) To obtain these probabilities, the relevant enion probabilities are combined in accordance with the assumption of independence. Alternatively, the independence assumption may be side-stepped altogether, as explained at the end of section 4.44. Provided that the enion probabilities are microconstant, the macrodynamic probabilities are also microconstant (section 3.4).

Macrodynamic transition probabilities depend only on the values of macrovariables. Such probabilities can be used to construct macrolevel laws that contain only the macrovariables on which the probabilities depend. The laws so constructed are simple in the sense defined in section 1.12.

There are certain circumstances in which the macrolevel laws will be simple in other ways as well. The two most important varieties of surplus simplicity are mathematical simplicity and determinism.

First, mathematical simplicity. In some systems, the macrostate transition probabilities depend on the macrostate in simple ways; that is, the transition probabilities are mathematically simple functions of the macrovariables. This is likely, for example, in a system where the rate with which the relevant enion events (say, rabbit deaths) occur depends in a simple way on the frequency of encounters between certain enions (say, rabbits and foxes). Because the enions in the sort of complex system under discussion wander randomly around the system, the expected number of encounters will, in the most straightforward cases, be directly proportional to the relevant population. For example, the expected number of rabbit/fox encounters, and so the expected number of rabbits eaten, may be directly proportional to the number of foxes. The result is a mathematically simple law.

Second, determinism. In some systems, the laws constructed from the transition probabilities are quasi-deterministic; that is, they approximate deterministic laws. There is a tendency towards quasi-deterministic laws in any complex system susceptible of enion probability analysis. The reason is as follows. The future values of macrovariables are determined by aggregating the relevant enion probabilities. Because of the independence of enion probabilities, the law of large numbers applies. The law entails that, in systems with many enions, the future value of a macrovariable is very likely to be approximately equal to its expected value. Thus the system will evolve with a probability near one to a macrostate with macrovariables approximately equal to the expected values. Single macrodynamic steps, then, are quasi-deterministic. Now, provided that the small amount of indeterminacy in each quasi-deterministic step does not make much of a difference to what happens further down the line (and this assumption will not always hold),[15] the macrolevel laws governing sequences of such steps will be quasi-deterministic.

When conditions are favorable, then, a complex system may be governed by a law that is simple in three distinct ways: first, it involves only macrovariables; second, it is mathematically simple; and third, it is deterministic.

4.62 Structured Populations

When enion probabilities depend in part on microlevel information, there is no technique that is guaranteed to aggregate them into a simple macrodynamic law. In many cases, however, the aggregation is straightforward, sometimes surprisingly so. In other cases, aggregation is not straightforward, but is nevertheless possible. (For an interesting discussion of aggregation strategies that complements my own, see Auyang 1998, part I.) In this and the following sections, I discuss three strategies for aggregation.

The probability of a rabbit's surviving a month varies with the rabbit's health (section 4.43). Thus there is no microconstant probability of a rabbit's surviving a month. What there may be is a microconstant probability that a rabbit with a given level of health survives a month. Call this a *microconditional probability*, meaning simply a microconstant enion probability that is conditional on certain microlevel information.

The task at hand is to aggregate microconditional probabilities in such a way that the microlevel dependence falls out. This and the succeeding sections consider, in turn, three approaches to the problem.

The first strategy accepts what seems to be an inevitable consequence of the use of microconditional probabilities: microlevel information will turn up in the macrolevel law. The hope is that it will turn up in statistical form, as information about the *structure* of the population, for example, as a set of variables representing the health profile of the rabbits, that is, the number of rabbits falling into each of a small number of health-delineated subpopulations. Then the problem is solved—the aggregation is achieved—at the cost of a fairly small increase in the complexity of the macrolevel laws, resulting from the addition of the macrovariables representing the population structure.

When an enion probability is sensitive to microlevel information about the enion itself, but not to microlevel information about other enions, it is reasonable to expect the structured population approach to succeed. In such a case, the way that the size of each subpopulation changes is determined by enion probabilities that are sensitive only to the value of one microvariable, the microvariable that characterizes the subpopulation—a certain level of health, say—and not to further information about the identity of the individuals that make up the subpopulation. As a result, the fate of every enion in a subpopulation will be determined by the value of the same enion probability, for example, the probability of survival given a certain level of health. To take into account the dependence of survival on health, then, the size of each health-delineated subpopulation must be tracked separately, but otherwise, everything can be framed at the macrolevel.

The structured population approach successfully deals also with the case of the walled ecosystem discussed in section 4.42. Because the probability of a rabbit's death in the walled system depends only on the number of rabbits and foxes on the same side of the wall, a macrolevel population law can be derived from the microconditional probabilities that contains only macrovariables representing the numbers of rabbits and foxes on each side of the wall.

What if enion probabilities are sensitive to microlevel information about *other* enions? Then things become more complicated. However, this sort of sensitivity is comparatively rare. What is more common is a sensitivity not to information about other individual enions but to the structure of the relevant population. An example is the probability of rabbit survival, which (one might suppose) varies with the health profile of the fox population, but not with the health of any particular fox, since a rabbit is as likely to encounter one member of a given subpopulation as any other. Thus, rabbit probabilities depend only on statistics concerning the fox population structure, which may be incorporated into the macrolevel laws just as above.

Laws that involve the structure of populations are well known. They are often found in more sophisticated ecological models, and are essential in models of natural selection, where the changing structure of the population, from the predominance of a less fit to that of a more fit trait, is the object of interest. The structured population approach, then, is often all that is needed to explain the simple behavior of complex systems in cases where enion probabilities are not entirely independent of microlevel information.

Two questions, however, arise. First, complex systems containing microconditional enion probabilities sometimes conform rather closely to macrolevel laws that make no reference at all to the relevant microvariable. How is this possible? Second, the structured population approach may not work out in cases where enions are more likely to interact with some members of a subpopulation than with others. These two questions are taken up in the next two sections.

4.63 Averaging

Although the probability of rabbit survival is sensitive to the health of both rabbits and foxes, good macrolevel laws of population ecology can be stated that do not refer to health. It is possible, using the *average* enion probability of rabbit survival when aggregating enion probabilities, to obtain laws about the population of rabbits that are roughly correct. This is an unexpected aid to the use of EPA, but how and when does it work?

Consider first a case where all that is desired is to predict the number of rabbits that survive a single month. Then if both (a) the structured population strategy is applicable, and (b) the average probability of survival is understood as the average of the microconditional probabilities weighted by the size of the relevant subpopulations, it is a matter of mathematics that the average probability will predict the number of survivors just as well as the microconditional probabilities themselves.

Now suppose that what is desired is to predict the number of rabbits that survive a period of several months, using, as before, the one-month survival probability. The average probability will predict survival in the first month well. To predict survival for the second month, the probability to use is not the average weighted according to the population structure at the beginning of the first month, but the average weighted according to the population structure at the *end* of the first month. Because the model in use does not represent population structure, however, that information is lost. Thus the original average must be used. Clearly, this technique will be successful if (and in most circumstances, only if) the original average is a good estimate of the later averages, which will be true if the health structure of the population is not changing,[16] that is, if the rabbit population is in equilibrium with respect to health (but not necessarily with respect to everything—population, for example, may cycle without affecting health structure).

A sufficient condition for the success of the averaging strategy, then, is that both (a) the structured population strategy can be applied, and (b) the relevant population structure is in equilibrium. The strategy can reasonably be applied, then, in situations where equilibrium can reasonably be expected.

The averaging strategy may also be used, note, in cases where enion probabilities are sensitive to microlevel information about other enions, provided that all relevant population structures are in equilibrium. For example, in the case where the probability of rabbit survival is sensitive to the health structure of the fox population, a weighted average probability can be used provided that the health structure of the fox population is in equilibrium.

4.64 Advanced Techniques

The structured population approach fails when both (a) an enion probability is sensitive to microlevel information about other enions, and (b) the enion concerned is more likely to interact with some enions within a subpopulation than with others.

A well-known class of cases in which these conditions hold is afforded by the first-order phase transitions. I will discuss one member of this class, the spontaneous magnetization of iron. At high temperatures, iron does not have its characteristic magnetic properties, because the magnetic moments of its atoms are not coordinated. If the iron is cooled slowly enough, however, the atoms come to have their magnetic moments aligned, and the material as a whole exerts a definite magnetic field: it becomes magnetized.

Consider a structured population approach to spontaneous magnetization. The obvious choice of subpopulations partitions the population of iron atoms according to the orientation of their magnetic moments. If there are just two possible orientations, up and down,[17] then there will be two subpopulations, one containing all upwards-oriented atoms and the other containing all downwards-oriented atoms. Now, the orientation of any particular atom depends almost entirely on the orientations of its immediate neighbors. This means that any given atom interacts much more strongly with some members of a subpopulation, namely, its neighbors, than with any others. Statistics concerning the subpopulations, then—the total numbers of *up* and *down* atoms—will not always reflect accurately the influences on a given atom. For this reason, the subpopulation approach, which is, in this case, more or less what physicists call the mean field approach,[18] although it does give good qualitative results, does not model spontaneous magnetization very accurately (Yeomans 1992).

Can the structured population approach be saved? One would have to treat the neighbors of each atom as a separate subpopulation, or more precisely, as two subpopulations, the *up* neighbors and the *down* neighbors.[19] On this scheme, then, there would be twice as many subpopulations as atoms. Obviously, the resulting "macrolevel" law will be almost as complicated, perhaps even more complicated, than the system's microlevel law.

What is required is a mathematical technique for dealing with this kind of population structure. One method, developed by Lars Onsager, solves the problem in a special, two-dimensional case. However, Onsager's method is not easily generalized. A far more widely applicable technique is *renormalization* (Wilson 1979). The renormalization approach in effect takes a model with subpopulations at the neighborhood level, and transforms it into another model also with subpopulations at the neighborhood level, but with only a handful of atoms, hence only a handful of subpopulations, in such a way that the transformed model behaves in the same way as the original model. Any macrolevel law true of the transformed model, then, is true of the original system, but the transformed model is far more tractable: because the number of subpopulations in the transformed model is small, aggregation can be performed relatively easily. It turns out that the transformed model, and so the original system, obey macrolevel laws that do not depend on fine population structure at all.[20]

Renormalization is a sophisticated mathematical technique for aggregating large numbers of otherwise recondite microconditional probabilities. It and any other such techniques—Onsager's, for example—serve to further extend the reach of EPA.

4.7 Grand Conditions for Simple Macrolevel Behavior

The generic explanation of simple behavior in complex systems is now complete. Let me conclude by bringing together in one place some of the more important assumptions made in the explanation. These assumptions are found chiefly in sections 4.3 through 4.6. What follows is not intended to be rigorous or exhaustive; my aim is to cover only what I consider to be the most notable elements of the explanation.

The very first requirement is that the simple behavior to be explained must be an aggregate behavior of many parts of the system, that is, a macrolevel law about the statistics of enion behavior.

The second requirement is that the many parts of the system whose aggregate behavior is to be explained must behave like enions. In particular, their dynamics must be the result of a series of causally isolated interactions, each interaction involving at most a few other enions (section 4.33).

Third, these interactions must be able to be represented as probabilistic, microconstant experiments (section 4.32).

Fourth, the interactions must cause the microvariables involved to change in a sufficiently inflationary and microlinear way (section 4.5).

Finally, the effect of the microdynamic probabilities on the microvariables must be sufficient to wash out the effects of starting at different regions within any given macrostate (section 4.4). In particular, the system must not contain what I have called *fences*, or if it does, one of the techniques for dealing with fences described in section 4.6 must be applicable.

A natural worry about a list of conditions such as this, intended to be satisfied only approximately, is that minor violations of individual conditions will add up to a major deviation from the expected behavior. For example, small departures from the independence requirements in the case of individual enions, the reader might worry, could bring about a cumulatively large departure from the macrolevel behavior that, given an assumption of independence, one would predict.

In order to explore this worry further, let me identify the most important sources of small departures from independence. These are:

1. A failure of IC-evolution to maintain a macroperiodic conditional distribution over a system's microvariables, and
2. The lack of independence among the fates of enions involved in the same microdynamic trial (section 4.44).

This is not an exhaustive list—the expanding toolbox of techniques for dealing with failures of microconstancy discussed in section 4.6 alone ensures this—but it covers the causes of the most significant problems that arise for independence from the very nature of the technique for vindicating EPA advocated in this chapter, as opposed to problems that arise for some particular kind of system because of a feature present only in that system.

I want to persuade the reader that there is no insidious synergy between (1) and (2), that is, roughly, that the departures from independence caused by (1) and (2) acting together are no greater than the sum of the departures caused by their separate action. The argument for this claim is simply that the two act in quite different ways. The effect of (1) is to create enion probabilities that differ slightly from the corresponding strike ratios. The effect of (2) has nothing to do with an enion probability's relation to its strike ratio: as can be seen from the discussion in section 4.44, the size of the effect depends only on the population size and the combinatorics of the complex system. Thus the effect of (2) will be independent of the effect of (1).

This argument does not, of course, show that departures from independence due to (1) and departures due to (2) might not combine to have a greater effect than each would have on its own. That is always a possibility, although there is no reason why the departures should be in the same direction rather than in opposite directions. (For a discussion of a particular case, see section 4.85.) What can be concluded is that, provided the departures due to (1) and (2) are very small, their combined effect will not be very large.

4.8 Statistical Physics

Sections 4.4 and 4.5 of this chapter developed a schema for an argument establishing the microconstancy and independence of a complex system's enion probabilities. In the next two sections I apply this schema to two particular complex systems, a gas in a box and a predator/prey ecosystem, paying special attention, as earlier promised, to the foundations of microdynamic independence. I make some remarks about social systems in section 5.4.

4.81 A Gas in a Box

My principal contention in what follows is that the enion probabilities of the kinetic theory of gases—in particular, the probability distributions over position and velocity known collectively as the Maxwell-Boltzmann distribution—may be understood as microconstant, independent complex probabilities. To make a rigorous case for this proposal would require another book; here I confine myself to making a prima facie case. If my conclusions can be generalized sufficiently far, the approach taken to the kinetic theory of gases will also provide an understanding of other branches of statistical physics, and so of the probabilistic element of formal, by which I mean Gibbsian, statistical mechanics.

The Maxwell-Boltzmann distribution consists of a uniform distribution over a molecule's position and a Gaussian distribution over the components of its velocity. Traditional derivations of the Maxwell-Boltzmann probabilities simply assume an inherently probabilistic element to the behavior of gases, the so-called assumption of molecular chaos,[21] which is then shown to lead to the Maxwell-Boltzmann distribution. My aim is to derive something akin to the assumption of molecular chaos in the form of a microdynamic probability distribution, and to use this distribution to argue, according to the schema presented in section 4.4, for the microconstancy and independence of the Maxwell-Boltzmann probabilities.

The properties of the Maxwell-Boltzmann probabilities, then, are the official end points of the inquiry, but as the reader will, I hope, come to see, the journey is more important than the destination: the same considerations used to establish the microconstancy and independence of the Maxwell-Boltzmann probabilities can, I believe, provide an understanding of all the dynamic properties of gases that fall within the ambit of kinetic theory, in particular, the tendency to equilibrium.

There is one obvious difficulty with the project I have just outlined: there is no hope of applying the methods developed in this study to a system in which there are no microdynamic probabilities, for example, a system in a deterministic world whose microlevel is the fundamental level, or near enough. The systems of kinetic theory fit this description rather well. The molecules of a classical gas in a box interact, for example, according to deterministic laws. How, then, to find room for microdynamic probabilities in a gas?

My answer is to create a level of description at which interaction is not deterministic. At this level, which I will call the *Boltzmann level*, positions and velocities are specified only approximately. The information in the specification is fine enough to determine which molecules are likely to interact, but coarse enough that the result of an interaction is not completely determined, not even at the Boltzmann level. (The system of description is what is called a coarse-graining.) The Boltzmann level, then, will serve as the microlevel for a gas.

Interactions between molecules may be conceived of as trials on a probabilistic experiment having ic-variables that range over values of position and velocity lying within a single Boltzmann-level microstate. The probability of a given outcome on such an experiment is the probability of that outcome conditional on the system's being in the relevant microstate. I will not choose any particular coarse-graining to serve as the Boltzmann level. Rather, I simply assume that some appropriate coarse-graining exists; later sections (4.85 and 4.86) provide some clues as to what level of granularity might be appropriate.

Throughout this section, I will deal with a rather idealized model of a gas. As well as ignoring quantum mechanics, I will assume that gas molecules are spherical, that they interact like billiard balls, and that they therefore have no effect on one another except when they collide. This is what is sometimes called the hard-sphere collision model for gases.

I will claim in section 4.88 that the hard sphere idealization does no harm to the argument, which can be expected to go through provided that real intermolecular interaction is at least approximately similar to the interaction of hard spheres. This is not, of course, true for gases that are not monatomic, but the approach described here for spherical molecules is, I will argue, reasonably easy to adapt to the non-spherical case.

One final idealization I make is to consider only a two-dimensional gas, made up of circles bouncing around a plane rather than spheres bouncing around three-dimensional space. It should be clear that everything I say is easily extended, upon addition of the relevant variables, to a three-dimensional gas.

4.82 Relative Angle of Impact

To show that the Maxwell-Boltzmann probabilities are microconstant and independent according to the scheme laid out in section 4.4, I must show that the microdynamic probabilities of kinetic theory—the probabilities governing collisions that appear when a gas is viewed, as it were, through the soft focus lens of the Boltzmann level—are independent.

There are two sets of sufficient conditions for microdynamic independence, one set for the case in which microdynamic probabilities are microconstant, one (relegated to website section 4.5A) for the case in which they are not. The very first issue to decide, then, is whether microdynamic experiments in a gas are microconstant.

The microconstancy of an experiment depends in part on the designated outcome; what I do in this section is to suggest a set of designated outcomes conducive to a microconstant microdynamics for a gas. The designated outcomes of the microdynamic experiments should be, I suggest, the coarse-grained values of a variable that I will call the *relative angle of impact* of a collision between two gas molecules. The relative angle of impact is the angle between the following two vectors: (a) the relative velocity of the two colliding molecules, and (b) the direction of the line connecting the centers of the molecules at the moment of impact. In the case of a two-dimensional gas, the relative impact angle is represented by a single real-valued variable taking on values between −90 and 90 degrees; it is the angle θ in figure 4.5.

The relative angle of impact plays a key intermediary role in the microdynamics of gas molecules. It is determined by the positions and velocities of two colliding gas molecules the moment before they collide, and it then determines the change in the velocities of the two molecules at the moment of the collision, and thus the ensuing change in their position after the collision.

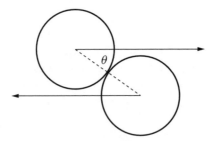

Figure 4.5 The relative angle of impact θ of a collision between two gas molecules. The arrows represent the relative direction of travel of the two molecules, which is, in this case, the horizontal. The dashed line connects the centers of the molecules at the time of impact. The angle θ between the dashed and solid lines is the relative angle of impact.

4.83 The Argument for Microconstancy and Independence

Consider a microdynamic experiment that begins with a collision between one gas molecule M and another, and ends with the very next collision involving M. The experiment spans, then, the interval between M's i^{th} collision and its $(i + 1)^{th}$ collision. Take as the designated outcomes of the experiment the different possible values of the relative angle of impact for the $(i + 1)^{th}$ collision, represented at the Boltzmann level. (Because the angle is coarse-grained, the possible outcomes make up a finite set.) The IC-variables for the experiment will include the position and velocity of M, the position and velocity of M's partner in the ith collision, and the position and velocity of M's partner in the $(i + 1)^{th}$ collision.[22] I will also consider as an IC-variable the relative angle of impact for the ith collision.[23] A microdynamic experiment is, by definition, conditional on the system's being in some microstate or other, so that the IC-variables for any particular experiment are restricted to values consistent with a given Boltzmann-level microstate. This microstate dependence will fall out of the picture shortly, making the argument much simpler than it otherwise would be.

The case for the microconstancy and independence of the enion probabilities of kinetic theory begins with the following two claims, made in section 4.84:

1. The evolution functions for microdynamic experiments of the sort I have described are microconstant.
2. The microdynamic experiments have a strike set of IC-variables with just one member, the relative angle of impact for the collision with which they begin (in the description above, the relative angle of impact for the ith collision). Thus, all the other IC-variables of the experiments—the positions and velocities of all the molecules involved in the collisions— are eliminable in favor of the relative angle of impact.

In section 4.85, I then show that

3. A macroperiodic conditional distribution is maintained over the relative angle of impact from trial to trial.

From the reasoning in section 3.76 concerning chained trials on microconstant experiments, and from the fact that the independence properties of only strike sets of IC-variables matter (section 3.45 and theorem 3.4), it follows that the relative angles of impact for a series of collisions involving a given molecule are stochastically independent.

At this stage I will have shown something else as well, namely, that there is a single probability distribution imposed over the impact angle for any collision, thus for any microdynamic experiment, regardless of the system's microstate. (It is shown in figure 4.10.) Therefore, the relative impact angles are both independent and identically distributed.

Finally, in section 4.86, I propose that:

4. If the relative angles of impact for all collisions in a gas are stochastically independent and distributed as I have argued, then all the microvariables of the system, that is, the positions and velocities of all the gas molecules, will take random walks of the sort sufficient, according to section 4.42, for the microconstancy of enion probabilities.

5. For the same reason, the conditions stated in section 4.44 for the independence of enion probabilities will be satisfied.

I conclude that the enion probabilities of kinetic theory, and in particular the probabilities that make up the Maxwell-Boltzmann distribution, are microconstant and independent. The argument, then, has five steps, spanning sections 4.84, 4.85, and 4.86.

It will emerge in section 4.9 that this argument is an example of a generic argument that turns on the notion, mentioned at the beginning of section 4.5, of a *randomizing variable*. In the case of the kinetic theory of gases, the randomizing variable is the relative impact angle.

Because the outcome of a microdynamic experiment, a coarse-grained impact angle, determines a range of IC-values for the next experiment, the same underlying mechanics of collisions will determine both the properties of the evolution function, which are the subject of step (2) of the overall argument, and the properties of the IC-evolution function, which are the subject of step (3). Consequently, there is considerable overlap between the argument establishing step (2) and the argument establishing step (3) (see also website section 3.7B). I deal with the overlap by postponing the more technical aspects of the discussion to the section concerning step (3), that is, section 4.85.

4.84 Microconstancy in Gas Microdynamics

My first task—step (1) of the main argument—is to show that the evolution functions for relative angles of impact are microconstant. Let X be a microdynamic experiment of the sort described above, spanning the time between two collisions and having the different possible relative angles of impact for the later collision as its designated outcomes. I must show that there is a micro-sized partition of the IC-variable space for X such that every partition member has the same strike ratio for each different outcome, that is, for each different angle of impact for the later collision.

To this end, consider the relation, for a given molecule M, between the IC-variables for X—that is, the initial conditions of M's i^{th} collision—and the relative angle of impact for M's $(i + 1)^{th}$ collision. The IC-variables describing the ith collision are restricted to those realizing a particular microstate, but because the subsequent trajectory of M depends in a very sensitive way on the details of the collision, even within the confines of this microstate, the collision could send M in a number of different directions. Thus, M's $(i + 1)^{th}$ collision could involve any of a number of different molecules.[24]

The role of X's IC-variables in determining the facts about the $(i + 1)^{th}$ collision can be broken into two parts: that of determining the molecule with which M will collide, and that of determining the relative angle of impact of the collision. Clearly, it is relatively high-level information about the trajectory of M and other molecules in the system that determines the identity of M's collision partner, and lower-level information that determines the relative angle of impact with this partner. In other words, it is high-level information about X's IC-variables that determines the identity of the partner, and low-level information about X's IC-variables that determines the angle of impact.

Now partition X's IC-variable space into regions each bringing about an $(i + 1)^{th}$ collision with a different molecule. Call the partition the *collision partition* for X. The collision partition is, I claim, a micro-sized constant ratio partition for the $(i + 1)^{th}$ collision's relative angle of impact.

The claim has two parts: that the collision partition is micro-sized, and that the collision partition is a constant ratio partition. The collision partition is micro-sized because, as stated above, even very small changes in X's IC-variables will cause significant changes in the trajectories of the molecules after the ith collision, and so will very likely change the identity of the molecule involved in the $(i + 1)^{th}$ collision.[25]

The collision partition is a constant ratio partition if different possible relative angles of impact for the $(i + 1)^{th}$ trial have the same strike ratio within each member of the partition. This is in fact the case, for the reason that the relative angle of impact depends on the relative position of a set of IC-values within the partition member in which it falls in qualitatively the same way for every member of the collision partition.[26]

I will, in section 4.85, discuss the facts that stand behind this assertion (see especially note 36), but until then, I ask the reader to accept it, temporarily, without justification.[27]

A much stronger version of the claim is also true. Not only are the strike ratios the same for any member of the collision partition for a given experiment X, they are also the same for any member of *any* collision partition. That is, any experiment that spans the interval between a molecule's i^{th} collision and its $(i + 1)^{th}$ collision will have a collision partition with the very same strike ratios. For the purposes of determining microdynamic probability distributions, then, we need to consider for practical purposes only one experiment of this sort: in a sense, X can be taken as *the* canonical experiment spanning any molecule's i^{th} and $(i + 1)^{th}$ collisions. This concludes the first step of the main argument: the experiment X is microconstant.

The second step is to show that all the IC-variables of X are eliminable in favor of the relative angle of impact of the ith collision. That is, for any assignment of fixed values to the relative positions and velocities of the two molecules involved in the ith collision, and to the positions and velocities of any other relevant molecules, the strike ratio for a given relative angle of impact for the $(i + 1)^{th}$ collision is the same.[28] At this stage, I present only a rather loose argument for this claim, based on an appeal to geometric intuition: as long as the relative angle of impact for the ith collision can vary freely within the parameters allowed by the specification of a Boltzmann-level microstate, it seems impossible to fix the relative positions and velocities for the collision so as to make some particular angle of impact for the $(i + 1)^{th}$ collision more likely than it would be otherwise. For a better argument, see section 4.85, especially note 36.

Let me pause to remind the reader of what is significant about a strike set: provided that there is a macroperiodic density over a strike set of IC-variables for an experiment, the probabilities for the designated outcomes are equal to their strike ratios. Experiments of the same type, then, will have the same probability distribution over their outcomes provided that they all have a macroperiodic density over a strike set of their IC-variables.

Now apply this result to gases. From the fact that the relative impact angle for the ith collision is a strike set for the canonical experiment X, it follows that, provided that there is a macroperiodic distribution over the relative angle of impact for a molecule's very first collision, the probability distribution over the relative angle of impact for the second collision will be the same, whatever the other facts of the case. If the probability distribution produced over the impact angle for the second collision is itself macroperiodic, the probability distribution over the impact angle for the third collision will be the same. And so on: the same, macroperiodic distribution will describe the relative impact angle for any collision.[29]

From here it is a short step to the independence of the impact angles for different collisions. In order to establish independence, it is necessary to consider the impact angle distribution for a collision conditional on all previous angles of impact. If it can be shown that this conditional distribution remains always macroperiodic, it follows from the reasoning in the last paragraph that the distribution remains always the same. Thus the distribution over the angle of impact for any collision in a gas, conditional on any previous impact angles, will be the same, and so angles of impact will be stochastically independent of each other, as desired. To establish that the conditional density is always macroperiodic is the next step of the argument.

Before I turn to this claim, however, let me take care of a loose end. I have been deliberately vague about the identity of the ic-variables for the experiment X that spans the period between the molecule M's i^{th} and $(i + 1)^{th}$ collisions. Among the ic-variables of X are, of course, the position and velocity of M and of the molecules that M strikes in its ith and $(i + 1)^{th}$ collisions. But a little thought shows that these are not the only ic-variables of X. The ic-variables of X must include the position and velocity of every molecule that M might possibly strike in its $(i + 1)^{th}$ collision, not just the position and velocity of the molecule that it actually strikes. The microdynamic experiment X, then, has many ic-variables. The reader is now in a position to see, I hope, that these ic-variables make no difference to the argument that I am presenting here, since they are all eliminable in favor of the relative angle of impact for the ith collision.

What I am claiming, then, is the following: although it is possible, by rearranging the molecules around M, to make a considerable difference to the probability that one rather than another will be involved in the $(i + 1)^{th}$ collision, it is impossible, as long as the angle of impact for the ith collision is macroperiodically distributed, to probabilify or deprobabilify, by such a rearrangement, any particular angle of impact for the $(i + 1)^{th}$ collision. All ic-variables of the collision experiment other than the relative impact angle may, as a consequence, be eliminated, enormously simplifying the treatment of the experiment.

4.85 Independence in Gas Microdynamics

I now show that the distribution over the relative angle of impact for a collision, conditional on the impact angle for all previous collisions in the gas, is
always macroperiodic. This, in conjunction with microconstancy, is sufficient,
as explained above, for the stochastic independence of any two impact angles.
For the reasons stated in section 4.5, sufficient conditions for the maintenance
of a macroperiodic conditional probability distribution over a variable are:

1. The ic-evolution of the variable is weakly inflationary to a sufficient
   degree,
2. The ic-evolution of the variable is sufficiently microlinear,
3. The distribution of initial values for the variable is itself macroperiodic.

I will simply assume that condition (3) holds, that is, that the microvariables
of a gas are typically distributed in such a way that the distribution of impact
angles for the first wave of collisions is macroperiodic.[30] My focus will be on
conditions (1) and (2).

Inflation first. The ic-evolution of relative impact angle is inflationary if
small differences in the relative impact angle for the ith collision involving a
molecule M bring about larger differences in the relative impact angle for M's
$(i + 1)^{th}$ collision. It is fairly obvious that this is so, but let me spell out the
argument.

Consider two almost identical collisions, one involving a molecule M and
the other involving a molecule N, with slightly different impact angles. The
small difference in the impact angles will bring about a commensurate difference in the direction and speed of M from that of N after the collisions.
From this point to the time of each molecule's next collision there will be no
change in either M's or N's velocities. Because of the difference in their directions of travel, there will be an increasing difference in their positions as the
time elapsed since the original collision increases. After a fairly short time, the
small difference in impact angle will have resulted in a reasonably large difference in the molecules' positions, as shown in figure 4.6. This reasonably large
difference in position will have an extremely large effect on the impact angles
of M's and N's next collisions.

Let me now put this subjunctively: had the impact angle of M's i^{th} collision
been slightly different, it would have had a rather different impact angle for
its $(i + 1)^{th}$ collision. Indeed, a small change in the impact angle of the ith
collision would have likely resulted in a completely different molecule's being

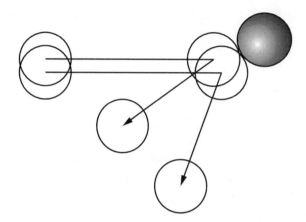

Figure 4.6 Collisions with convex surfaces bring about inflationary evolution. The initial positions of two spheres are represented by the overlapping circles. The arrows show the trajectories of the centers of mass of the two spheres as they collide with the sphere at the top right.

M's partner in the $(i + 1)^{\text{th}}$ collision. For this reason, the inflation of impact angle is *folding*, with one fold for each potential collision partner. The folding may disguise the effect of the inflation, but as I showed in section 3.7, folding inflation is just as effective as non-folding inflation in providing the stretch necessary to create a macroperiodic conditional probability distribution.

Microdynamic ic-evolution is undeniably inflationary, but is it sufficiently inflationary for independence? What is required for independence is, at a minimum, that the fundamental-level states corresponding to a single value for the Boltzmann-level impact angle of the ith collision spread out, due to inflation, to encompass every possible impact angle for the $(i + 1)^{\text{th}}$ collision. That is, no impact angle for the $(i + 1)^{\text{th}}$ collision should be inaccessible from any microstate corresponding to the ith collision.

This condition can be met by adjusting the coarse-graining of Boltzmann-level impact angles. If the angles are specified sufficiently coarsely, then whatever level of inflation is present between the ith and $(i + 1)^{\text{th}}$ collisions will suffice to ensure that all impact angles are possible outcomes of the $(i + 1)^{\text{th}}$ collision, no matter what the impact angle of the ith collision.[31] Thus condition (1) for microdynamic independence is satisfied.

The question of microlinearity is not so easily resolved, although I think it is fairly clear that the ic-evolution function mapping the relative angle of impact for a molecule M's i^{th} collision to the impact angle for M's $(i + 1)^{th}$ collision is, on the whole, smooth enough to qualify as microlinear. What I will do in the remainder of this section is to examine in some detail the relation between the relative angles of impact of two successive collisions for a particular two-dimensional case. My principal aim is to show that the IC-evolution of impact angle is microlinear (though with an exception to be discussed at the end of the section). Two other aims will be realized in the course of the discussion: (a) I will show that the qualitative claims made above about inflation are true, that is, that impact angle is inflated sufficiently, in the technical sense of *sufficiently* required for the maintenance of a macroperiodic conditional distribution over impact angle, and (b) I will show that an important claim made in the previous section, that the collision partition for a microdynamic experiment is a constant ratio partition, is true (see note 36).

The IC-evolution of impact angle can be divided into three phases. In the first phase, the impact angle for the ith collision determines the velocity of M immediately after the ith collision. In the second phase, the velocity and position of M after the ith collision determine its position at the time of the $(i + 1)^{th}$ collision. In the third phase, the position of M relative to its partner in the $(i + 1)^{th}$ collision determines the relative angle of impact for the $(i + 1)^{th}$ collision. I will amalgamate the treatment of the second and third phases.

I begin with the first phase. Let θ be the relative angle of impact of M's i^{th} collision. Take as the frame of reference, here and for the remainder of this section, the frame with respect to which the center of mass of the two molecules involved in the ith collision is stationary. Then the effect of the collision on M is the same as if M had hit a wall coincident with the line tangent to the two molecules at the point of impact, and therefore perpendicular to the line connecting the centers of the two molecules (see figure 4.7). The angle of reflection is equal to the angle of incidence, so the effect of the collision is to rotate M's direction of travel through an angle equal to twice the relative angle of impact θ.

Now suppose that M's new direction of travel sends it towards a molecule N. Here I make a major simplifying assumption: I assume that N is not moving relative to the frame of reference, which is, recall, the frame with respect to which the center of mass of M and the molecule with which it just collided is stationary. As a result, I can ignore the role of M's speed in determining its relative impact angle with N, considering only its direction of travel.[32]

Assuming that M and N are very many molecular diameters apart, M will collide with N only if its direction of travel lies in a very narrow band. What I

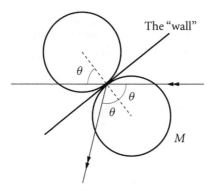

Figure 4.7 The effect of the relative angle of impact θ on M's direction of travel is to rotate the direction of travel by 2θ, when viewed in the frame of reference in which the colliding molecules' center of mass is stationary.

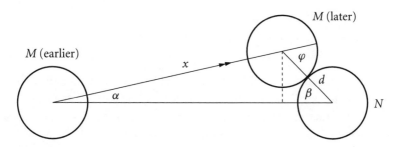

Figure 4.8 The effect of the velocity (more exactly, the effect of the direction of travel) of M on the relative angle of impact φ for M and N's collision. From left to right, the molecules are: M at the time of the ith collision, M at the time of the $(i + 1)^{\text{th}}$ collision, and N. The length of the dashed line is equal to both $x \sin \alpha$ and $d \sin \beta$.

want to do is look only at angles in that band, and to write down an expression that relates such an angle to the relative impact angle for M and N's collision.

I will measure M's angle of travel relative to a line connecting M and N's centers at the time of the ith collision (see figure 4.8). I call the angle of travel α. I will be concerned only with positive values of α; the treatment of negative values is of course entirely symmetrical.

Because N is not moving, the relative velocity of M and N is just the velocity of M, and is thus represented by a vector having direction α. The relative angle of impact for the $(i + 1)^{\text{th}}$ collision is the angle φ between this vector and the vector connecting M and N's centers at the time of the collision.

Now I calculate φ as a function of α (all references are to figure 4.8). Let x be the distance traveled by M between the ith and the $(i + 1)^{\text{th}}$ collision. Let d be the diameter of a molecule in the gas, and hence the distance between M and N's centers at the time of their collision. Then, where β is the angle labeled as such in figure 4.8,

$$x \sin \alpha = d \sin \beta$$

from which it follows that

$$\beta = \arcsin \left(\frac{x}{d} \sin \alpha \right)$$

and thus, since $\varphi = \alpha + \beta$,

$$\varphi = \alpha + \arcsin \left(\frac{x}{d} \sin \alpha \right).$$

Because the distance x traveled between collisions is much greater than the molecular diameter d, the coefficient x/d is very large and so the function is strongly inflationary. This is the quantification of IC-evolution's inflationary power promised above; given a reasonable choice for the Boltzmann level, the reader should be able to see, the inflation is of the degree required for independence.[33] (Readers who want the complete quantitative argument for the maintenance of macroperiodicity will find it later in this section.)

Now, since α is restricted to angles that will cause a collision between M and N, it is very small. This allows two simplifications: I drop the first occurrence of α and I replace $\sin \alpha$ with α.[34] Thus

$$\varphi = \arcsin \frac{x}{d} \alpha$$

$$= \arcsin \frac{2x}{d} \theta$$

for values of θ that bring about a collision between M and N.

The IC-evolution function for impact angle, then, is just the arc sine function, that is, the sine function turned on its side. The IC-evolution function for

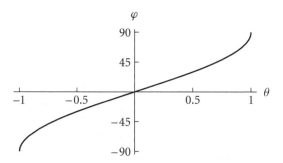

Figure 4.9 The relative impact angle IC-evolution function for the case in which $2x/d = 57$ (with the angles measured in degrees).

the case in which $2x/d = 57$ (assuming angles measured in degrees), and the zero points of θ and α coincide, is shown in figure 4.9.

Is the relation sufficiently microlinear? To answer this question, the partition with respect to which microlinearity ought to obtain must be specified. As explained in section 3.76, this partition is determined by the optimal constant ratio partition for the *next* trial, that is, for the trial in which the relative impact angle for the $(i + 1)^{\text{th}}$ collision determines the impact angle for the $(i + 2)^{\text{th}}$ trial. An optimal constant ratio partition for this next trial will, recall, be a partition into contiguous sets of IC-values each leading to an $(i + 2)^{\text{th}}$ collision with a different molecule. Suppose that this partition divides the $(i + 1)^{\text{th}}$ impact angle φ into 180 intervals each spanning 1 degree of angle. Then the IC-evolution function is microlinear if, when its range is divided into these 180 equal sections, corresponding to the members of the optimal constant ratio partition for the $(i + 1)^{\text{th}}$ trial (that is, the trial that culminates in the $(i + 2)^{\text{th}}$ collision), the function is approximately linear over the inverse image of each of these sections (see section 3.76). Clearly, this is so, except at the very extremes of the graph; more on this exception shortly.[35]

Now I will step back to sketch the big picture, that is, to show how the properties of molecular collisions discussed in this section ensure that a single probability distribution is maintained over the relative impact angle at all times, conditional on any sequence of impact angles for previous collisions.

Assume that the distribution over the impact angle of the ith collision is macroperiodic. That is, the distribution is approximately uniform over each member of the optimal constant ratio partition for X, the experiment determining the impact angle of the $(i + 1)^{\text{th}}$ collision. The distribution over the

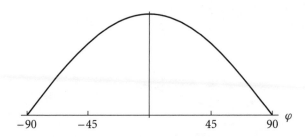

Figure 4.10 The density over the impact angle for M's $(i + 1)^{\text{th}}$ collision, conditional on the Boltzmann-level impact angle for its ith collision, whatever that angle may have been. It is, approximately, the cosine function.

relative angle of impact for the $(i + 1)^{\text{th}}$ collision, conditional on the impact angle for the ith collision, is therefore this uniform distribution transformed by the impact angle's IC-evolution function.

The transformed distribution will have a density, it turns out, of the approximate form:

$$q(\theta) = \cos \theta$$

which is shown in figure 4.10. Happily, the cosine distribution is observed, in experiments conducted by the author, to be very close to the actual distribution over impact angle in a simulation of a two-dimensional gas. It is also the distribution that is predicted by the Maxwell-Boltzmann statistics themselves. Thus it seems that the simplifications and approximations made above in the derivation of the cosine density did not result in any real distortion of the important facts about IC-evolution.[36]

Let me now return to the question of the microlinearity of impact angle IC-evolution, and in particular, to my earlier comment that the impact angle IC-evolution function does not look microlinear for extreme values of the impact angle. Is IC-evolution nevertheless microlinear enough?

This is equivalent to the question whether the cosine density is sufficiently macroperiodic.[37] For the purposes of assessing macroperiodicity, recall, the relevant partition is an optimal constant ratio partition, that is, a partition of the impact angle space into regions each bringing about a collision with a different molecule. In a gas such as nitrogen at atmospheric pressure, there are something like 120,000 molecules within probable collision distance,[38] so this is a very fine partition, making macroperiodicity easy to achieve.

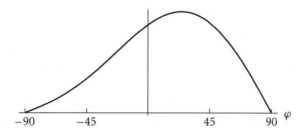

Figure 4.11 Impact angle density for the $(i + 1)^{\text{th}}$ collision conditional on a very high impact angle for the ith collision.

The reader will note, however, that the cosine distribution is not macroperiodic by any standard at the very extremes. Within a range spanning very large impact angles, for example, the range from 89 to 90 degrees, the lower values are considerably more probable than the higher values. An angle between 89 and 89.5 degrees is about three times more likely than one between 89.5 and 90 degrees.

Now I will show that this constitutes a real problem for my argument. Suppose that any angle between 89 and 90 degrees, considered as an impact angle for the ith collision, brings about an $(i + 1)^{\text{th}}$ collision with the same molecule, with an angle of 89.5 degrees bringing about a head-on collision, that is, a collision with a relative impact angle of 0. Any impact angle less than 89.5 degrees, then, will result in the $(i + 1)^{\text{th}}$ collision having a positive impact angle; any angle greater than 89.5 in a negative impact angle. Since the lower values for the ith impact angle are more probable, a positive $(i + 1)^{\text{th}}$ impact angle is more probable. The probability distribution over the impact angle of the $(i + 1)^{\text{th}}$ collision, conditional on a high impact angle for the ith collision, is qualitatively different, then, from the distribution with the cosine density; see figure 4.11. It follows that the $(i + 1)^{\text{th}}$ impact angle is not always independent of the ith impact angle.

There are several different ways to deal with this worry effectively. I will briefly sketch three of these, for the sake of illustrating the diversity of techniques that may be brought to bear on a problem such as this. Let me first, however, take care of a preliminary matter. Observe that the skewed conditional density shown in figure 4.11 is, like the cosine density, macroperiodic everywhere except the extremes (relative to a reasonably fine partition). Consequently, the effect of the impact angle IC-evolution function on any small

segment of the graph, other than a segment at the extremes, will be to produce the cosine density once again. It follows that the density for the $(i + 2)^{\text{th}}$
impact angle, conditional on the ith angle having been very high and on the
$(i + 1)^{\text{th}}$ angle having been anything but very high or very low, is the cosine
density. The density over impact angle conditional on as many previous impact angles as you like, then, will depart significantly from the cosine density
only if the immediately preceding impact angle is very high. This considerably
simplifies the problem of dealing with departures from independence.

Let me now present the three solutions to the problem. First, it is always
possible to individuate the microdynamic experiments more finely, so that a
collision following on from a collision with a very high relative impact angle counts as the result of a different kind of microdynamic experiment than
other collisions. The dependence is thereby built into the definition of the experiment, thus removing it as an obstacle to microdynamic independence. The
disadvantage of this approach is the proliferation of microdynamic probability
distributions, resulting in a somewhat more complicated treatment of enion
probabilities. Recall, however, that the sole function of the microdynamic distributions in my argument is to randomize a system's microvariables over the
duration of a trial on an enion probabilistic experiment, as described in section 4.42; for this purpose, a variety of different independent distributions
need be no less effective than a single distribution. This is just as well, since for
some systems, such as a predator/prey ecosystem, and even, perhaps, certain
complicated gases (section 4.88), there are many qualitatively different kinds
of microdynamic interactions and so many different microdynamic distributions.

The second approach relies on coarsening rather than refining the description of a certain element of a gas's microdynamics: instead of following the
evolution of the relative impact angle, a quantity that varies from −90 to 90
degrees, this approach follows the evolution of the *magnitude* of the relative
impact angle, a quantity that varies from 0 to 90 degrees. Two claims are then
made about this somewhat more coarse-grained variable:

1. The dependence problem stated above does not arise. The distribution
   over the impact angle magnitude for the $(i + 1)^{\text{th}}$ collision, conditional
   on the magnitude for the ith collision, is the cosine distribution even for
   very high i^{th} collision magnitudes.
2. The IID distribution of impact angle magnitude has enough randomizing power to ensure, in the long term, that the appropriate

distribution—the Maxwell-Boltzmann distribution—is imposed over the microvariables of a gas in a box. For example, from the assumption that the conditional distribution of impact angle magnitudes for every collision is the cosine distribution, it follows that the probability distribution over the position of a particular molecule, after a sufficient period of time has elapsed, is uniform.

The justification for (1) is roughly as follows: taking the magnitude of the impact angle folds the distribution in half around the point where the impact angle is zero. As a result, the imbalances in the probability distribution are rectified—the bias towards high values on one side makes up for the bias towards low values on the other side—and the distribution's symmetry is restored.[39] Claim (2) is more difficult to justify; in fact, it is, I think, strictly speaking false. A weaker version of the claim, however, sufficient for the randomization of the microvariables, is true. The argument for all of this is somewhat complicated, and presupposes material from section 4.86; in the interests of brevity it is omitted.

The third approach to the problem provides my preferred solution. In the presentation of the problem above, I assumed that an impact angle of exactly 89.5 degrees for the molecule M's i^{th} collision results in an impact angle of 0 degrees for its $(i + 1)^{th}$ collision, its collision with the molecule N. This corresponds to a very precise assumption about the position of N. But for the purpose of determining microdynamic probabilities, microvariables such as N's position ought only to be specified at the Boltzmann level. This introduces a small amount of fuzziness into N's position, and the positions of the molecules around it. Once the fuzziness is taken into account, the lopsidedness of the conditional distribution disappears.[40]

Suppose, for example, that a full specification of N's Boltzmann position constrains its precise position just enough to entail that *some* angle of impact for M's i^{th} collision between 89 and 90 degrees results in a head-on collision between M and N. In the case where, say, it is an 89 degree impact angle that results in a head-on collision, the more probable impact angles between 89 and 90 degrees—those between 89 and 89.5 degrees—will result in a negative impact angle rather than a positive impact angle. The less probable angles, those between 89.5 and 90 degrees, will result in a collision, not with the "negative" side of N, but with, say, the "positive" side of N's next-door neighbor, resulting in a positive impact angle. Thus in this particular configuration, it is negative rather than positive impact angles that will have their probability boosted by the departure from macroperiodicity between 89 and 90 degrees in the distribution of the ith impact angle. In other configurations having the same Boltzmann-level description, such as the configuration originally considered above, positive impact angles come out as more likely. The probability distribution over impact angles conditional on the Boltzmann-level position of N is the average over the distributions for all of these different configurations.[41] Some are lopsided one way, some the other. The asymmetries will cancel one another out.

The net effect of specifying the positions of M's collision partners only vaguely, then, is to render positive impact angles for the $(i + 1)^{th}$ collision, conditional on a very high impact angle for the ith collision, no more likely after all than negative impact angles. The symmetry is restored to the conditional distribution, and restored in such a way that it is just the distribution with the cosine density. The truth of my original assertion, that impact angles for all collisions have the same, independent probability distribution, is thus preserved.[42]

4.86 A Hard Sphere's Random Walk

I have argued that the relative angles of impact for any two molecular collisions in a gas are independent. What remains—steps (4) and (5) of the main argument outlined in section 4.83—is to show that this is a sufficient basis for the microconstancy and independence of the Maxwell-Boltzmann probabilities.

Perhaps the surest way to do this is to take the facts about the impact angle distribution derived above and to use the argument schema presented in section 4.42 to show that the distribution over the gas molecules' positions and velocities is the Maxwell-Boltzmann distribution. The Maxwell-Boltzmann probabilities would be derived, then, by showing that they are the result of each molecule's taking a random walk through position and velocity space, the randomness being supplied by the impact angle distribution. Just because the derivation conforms to the random walk schema of section 4.42, it will follow that the enion probabilities assigned by the Maxwell-Boltzmann distribution are microconstant and independent.

A derivation of this sort can be divided into two parts: a philosophical or conceptual part, in which it is shown how it is possible to regard the impact angle distribution as inducing a random walk through position and velocity space, and a mathematical part, in which it is shown that the particular random walk induced by the impact angle distribution leads to the Maxwell-Boltzmann distribution.

In this section, I will provide the philosophical part of the argument, but not the mathematical part. Mathematical techniques for deriving the Maxwell-Boltzmann probabilities, given an appropriate assumption of "molecular chaos," are well known from the work of Maxwell and Boltzmann themselves, and need not be rehearsed here.[43] My goal, then, is to set up and justify the assumptions required for the mathematical derivation to go through, and in particular, to provide a way to think about the relationship between the IID distribution over relative impact angle and gas microdynamics as a whole.

I suggest that the microdynamics of a gas molecule be conceived of as having two phases:

1. The probabilistic phase: This phase lasts only for the moment in which two colliding molecules touch. It is initiated by the collision of any two molecules, and it instantaneously determines a relative angle of impact for the collision. It is probabilistic because the impact angle is determined by the IID impact angle distribution. Since this distribution is the same for any collision, the Boltzmann-level positions and velocities of the colliding molecules have no influence on the dynamics of the probabilistic phase.

2. The deterministic phase: This phase lasts for the entire period between any two of a molecule's collisions. It consists of two parts. In the first part, the impact angle for a collision instantaneously determines, at the moment of the collision, new Boltzmann-level velocities for the colliding molecules. In the second part, the velocities determine the change in Boltzmann-level position between one collision and the next. Both phases operate in accordance with the deterministic laws of motion.

It is important that all variables—impact angle, velocity, and position—are described at the Boltzmann level; only then can the deterministic and probabilistic phases coexist as parts of a single, consistent scheme for describing gas microdynamics.

On the picture I have suggested, there is a one-way flow of influence from the impact angles of the collisions to the positions and velocities of the colliding molecules. The positions and the velocities of two colliding molecules do not probabilistically influence their impact angle, but the impact angle very much affects the positions and velocities. The influence comes in two steps: at the moment of the collision, the angle profoundly affects the velocities of the molecules, and then, during the time between that collision and the next, the changed velocities profoundly affect the positions of the molecules.

Because the influence is both profound and unidirectional, IID impact angles are able to randomize the position and velocity of every gas molecule, and thus every one of a gas's microvariables. A single collision will not completely randomize a molecule's position and velocity, in the sense that it will not create a position or velocity distribution that is completely independent of the molecule's previous state, but it will constitute a step in a random walk that will, over time, bring about a complete randomization. This is just what is required by the argument for the microconstancy of enion probabilities in section 4.42.

Let me say something more about the philosophical underpinnings of my treatment of gas microdynamics. The picture I have presented requires, I suggest, a certain kind of doublethink. From the perspective of fundamental physics, the velocities and positions are of course determining the impact angles, just as much as the impact angles determine the velocities and positions, and the one operation is just as deterministic as the other. But the arguments put forward in sections 4.84 and 4.85 license us to ignore, though not to deny, these facts, and to treat an impact angle as probabilistic and as independent of the preceding positions and velocities. We can do this because impact angle is indeed *stochastically independent* of the *Boltzmann-level* facts about position and velocity. This is all that is needed for EPA; one must be careful, however, not to slip into thinking, falsely, that impact angle is *causally independent* of Boltzmann-level position and velocity or that it is stochastically independent of *fine-grained* position and velocity—or worst of all, because most manifestly false, that it is causally independent of fine-grained position and velocity.

Within the boundaries marked out by these admonitions, it is entirely legitimate to treat the probability distribution over impact angle as floating free of the microvariables of the gas, and as stepping into the system just to give pairs of molecules a little random kick whenever they collide. On this view, the system's microdynamics is as a rule deterministic; the deterministic rule is, however, suspended during a collision, at which point a probabilistic law rules instead.

I have one important remark about the details of this picture. Collisions aside, the dynamics of a system described at the Boltzmann level can never truly be deterministic. Changes in position, even changes in Boltzmann-level position, will eventually come to depend on facts about velocity that reside at too low a level to appear in a Boltzmann-level description of initial conditions.

The indeterminism can be captured probabilistically by assuming that there is a uniform probability distribution over the different fundamental level (fully precise) values of position and velocity within each Boltzmann point, that is, within each region of fundamental state space corresponding to a single set of initial conditions specified at the Boltzmann level. (The assumption is, of course, one of macroperiodicity relative to the partition that induces the Boltzmann coarse-graining.) As long as there are no collisions, the probability distribution will evolve according to deterministic laws, and so in a fairly orderly way. This is what I am calling "deterministic" microdynamics.

How to tell when a collision takes place? Boltzmann-level information will not deliver a definite verdict on the occurrence of a collision, but it will deliver a probability for a collision.

This concludes step (4) of the main argument, establishing the microconstancy of a gas's enion probabilities, the Maxwell-Boltzmann probabilities. All that remains is step (5), the argument establishing the independence of the Maxwell-Boltzmann probabilities. Independence follows immediately, as explained in section 4.44, from the microconstancy of the enion probabilities and the independence of the outcomes of the microdynamic experiments, that is, the independence of the impact angles.

4.87 Comparison with the Ergodic Approach

The approach to understanding the probabilistic element of statistical physics presented above is similar in some ways and different in many ways from other attempts to provide a foundation for statistical physics. (An especially useful resource on the foundational issues is Sklar 1993.)

It is different most of all because, whereas other approaches attempt to provide a foundation for the probabilities of formal, or Gibbsian, statistical mechanics, my approach provides a foundation for the probabilities of kinetic theory. (I envisage that the probabilities of statistical mechanics will be derived from those of kinetic theory.) This leads to an emphasis on the properties of the evolution functions of encounters between individual molecules, rather than on the properties of the evolution function of a system as a whole.

My approach is most similar, as noted in section 1.25, to the attempt to find a foundation for statistical physics in modern ergodic theory. This is especially true of the work in ergodic theory showing that certain physical systems are K or Bernoulli systems. Some of the mathematical properties to which I appeal have obvious analogs in the modern ergodic approach; inflationary evolution, for example, is important on both kinds of account (as I pointed out earlier, in note 25 of chapter 3).

I will mention three differences between my approach and the ergodic approach. First, like most attempts to provide a foundation for statistical physics, the ergodic approach is concerned with the independence of successive steps of the evolution of a system as a whole rather than the independence of the steps of the evolution of individual enions. This difference was previously discussed in section 1.25.

Second, the ergodic approach looks for a more extreme form of independence than I do. I look for enough independence to regard evolution as a Markov chain, so that the probability distribution over an enion's next microdynamic step depends on its current microstate; the ergodicists hope to show that the next step of a system is completely independent of its current state.[44] On my treatment of statistical physics, then, a system takes a random walk through state space, each new location depending in part on the last, whereas on the ergodicists' view, the system jumps from place to place as though each new location were the outcome of a single probabilistic experiment—as though each new location were determined by, say, a drawing from an urn.

Taken together, these two characteristics of the ergodic approach make it quite unsuitable for providing a general explanation of the simple behavior of complex systems, for the reason that a system whose evolution has the independence properties pursued by the ergodicists is capable of satisfying only a very limited number of kinds of macrolevel laws. Periodic, or cycling, behavior, for example, is not possible.[45]

A third feature of the ergodic approach is that, while the theorems of the approach give sufficient conditions for the existence of a partition—that is, a level of description—relative to which a system satisfying the conditions will exhibit the desired level of randomness in its evolution, the conditions do not guarantee that the level of description is the same as the level employed in EPA. Thus the ergodic approach may fail to explain at all why EPA is successful. This worry about the ergodic approach is raised by Sklar (1993, 240–241), and explained at length by Winnie (1998).

4.88 Beyond Billiard Balls

The treatment of statistical physics outlined in sections 4.84, 4.85, and 4.86 above suggests that the probabilistic aspect of kinetic theory, and thus the simple behavior of gases, can be understood using the approach developed in this study. If I am right, the understanding will be the most satisfying possible: the probabilistic behavior of gases will rest not on certain very particular mathematical properties of the laws of fundamental physics, but on very general facts about the nature of collisions.

Let me enumerate some of the ways in which the approach to understanding kinetic theory suggested here would have to be generalized in order to provide a fully satisfying account of the foundations of statistical physics.

First, real gas molecules do not literally bounce off one another, but repel one another when they come close with very strong short-range forces (though the net effect is not too different from a bounce). The scope of the treatment of molecular interaction must be widened, then, so as to cover repulsive bouncing as well as literal bouncing. There is every reason to think that the various properties asserted above of impact angle ic-evolution will continue to hold under such a widening. In particular, ic-evolution will be no less inflationary and no less (perhaps more) microlinear.

Second, most gas molecules are not spherical. This does not much affect the scattering effects of their collisions, but it does introduce other variables into the mix: molecular rotation and vibration. I expect that what I have said about the relation of position and velocity to impact angle is also true of these other variables, namely, that they are randomized over time by randomly distributed impact angles, and that they can be eliminated in favor of the impact angle variable in the microdynamic probabilistic experiment, described in section 4.84, by which one impact angle determines the next.

In more complex cases there may be more than one kind of collision, hence more than one kind of microdynamic probabilistic experiment, each imposing its own probability distribution over impact angles. As noted elsewhere, this will not interfere with randomization.

Third, statistical physics treats not only gases but liquids and solids. Here the generalization is far more difficult. But I would recommend the same approach as in the case of gases: to find a randomizer such as relative angle of impact and to show that the values of the randomizer for every interaction are independent and identically distributed. Identical distribution is not a requirement, but it does simplify the argument.

Fourth, real molecules attract one another very slightly. They therefore experience a small long-term coupling. I would hope that this coupling can be shown not to interfere with the properties of the relative angle of impact and its ic-evolution assumed in my argument. Certainly, it does not seem that a small coupling would affect inflation or microlinearity; if so, the coupling does not compromise independence.

One theoretical obstacle to the success of the argument I have presented here, due in part to the existence of attractive forces, is a result known as the KAM theorem. The theorem entails that, in certain kinds of dynamical systems, there exist regions of initial conditions from which a system can never depart. That is, if a system starts out in such a region, it will never leave. This is not a result of fencing in the sense of section 4.42, since the relevant enion probabilities taken one by one will not prohibit the system's departure from the region, but is rather a result of the less than total independence of enion probabilities. The existence of such regions has been known since the inception of kinetic theory; what is interesting about the KAM theorem is that it shows that they are not infinitesimally small.

It is not known whether the KAM theorem applies to real systems in statistical physics, but the consensus is that it likely does. If so, the important question to ask concerns the size of the regions of initial conditions within which systems remain trapped. If these regions are very small, they can be ignored, like other *bad regions* of initial conditions (see approximation 2.2.2). If not, not.

The KAM theorem is proved by showing that, in the right circumstances, the influences of the kinds of interactions that one would expect to lead to inflationary evolution cancel each other out, so that inflation does not occur. The received wisdom is that, the more independent and potentially inflationary influences are acting at one time, the more delicately things have to be arranged for these influences to cancel out. Thus, the more enions in a system, the smaller the regions of initial conditions leading to non-inflationary evolution. If this reasoning is correct, then in a gas where there are about 10^{25} molecules (as there are in a cubic meter of nitrogen at normal pressure and temperature), the regions will be unspeakably small, and therefore may be passed over in silence. Unfortunately, no one knows whether the regions are as small as expected. Computer simulations, however, give grounds for optimism (Sklar 1993, 174).

A final observation: because of inflation, very small changes in the state of a system of colliding molecules will have a large effect on its microlevel evolution. There is great scope here for simple probabilities at the fundamental level to influence evolution, so great, in fact, that if such probabilities exist, it may well turn out that the enion probabilities of kinetic theory are simplex, as suggested by Albert (2000). (I emphasize the conditional form of this claim, given both the difficulties in formulating a version of quantum mechanics that invokes simple probabilities (Albert 1992) and my own suspicions about the existence of simple probability, for which see section 5.6.) If the probabilities of statistical physics are simplex, the microdynamic probabilities of ecology may be simplex too, for reasons given in section 4.9, in which case the enion probabilities of ecology would be simplex. It is far from impossible, then, that quantum probabilities could percolate up to the level of medium-sized objects.

4.9 Population Ecology

4.91 Randomizing Variables

My treatment of population ecology will be similar to my treatment of statistical physics in the following way. In the treatment of statistical physics, I took one variable describing a property of microdynamic enion interaction, the relative angle of impact, and argued that

1. Values of this variable for different enion interactions are stochastically independent, and
2. The independent values for different interactions are sufficient to randomize all the system's microvariables in the manner required to establish the microconstancy and independence of enion probabilities.[46]

I will call a variable for which (1) and (2) are true a **randomizing variable** for a system. Note that, although I argued in section 4.8 that relative impact angle is identically distributed for every collision, this property, unlike independence, is not a part of the definition of a randomizing variable. The identical distribution makes (2) on the whole easier to establish, but it is not at all necessary for (2).

The probability distribution over a randomizing variable can be seen as the ultimate source of a system's probabilistic behavior. It represents the system's stock of primal randomness, of which the probabilistic distributions over a system's other microvariables and hence over its macrovariables are, in some sense, just a reflection.

My strategy in this section is to find microdynamic variables that might serve as randomizing variables for a predator/prey ecosystem. Unlike a gas in a box, an ecosystem has a diverse set of enion interactions from which to choose potential randomizing variables. At least two kinds of microdynamic events seem especially promising sources: interactions between predators and prey, and individual organisms' foraging decisions, that is, their decisions as to where to look for food. Quite likely, in my opinion, the outcomes of both predator/prey encounters and foraging decisions will serve as randomizing variables for an ecosystem.

This raises the question whether the effect of either one of these variables acting alone would be enough to randomize the microvariables, or whether they have sufficient randomizing power only when acting together. This is not an issue that will be much discussed in what follows. Indeed, I will use the expression *randomizing variable* rather loosely to refer both to variables that are sufficient in themselves to randomize a system, and to variables that are a part of a set of variables that are sufficient only when acting together to randomize a system.

There are two very important problems that arise in the search for an ecosystem's randomizing variables. The first is a practical problem: we know very little about the detailed mechanics either of predator/prey interactions or of foraging decisions. Thus it is difficult to answer a question such as whether ic-evolution in predator/prey interactions is microlinear. There is a small amount of work on apparently probabilistic aspects of predator/prey interactions—see, in particular, Driver and Humphries (1988)—but more is known about foraging decisions, so I will focus on foraging in what follows.

The second problem is more theoretical. The values of a randomizing variable must be independent, and the techniques for demonstrating independence investigated in chapter three and applied to complex systems in section 4.5 require that ic-evolution of the variable be inflationary. But almost every microvariable in an ecosystem undergoes, under some circumstances, quite deflationary evolution. Deflation has a tendency to eliminate, or at least to suppress, any randomness in a variable. I call this the *homeostasis problem*. The next section discusses the problem at some length.

4.92 The Homeostasis Problem

The term *homeostasis* is borrowed from physiology, where it denotes a process by which a system maintains some variable or variables at a predetermined target value. The value of a homeostatically maintained variable, such as mammalian body temperature, will fluctuate because of environmental exigencies; what a homeostatic mechanism does is to compensate for these fluctuations, bringing the variable back to the target value. Movements of the maintained variable, then, are of two sorts: *excursions* from the target value brought about by external factors, and *returns* to the target value brought about by the homeostatic mechanism. The excursions, being fluctuations that are at least in part environmentally driven, may be probabilistically patterned, and so may cause probabilistically patterned changes in the maintained variable. But these changes can only be temporary; the return will eliminate the effect of the fluctuation and thereby render it irrelevant to the future time evolution of the variable.

One can draw an analogy between the regulation of body temperature and a rabbit's regulation, as it were, of its position. Let me consider a deliberately oversimplified case. Say that the rabbit decides every morning where to forage. The rabbit then makes its way to the chosen foraging location, forages away for the rest of the day, and returns at night to its warren. The journey to the foraging spot is an excursion; the journey home is a return that, it seems, wipes out any randomizing effects that the rabbit's foraging decision might have had on its position. This is not, strictly speaking, a case of homeostasis, as the rabbit is not trying to stay inside its warren at all times, but it is useful, for my purposes here, to use the term *homeostasis* in a broad sense that encompasses cases such as this.

There are two kinds of difficulties raised by the homeostatic character of the change in rabbit position. First, if rabbit position periodically returns to a certain value (in this case, *home*, each night), it might be thought that position cannot truly be randomized as required by the argument for microconstancy in section 4.42. But this is not correct. Although rabbit position at night may not be randomized, this does not mean that rabbit position during the day is not randomized, and it is rabbit position during the day that makes a difference to designated outcomes such as rabbit death. To randomize rabbit death, then, it is sufficient to randomize rabbit position during the day, and here periodic homeostasis is no obstacle. (See, however, section 5.4 for further worries along these lines in the case of social systems.)

The second difficulty raised by periodic homeostasis is the more serious one. It seems that, as a result of homeostasis, rabbit position cannot be a randomizing variable in the rabbit/fox ecosystem. The reason is that the position variable is the same at the beginning of one day as it is at the beginning of the previous day. If it is this variable that decides, say, where the rabbit forages on a given day, then far from being independent, these decisions will always be the same. Put technically, the problem is that any inflation of the position variable that occurs during the day is undone by a corresponding deflation that occurs each evening as the rabbit heads home, leaving the position variable in more or less the same state as it was in the morning.

Suppose that the movements of every organism in the ecosystem are homeostatic. Then, it seems, none of the system's position microvariables can serve as a randomizing variable. Worse, no temporary states of affairs that are determined by positions—analogs of relative impact angle in a gas such as, say, predator/prey encounter angle—could be randomizing variables, either. But what else is left?

In the remainder of this section I discuss some possible solutions to the second homeostasis problem, that is, some possible sources of randomization in a system in which the major microvariables periodically undergo the sort of deflation that I am calling homeostatic. I single out one of these as a promising source of randomization in a large class of ecosystems; this idea is then developed in sections 4.93 and 4.94.

The first approach to the homeostasis problem is the most direct; it seeks to rehabilitate the homeostatic variables as randomizers. Stated in terms of the rabbit case described above, the idea is that the return of the position variable to its former state is only superficial. Although the high-level information about the rabbit's position is the same from day to day—each morning, the rabbit is in the warren—the lower-level information, according to this approach, changes, perhaps by inflation of the lower-level information from the day before. One need only add that it is the lower-level information that determines the rabbit's foraging decisions and so on, and position is once again a potential randomizing variable.

Although this solution to the problem may be successful for some systems, it requires certain rather strained assumptions about the ecosystem. In effect, what is being said is that the source of randomness in a rabbit's daily wanderings is the low-level information about its position at night, which is to say, the precise details about its sleeping position and so on. While we do not, I think, know enough about rabbit ethology to rule out this possibility (see, for example, the end of section 4.93), I would not want the credibility of my vindication of EPA to depend on it.

The second approach explores the possibility that, although no homeostatically deflated variable can be randomizing on its own, a set of such variables might be randomizing provided that they are not all deflated at the same time.

Let me begin with an example. Suppose that, every morning, rabbits make decisions as to where to forage by scouting out the positions of foxes and trying to avoid them. Suppose that foxes decide where to forage in the morning based on the observed positions of rabbits the previous afternoon. Then the rabbits' positions on one afternoon determine (by way of the foxes' memory mechanisms) the foxes' positions the next morning, which determine the rabbits' positions the same afternoon. In this way, the influence of the rabbits' positions on a given day may still be felt many days later. It does not matter that the rabbits themselves forget their positions and head back to the same place every night; the influence of their positions lives on overnight in the foxes' memory. We can also allow that the foxes forget all about the rabbits' positions between the morning and the afternoon, since by then the memories will have done their work by influencing the foxes' positions and thus the rabbits' positions for the whole day.

To summarize: the rabbits' positions on one afternoon will depend on their positions the same morning, which will depend on the foxes' positions that morning, which will depend on the foxes' memories, which will depend on the rabbits' positions the previous afternoon. Through this chain of influence, the rabbits' positions on one day influence their positions on the next day, despite the period of intermediate homeostatic deflation. Information about earlier positions of rabbits, then, is handed on from variable to variable in such a way that, by the time the value of one variable is "erased" by some periodic homeostatic deflation, it has left its mark in some other causally efficacious part of the system.

More generally, although the value of every variable ζ may be at some time homeostatically erased, in the sort of case illustrated by the example there is always some other variable that (a) has its value determined by ζ before ζ is erased, and (b) either directly or indirectly determines the value of ζ after ζ's erasure. In this way, values of ζ before erasure manage to determine, indirectly, values of ζ after erasure. There is no reason why this indirect determination cannot be inflationary, microlinear, and so on.

Another example: suppose that we have a gas in a box. One by one, we pluck molecules out of the gas and reintroduce them in a single, predetermined way. In this way, the microvariables describing the selected molecule are erased, and replaced with values that in no way reflect the state of the gas at the time of the selection. Even if we repeat this process for every molecule in the gas, we can reasonably expect that the gas will continue to behave randomly, that is, in accordance with the probabilities of kinetic theory. While a molecule is out of the box, the remainder of the molecules carry the system's stock of randomness. That each of these molecules has had at some time in the past, or will have at some time in the future, its own microvariables erased, in no way compromises its power to carry the randomness in the present.

The moral, then, is this. In a process that homeostatically maintains the value of a certain variable, the return erases the effect that the excursion has *on that particular variable*. Over the duration of an excursion, however, the wandering variable will have effects on other variables. The return from an excursion cannot normally undo those effects. Thus, even though an excursion itself is always temporary, its effects may endure.

This second approach to the problem of homeostasis may, I think, turn out to be quite important, but its application to any particular case does require quite detailed knowledge of the workings of the system in question. In the case of the ecosystem, in particular, it requires considerable knowledge of the intricacies of rabbit and fox cognition.

The third and, of course, most obvious solution to the homeostasis problem, is to find some variable that is *not* homeostatically deflated and to show that it is capable of randomizing the system. This is what I will find myself doing by the end of this discussion of population ecology.

Here is the route I propose to take:

1. I focus on foraging decisions as randomizers, that is, I take the outcomes of foraging decisions as my randomizing variables.
2. I show that a certain aspect of foraging decision making among higher animals is probabilistically patterned, and suggest that this probabilistic element is sufficient to randomize the evolution of an ecosystem's microvariables.
3. I present reasons for thinking that the source of randomness in this probabilistically patterned element is something internal to the deciding organism, rather than an environmental cue.
4. I search for a possible neurobiological root of this randomness.

4.93 Probabilistic Foraging Behavior

Consider two examples of probabilistic foraging behavior. The first is that of the desert ants of the genera *Cataglyphis* and *Ocymyrmex*. Rudiger Wehner showed that the movements of foraging ants could be described by a probabilistic model in which each ant conforms to the following rule, as described by Hölldobler and Wilson (1990):

> *Continue to forage in the direction of the preceding foraging trip whenever the trip proved successful. Otherwise abandon that direction and select a new one at random, but decrease the probability of doing so as the number of previously successful runs increases. (p. 385)*

It is worth noting that these ants, in the course of their adventures, move in an extremely irregular pattern, often turning the equivalent of many full circles in a second. This furious motion must inflate small navigational errors since, Hölldobler and Wilson write, even when following up a successful expedition, the ants often deviate sufficiently from their earlier path to find new items of food. Other varieties of ants with quite different foraging strategies also wander—in a random way—sufficiently far from the straight and narrow to encounter new food sources (p. 385). So there are two random elements in ant foraging: the probabilistic decision rule and the probabilistic deviations from the course set by that rule.

Turning to more complex organisms, it is a well-established fact that many higher organisms, including humans, exhibit a behavior known as *probability matching*, which involves making a kind of randomized foraging decision. The behavior naturally arises in a situation where several resources (usually food sources) of varying abundance are available to a group of animals. The animals spread themselves out so that the concentration of creatures at any one source matches the proportion of the total available resource accounted for by that source. Twice as many bread crumbs, for example, attract twice as many birds.

Experiments show that the mechanism primarily responsible for this phenomenon is a probabilistic foraging rule: choose between resources probabilistically, assigning a probability of choosing any source equal to the proportion of food delivered by that source.[47] Gallistel (1990) provides a survey of recent work in this area, including a description of probability matching among undergraduates in a Yale psychology class (p. 351).[48]

There is good empirical evidence, then, that many organisms' foraging decisions are randomized by some sort of probabilistic mechanism.[49] Trials on this mechanism seem to be fully independent of one another, and of trials on the probabilistic mechanisms that determine the foraging decisions of other organisms.[50]

If this observation is to be used to vindicate the application of EPA to certain ecosystems, and so to explain the simple behavior of those ecosystems, two questions must be answered:

1. Are probabilistically patterned foraging decisions sufficient to randomize an ecosystem's microvariables in the way required for the microconstancy and independence of the ecosystem's enion probabilities?
2. What underlying facts explain the probabilistic patterns found in foraging decisions?

As for the first question, it seems to me that the answer is almost certainly yes, since foraging decisions exert such a great influence on the positions of organisms within an ecosystem, and these positions in turn determine so much else—the relative positions of predators and prey, for example—that is important to ecological microdynamics. A stronger argument would have to examine particular microlevel models of a predator/prey ecosystem, something that I will not do here.

It is the second question I will take up in what follows, focusing exclusively on the psychological underpinnings of probability matching. What especially interests me is the source of the probabilistic element of probability matching. Is it some external, environmental cue that causes the probability matcher to opt for one resource rather than another? Or is it some internal process, the neurobiological equivalent of a coin toss?

The phenomenon of probability matching was discovered by behaviorists, who subjected animals to repeated experiments in which the animal's surroundings were very sparse and very tightly controlled—a maze, a cage, or whatever. Even in these experiments, where the animal's environment is almost identical from trial to trial, decisions are probabilistically patterned. This strongly suggests one of two possibilities:

1. The source of the randomness is extremely small variations in the environment, that is, very low level information about the environment, or
2. The randomness has an internal source.

Either possibility is congenial to my project. The first possibility fits well with the first approach to the homeostasis problem discussed in section 4.92, which requires that very low level facts about a rabbit's position within its warren influence its foraging decisions. The second possibility fits well with the third approach to the homeostasis problem, by suggesting the existence of an internal randomizing variable that, unlike position, is not obviously deflated by periodic homeostasis.

It is this second line of inquiry that I will follow in the remainder of the section, exploring the possibility that the probabilistic patterns in foraging decision making have their basis in some neurobiological process, and that this process involves some variable or variables that are continuously inflated in the sort of way that provides a foundation for the stochastic independence of foraging decisions.

Because more or less nothing is known about the neurobiology of foraging decision making, I will be unable to complete this investigation. What I can do is to show that there exists in the brain a certain amount of randomness that might be harnessed by a decision maker to implement the probability matching rule. Whether this randomness is in fact so used, I cannot say.

4.94 The Neurobiology of Randomized Decision Making

There are many probabilistically patterned phenomena in the brain (Tuckwell 1989), but the currently most salient probabilistic patterns in neurobiology are the patterns with which neurons fire, or "spike." It is well known that neurons spike in response to sufficiently many of the right kind of inputs from other neurons, and that it is through spiking that neurons transfer signals to other neurons, and hence that thought occurs. What is less well known is that the timing of most neurons' spiking is quite irregular. It has been shown that the spike rate in at least certain kinds of cortical neurons has, approximately, a Poisson distribution (Softky and Koch 1993). The spiking, then, is probabilistically patterned.

A Poisson distribution has a single parameter, the mean rate of the occurrence of the designated outcome, that outcome being, in the case under discussion, the event of a single neural spike. The pattern of the spikes in a single Poisson-patterned spike train, then, may be conceptually divided into two parts: the mean rate of spiking, and fluctuations around this mean rate. These correspond to, respectively, the long-term order and the short-term disorder in the pattern. I suggest that the short-term disorder could be harnessed by a foraging decision maker to implement the probability matching rule.

Concerning this suggestion, three questions ought to be asked:

1. Are the fluctuations in spike rate as random as they seem? Or are they correlated with other behaviorally significant elements of neural activity? For my purposes, a lack of correlation is what is desired.
2. Can the fluctuations be harnessed to do cognitive work? Can a foraging decision maker be built that takes advantage of the fluctuations to implement the probability matching rule?
3. What is the source of the fluctuations? Can the source be shown to have the kinds of properties that will make fluctuations at different times, and hence foraging decisions at different times, stochastically independent?

Answering each of these questions will involve some engagement with the relevant neuroscientific research. Before I begin, let me remind the reader that I am exploring just one neurobiological avenue to randomness in what follows; others exist, and should also, at some stage, be explored—but not here.

The first question, concerning the randomness of the fluctuations, can be answered definitively only when a major contemporary debate in neuroscience is settled, namely, the debate concerning the nature of the *neural code*, that is, the debate concerning the role of the different properties of a spike train in conveying information from one neuron to another. For a recent, concise overview of this and related questions, the reader may consult deCharms and Zador (2000). I will sketch three views about the code.

The first, traditional view is that all information in the spike train is conveyed by the mean spiking rate. The fluctuations, then, convey no information; they are just *noise*, in the information-theoretic sense. This is called the *rate coding* theory. For a recent defense of the rate coding view, and references to much of the literature, see Shadlen and Newsome (1994, 1998).

According to the second, more recent view, called the *temporal coding* theory, the fluctuations are an important part of the neural code, and indeed, are not properly regarded as fluctuations at all. Rather, each spike in the train is a precisely timed, independent message. Thus the exact length of the intervals between spikes conveys information over and above the mean spiking rate; the mean spiking rate, far from being the message, may be only a by-product of the coding scheme. This kind of temporal coding theory is advanced by Softky and Koch (1993, 348).[51] If it is correct, then the "fluctuations" in a spike train are presumably correlated with other information processing going on in the brain, and may well make poor randomizers.

One might well wonder how, if the temporal coding view is correct, the spike train comes to have the form of a Poisson distribution. Advocates of the temporal coding view observe that, if (a) there is one spike per message, (b) the probability of any given message's being sent is the same, and (c) the messages are stochastically independent (rather strong assumptions, I note in passing), then spikes will have a Poisson distribution. In effect, the probabilistic patterns in the spike train reflect probabilistic patterns in whatever facts the code is representing—presumably, in some cases, probabilistic patterns in the external environment.

The third, and most recent, view is in some ways a synthesis of the rate and temporal coding views; I will refer to it as the *synthesizing* theory. In proposing the synthesizing theory, Rieke et al. (1997) agree with the rate coding view that the message in a spike train may be contained in the expected mean firing rate of the neuron emitting the train. Note the emphasis, now, that it is the *expected* rate that is important.

On either the rate coding or the synthesizing view, neurons receiving a spike train must analyze it so as to estimate, from the actual pattern of spikes, the expected mean firing rate of the emitting neuron. The synthesizing view differs from the rate coding view advocated by Shadlen and Newsome on the question as to how this estimate is made. Shadlen and Newsome hold that a receptor neuron estimates the expected firing rate by calculating the actual firing rate, and then setting the estimate of the expected firing rate equal to the actual firing rate. Rieke et al. hold that a receptor neuron estimates the rate by looking not just at the actual mean firing rate but also at the precise timing of the spikes. This becomes very important when the messages are short, since short messages mean that very few spikes are emitted at a given mean firing rate before shifting to a new rate.

The synthesizing view combines the rate and temporal coding views in the following sense: it agrees with the rate coding approach that a message is encoded in an expected mean firing rate, but with the temporal coding approach that the decoding neuron must look at the precise timing of individual spikes to recover the message. Decoding has, as it were, two steps: from the precise timing of the spikes to the expected firing rate, and then from the expected firing rate to the message.

What is important, given my purposes, is that, on the synthesizing view, there is no reason to expect fluctuations in the spiking rate to be correlated with any information being processed elsewhere in the brain. The fluctuations are, as on the rate coding view, merely noise, and as such, would make suitable randomizers.

I will proceed by assuming that either the rate coding or the synthesizing view is correct, and therefore that the fluctuations in the spike train are a source of high-quality randomness. The next question to ask is whether this randomness could be harnessed so as to implement the probability matching rule neurobiologically.

The simplest way to answer this question is to show how it can be done.[52] Suppose that an organism is considering which of two food sources to exploit. Assume that there are two neurons in the organism's brain, which I call the *outputs*, corresponding to the two food sources, and that whichever of the outputs fires first determines which food source will be exploited. Assume also that there are two other neurons, which I call the *inputs*, also corresponding to the two food sources. Let each of the inputs fire at a mean rate proportional to the size of the food source to which it corresponds. If one source is twice the size of the other, then, the input for that source will on average fire twice as fast—only on average, however, due to the fluctuations that are a part of any Poisson process. Now simply link the corresponding inputs and outputs, and arrange things so that a single spike from an input is sufficient to trigger the output neuron. It is easy to show, given the assumption that spiking is a Poisson process, that this will implement the probability matching rule; in the case where one input has a mean firing rate twice that of the other, for example, the chance of its firing first is 2/3 and that of the other's firing first is 1/3.

Is this mechanism biologically realistic? The most contentious part of the picture I have suggested is the assumption that the outputs fire upon receiving a single input. Shadlen and Newsome, for example, vigorously deny that cortical neurons work in this way. Unfortunately, requiring more than one spike to trigger the outputs destroys the simple relationship between spiking rate and probability (the larger food source will be disproportionately favored). However, even Shadlen and Newsome concede that some neurons are sensitive to single spikes, and the consensus in the field seems to be shifting in the direction of single spike sensitivity (Rieke et al. 1997; Bair 1999).

Of course, the mechanism I have suggested may be completely fictional; the true implementation of probability matching might, for example, involve arrays of hundreds or thousands of neurons rather than just a handful. What I have shown is that there is no obstacle in principle to implementing the probability matching rule using the Poisson fluctuations in the spike train as the source of randomness. On the contrary, it is extraordinarily easy.

Let me now turn to the third question: supposing that either the rate coding or the synthesizing view is correct, what is the origin of the spike train fluctuations? Shadlen and Newsome (1998) suggest that the fluctuations in the spike train emitted by a neuron are largely a reflection of the fluctuations in the spike trains of the neurons for which it is a receptor. But this cannot, of course, be the whole story; the fluctuations must begin somewhere. And they seem to get started very early: the Poisson distribution shows up even in spike trains issuing from sensory organs (Rieke et al. 1997). One among several possible original sources of the fluctuations is the probabilistic behavior of chemical synapses, which I will make my final topic.

Most communication between neurons goes by way of a chemical synapse. When the emitting neuron fires, it releases a certain quantity of neurotransmitter, which contributes to the activation of the receptor neuron. It has long been known that firing does not always lead to the release of neurotransmitter, and that releases are probabilistically patterned; to be precise, they are patterned like the outcomes of a Bernoulli process such as a coin-tossing experiment. Thus the release probabilities are independent and identically distributed.[53] (For this and much more about synapses, see Cowan, Südhof, and Davies 2001.) I suggest, tentatively, that these IID probabilities are what cause the fluctuations in the spike train. It would follow that neurotransmitter release probabilities provide the necessary random element in the foraging decisions of organisms using the probability matching rule.[54]

(For the view that synaptic unreliability contributes to fluctuations, see Zador (1998); for the view that it may not be the sole source of fluctuations, see Stevens and Zador (1998).)

What, then, is the physical basis of the neurotransmitter release probabilities? Unfortunately, the answer to this question is not known. It seems possible, however, that it might have to do with statistical fluctuations in the rate of some chemical reaction in the chemical pathway leading to neurotransmitter release, a reaction involving perhaps only very small amounts of some crucial signaling substance. (The pathway leading to neurotransmitter release is poorly understood.) Since these fluctuations in the chemical reaction, should they occur, are due to phenomena that are the subject matter of statistical physics, such as temperature differentials and collision angles, this suggests that the probabilistic aspect of neurotransmitter release is due to the probabilistic aspect of statistical physics.

I argued in section 4.8 that the dynamics of statistical systems in physics satisfies the conditions for microdynamic independence. If every step of the rather speculative chain of explanation in this section is correct, then the rabbit/fox ecosystem's microdynamic probabilities derive their independence from the independence of the microdynamic probabilities of the statistical physics of the brain. Thus, the probability distribution over relative angle of impact may be the reservoir of randomness not only in statistical physics but in ecology too.

Bigger than Chaos

At the microlevel of a complex system there is chaos—a proliferation of enions, interacting in many and various ways. Because of this chaos, the behavior of complex systems might be thought to be unstable, impossible to describe simply, and quite unpredictable.

Yet the microlevel also contains the seeds of something much bigger than chaos: microconstant probability. Microconstant probability is bigger than chaos because it is indifferent to the microlevel details that exhibit chaos. But it is bigger, too, in the way that it takes the source of chaos—sensitivity to initial conditions—and creates microdynamic independence, the independence of enion probabilities, and ultimately the macrolevel probabilities that sculpt the simple dynamic lines of our world.

5

Implications for the Philosophy of the Higher-Level Sciences

There is one scientific theory that is not like the others: fundamental physics. By convention, however, fundamental physics is the paradigmatically normal science, and all other sciences are more or less "special" or at a "high level" of description. By *higher-level sciences*, then, I mean all sciences other than fundamental physics, with occasional emphasis on those sciences that are at a considerably higher level of description: biology, psychology, and the social sciences. Most higher-level sciences are partly or wholly about complex systems, that is, systems of many, somewhat autonomous, strongly interacting parts. This chapter examines the implications of my account of enion probability analysis for the philosophy of the higher-level sciences.

5.1 Reduction

5.11 Varieties of Interlevel Relations

Enion probability analysis provides an understanding of the way in which higher-level laws depend systematically on lower-level laws—for example, of the way in which laws of population ecology depend on laws of animal behavior. In a sense, then, EPA shows how higher-level sciences can be reduced to lower-level sciences, in those cases where the higher-level science describes a complex system.

I say *in a sense* because *reduction* means many things to many people, so much so that the term had perhaps better be abandoned. Consider three views about the dependence relations between the laws of higher- and lower-level sciences.

According to *metaphysical anarchy*, there are few dependence relations between higher- and lower-level laws, and where they exist, they are unsystematic and may go in either direction—the lower-level laws may depend in part on the higher as well as the higher on the lower. This view is articulated by Dupré (1993).

Particularist physicalism has a positive part and a negative part. The positive part: any process governed by a high-level law occurs as a result of one or more low-level processes. For example, the process resulting in the death of 5% of a particular group of rabbits—a process governed by a high-level law of population ecology—occurs because of low-level behavioral processes, in the sense that each of the deaths was brought about by a behavioral process. The negative part: there is usually no systematic relationship between the high-level processes that fall under a given law and the low-level processes on which they depend. The relationship between the high- and low-level processes in one instance of the high-level law may be different in every respect from the relationship in another. Thus, there is no systematic relationship between the high-level law and the relevant low-level laws. This is the picture presented in Fodor (1975).[1]

In *systematic physicalism*, as in particularist physicalism, any process governed by a high-level law occurs as a result of one or more low-level processes. But the relation between the two levels is systematic. That is, there is some kind of relationship between the processes falling under a given high-level law and the low-level processes on which they depend that is the same in every case. The most straightforward sort of systematicity arises when more or less the same kinds of low-level laws, combining in the same sort of way, are responsible for the unfolding of each high-level process governed by a given high-level law, and thus for the law itself.

Where enion probability analysis succeeds, it demonstrates the existence of a systematic relation between the macrodynamic and the microdynamic processes in a complex system. Thus, EPA demonstrates the existence of a systematic relation between a system's high-level laws and the lower-level laws governing the behavior of its parts. In so doing, EPA vindicates systematic physicalism.

The nature of and the reasons for the systematic dependence in particular cases are the subject of chapters one to four. I will not rehearse them here. In the remainder of this section I ask: where did the opponents of systematic physicalism go wrong? I will consider two arguments that have been raised against systematicity, making, along the way, some very general comments about the reasons that systematic relations between higher and lower levels can exist.

5.12 Multiple Realizability

The paradigm of systematicity is the case in which a higher-level law is a systematic consequence of a single set of lower-level laws. I have provided, in chapter four, two putative examples of such systematicity: the laws governing the population of rabbits and foxes depend systematically on the laws governing the behavior of individual rabbits and foxes, and the laws governing the statistics of gas molecules depend systematically on the laws governing the collisions of individual molecules.

Not every higher-level law is the consequence of a single set of lower-level laws, however, no matter how loosely *single set* is construed. This is because the properties that figure in some higher-level laws are *multiply realizable*. They can be instantiated by many different kinds of entities, kinds governed, at the lower level, by entirely different laws.[2]

Consider, for example, a very general law governing the population dynamics of a predator/prey system. There are many kinds of predators and prey, thus many lower levels to which the higher-level law will apply. Different instances of the higher-level law will belong, at the lower level, to quite different domains. One process falling under the higher-level law will be, at the lower level, a process involving the behavior of rabbits and foxes. Another will be a process involving the behavior of flies and spiders. Yet another might not be a biological process at all: if the predators and prey exist in an artificial life environment, the process might be a computation.

How, in such a case, can there be a systematic relationship between the higher and the lower level? My discussion of enion probability analysis provides the answer: the lower-level properties on which the simple behavior of complex systems depends are themselves multiply realizable.

For example, a predator/prey law may depend in part on the fact that the predators and prey in any system encounter each other in a randomly determined way. The randomness of the encounters may depend in turn on the independence of microdynamic probabilities, which may depend on the microlinearity of microdynamic ic-evolution. All of these properties—being made up of randomly determined encounters, having independent microdynamic probabilities, and having microlinear ic-evolution—are sufficiently abstract that they can be possessed by quite different processes, including both biological and computational processes. A property such as that of having microlinear ic-evolution is a low-level property, since it is a property of a system's microdynamics, but it is multiply realizable, since it can be shared by utterly different low-level processes. My explanation of the foundations of EPA in the preceding chapters shows, then, that higher-level laws depend on lower-level properties that any lower level can have.

Wherever there is a higher-level law, as opposed to a mere true generalization, concerning properties that are multiply realizable, I conjecture, it can be explained as a systematic consequence of multiply realizable properties of lower-level laws. In the case of complex systems, EPA shows the way.

5.13 The Failure of Classical Reductionism

The best-known variety of systematic physicalism is classical reductionism (Nagel 1979). The shortcomings of classical reductionism have naturally been regarded as casting doubt on systematic physicalism in general.

According to classical reductionism, the systematic relation between a higher-level law and the lower-level laws that it reduces to is the following: in conjunction with suitable *correspondence rules* (sometimes called *bridge laws*), the lower-level laws logically entail the higher-level laws. A correspondence rule is, in effect, a translation of the vocabulary of the higher level of description into the vocabulary of the lower level of description. Classical reductionism favors a translation in the form of a biconditional. Thus if Q is a predicate used at the higher level, reduction requires a translation of Q into the language of the lower level by way of a correspondence rule of the following form:

$$(x) : Q(x) \leftrightarrow P(x)$$

where P is a predicate, presumably syntactically complex, couched in the language of the lower level. Once the translations are given, everything that happens to a Q at the higher level can be seen, if the reduction is successful, to happen to a P at the lower level, and vice versa.

The most influential presentations of classical reduction did not absolutely require that correspondence rules take the form of biconditionals, but neither did they investigate other possibilities. As a result, the biconditional requirement is now considered an intrinsic part of classical reductionism. It is this version of classical reductionism that modern writers take to have failed; in what follows, then, I assume that classical reduction insists on correspondence rules of the biconditional form.

For a classical reduction to succeed, every predicate in the higher-level law must be translated. If the higher-level law mentions rabbits, for example, the term *rabbit* must be translated into the language of the lower level. If everything is to be reduced to fundamental physics, then along the way *rabbit* must be translated into the language of fundamental physics, which is to say, necessary and sufficient conditions must be given in the language of physics for an object's being a rabbit.

The necessity of providing such a translation has struck writers such as Dupré (1993) as imposing an impossible demand. Classical reductionism must fail, and so much the worse for systematic physicalism. The assumption is that classical reductionism is systematic physicalism's best hope. Exhibiting an alternative form of systematic physicalism—the form implicit in enion probability analysis—sidesteps this argument neatly enough. But it is illuminating to see why, exactly, classical reductionism fails.

As I have said, it is classical reductionism's requirement that all higher-level terms be translated into the language of physics that drags it down. To see how, consider an example. Suppose that one wants to systematically account for the law that 5% of rabbits in a given ecosystem will die in the course of a month. Reductionism says that a physical definition of *rabbit* is needed. Here are three questions that arise in formulating such a definition. First, rabbits evolved from something other than rabbits; call them proto-rabbits. At exactly what point do proto-rabbits become rabbits? Second, is an organism's history relevant to rabbithood? For example, if an exact molecular replica of a rabbit were created in the laboratory from old tires and corn syrup, would the organism count as a rabbit? Third, is there some specification of the layout of an animal's DNA to which all and only rabbits conform?

These questions may be merely hard, almost impossible, or utterly hopeless. But whatever else they are, they are clearly irrelevant to the problem of why 5% of rabbits in a certain kind of ecosystem (and not 1% or 90% or some complex function of the starting positions of all the rabbits) typically die in the course of a month. None of the organisms in the ecosystem falls into any of the gray areas of rabbithood that the questions concern, thus none of the questions need be resolved to understand, systematically, the mortality of the organisms.

It turns out, then, that classical reductionism asks for far more than it needs. It forces us to consider absolutely every physical property that is relevant to rabbithood, when only a subset of these properties make a difference to the high-level law that is to be explained.

How might the classical picture be amended so as to demand less without compromising the goal of giving systematic lower-level explanations of higher-level phenomena? I will divide the classical view into two parts:

1. A law is reduced by deducing it, with the help of correspondence rules, from lower-level laws and facts.
2. The correspondence rules used in the deduction must give necessary and sufficient conditions for the satisfaction of the terms appearing in the law to be reduced.

A conservative amendment of classical reductionism—perhaps not the best amendment, but the simplest—would retain (1) but relax (2), by asking not for a complete set of correspondence rules translating between higher-level and lower-level vocabulary, but only for those rules of higher/lower-level correspondence that are absolutely necessary to deduce the higher-level generalization in question from the lower-level theory.

What will these correspondence rules look like? Suppose that the goal is to give a lower-level explanation of a law of the form:

In environment E, all Qs have property A

where Qs might be rabbits and A might be the property of having a death rate of 5% per month. In order to deduce this law from the lower level, what one absolutely needs are lower-level properties P and B such that

1. All Qs in E have P
2. In E, everything with P has B
3. In E, anything with B has A

where (1) and (3) are the correspondence rules and (2) is the relevant lower-level law. In the example, this might involve finding a lower-level property P that all rabbits share—a necessary property of rabbits—and a lower-level property B that is sufficient for having a 5% death rate, which will involve finding a sufficient property for death.

Rules of the form of (1) and (3) are easier to come by than the correspondence rules demanded by classical reductionism in two ways. First, rules (1) and (3) give either necessary or sufficient conditions, but not conditions that are both necessary and sufficient. Giving a necessary condition for rabbithood is far easier than giving a necessary and sufficient condition; similarly, giving a sufficient condition for death is far easier than giving a necessary and sufficient condition. Even when the same predicate appears in both the antecedent and the consequent of the high-level law, one need find only a necessary and a sufficient condition for instantiating the predicate, potentially a much easier task than finding a condition that is both necessary and sufficient.

Second, the scope of rules (1) and (3) is limited to environments of type E, thus, the rules need give only locally reliable necessary or sufficient conditions. Again, this makes such rules much easier to find. Perhaps it is even possible to find locally reliable necessary *and* sufficient conditions for some higher-level terms, though I doubt that this achievement would be particularly useful in most interlevel explanations.

In summary, then, to achieve classical reductionism's goal of deducing higher-level generalizations from lower-level generalizations, one's correspondence rules need only be locally reliable rules that give either necessary or sufficient conditions for the relevant higher-level properties, rather than globally exceptionless rules giving necessary and sufficient conditions. So understood, the classical goal seems within reach; EPA then shows how a deduction of a high-level law might proceed.

5.2 Higher-Level Laws

The single most important philosophical question about the higher-level sciences is, I think: are there any higher-level laws? If the answer is yes, then the principal goal of the higher-level sciences must be to discover the nature of the high-level laws, if not in every detail then at least in outline. In particular, science ought to determine the kinds of variables that are related by the laws, and the form of these lawful relations.

I have shown that, in large classes of complex systems, there are simple laws linking variables at the macrolevel. In this section, I will look more closely at the nature of the laws whose existence has been established, and I will draw some morals for the structure of those higher-level sciences that concern complex systems.

5.21 Laws from Enion Probability Analysis

Let me summarize the method, developed in chapter four, by which one might demonstrate that a given complex system obeys a simple macrolevel law. The procedure can be divided into two steps, which I will call the *qualitative* stage and the *quantitative* stage.

In the *qualitative stage*, some rather general decisions are taken as to how to construct a probabilistic model of the system. Some of the more important choices to be made are the following:

1. A microlevel must be chosen. In the case of population ecology, I made the rather obvious choice of the behavioral level; in the case of kinetic theory, I invented the Boltzmann level.
2. A policy for individuating microdynamic experiments must be chosen. Much of the discussion of gases and ecosystems in sections 4.8 and 4.9 was concerned with this question. The decision about individuation will, more than any other, determine the availability of randomizing variables.
3. A certain set of enion statistics must be chosen as the macrolevel variables. This choice determines what kinds of information the enion probabilities are and are not allowed to depend on. Here, also, a decision may be made as to whether to take a structured populations or averaging approach to eliminate any microlevel dependencies that look to cause trouble, as described in section 4.6.
4. An approximate time scale must be chosen for the experiments to which enion probabilities are attached. In the case of the rabbit/fox ecosystem, for example, I suggested a trial duration of one month.

There is often more than one reasonable way to make these decisions. But there is also a strong constraint: the decisions must result in a model in which the enion probabilities satisfy the probabilistic supercondition, that is, in which they are independent of microlevel information.

The qualitative stage, properly executed, delivers up a suitable framework for making the arguments outlined in chapter four for the independence of microdynamic probabilities, the microconstancy of enion probabilities, the independence of enion probabilities, and so on. In order to establish the existence of some particular macrolevel law, some more particular information must be taken into account, for example:

1. The exact form of the probability distributions for each of the microdynamic experiments,
2. The details as to the deterministic aspect of microdynamic evolution,
3. The probability distributions over any exogenous variables, and
4. The structure of the background environment.

The *quantitative stage* attends to this information and applies whatever mathematical arguments are necessary to derive the macrolevel law.

Much of what I have to say in this section follows from the fact that the information taken into account in the quantitative stage may be far more system-specific than the information taken into account in the qualitative stage. That is, the decisions made at the qualitative stage will tend to be the same for a wide range of systems, but the information added at the quantitative stage may differ from system to system. The first consequence of this observation, discussed in the next section, is that the laws governing complex systems will tend to be laws of *narrow scope*.

5.22 Narrow-Scope Laws

I will begin by defining the *scope* of a law. Every law has certain *conditions of application*, which specify the kinds of systems concerning which a law makes assertions. The scope of a law is a measure of the number of systems that satisfy its conditions of application; the more systems, the wider the scope. Laws about tigers have narrower scope than laws about mammals, for example. The fundamental laws of physics have the widest possible scope: they are intended to apply to everything there is. Higher-level laws will tend to have narrower scope, but how much narrower? That is an open question. Discussions of, for example, biological laws often focus on laws of relatively wide scope, such as the laws of population genetics (Sober 1984), Mendel's laws (Waters 1998), and the competitive exclusion principle (Weber 1999). But there are any number of laws of narrow scope in biology. The population laws governing particular ecosystems are just one class of examples.

Note that scope should not be confused with exceptionlessness. A law of very wide scope may have many exceptions, while a law of very narrow scope may have very few. (Indeed, this is the normal state of affairs in the higher-level sciences.) A law does not fail outside its scope; it simply makes no pronouncement at all.

The scope of a law governing a complex system is determined by the scope of the decisions and information taken into account in the qualitative and quantitative stages described above. Because the quantitative stage incorporates more system-specific information than the qualitative stage, it is the information taken into account in the quantitative stage that will set the scope of a law.

In a science such as population ecology, the result will be, it seems, a profusion of narrow-scope laws. As environment, weather patterns, and most of all, the particular organisms making up a system change, so will the relevant enion probabilities change, resulting in different macrolevel laws for different ecosystems, and even for the same ecosystem at different times.[3]

There are certain curbs on this nomological profusion. First, some changes in an ecosystem, though they change the laws, will not change the mathematical structure of the laws. The result is what might be considered the same law but with different values for its parameters. In population ecology, laws of this sort are often encountered.

Second, some changes in a system will not change the laws at all, because they will not change the enion probabilities. Kinetic theory provides a striking example: substituting the molecules of one gas for another, even though the molecules might be quite differently shaped, will not alter the fact that the positions and velocities of the gases conform to the Maxwell-Boltzmann distribution. I assume, however, that the systems of population ecology are more typical of complex systems in general, and therefore that the simple macrolevel laws whose existence is explained in this study are predominantly narrow-scope laws.

Narrow-scope laws, though important, are of less interest, or at least, less use, to science than generalizations of wider scope. The remainder of this section therefore turns to the examination of wide-scope facts about complex systems.

5.23 Rules of Construction

The natural place to look for wide-scope generalizations about complex systems is in the assumptions made about systems during the qualitative, rather than the quantitative, stage of EPA (see section 5.21). But at the qualitative stage one finds not laws, or any other kind of generalization about systems, but methodological rules. Because these rules are used to construct models, call them the *rules of construction* for the systems that fall within their scope.

Let me say something more about the construction rules before returning to the topic of wide-scope generalizations. I have characterized the construction rules in a way that makes explicit reference to the techniques developed in this book for the study of complex systems. But the notion has application in any science in which model-building methods play an important role in scientific inquiry.

The practitioners of population ecology, for example, although they do not discuss, say, the individuation of microdynamic experiments, do have a toolbox of techniques for building models of population change in ecosystems. I count these techniques among the construction rules. Indeed, there is an obvious overlap between the model-building techniques of population ecology as it is practiced today, and the methodological rules used to build a model of an ecosystem for the purposes of explaining simple behavior in the way suggested by this study. Both methods require, for example, a choice of macrovariables.

On the whole, the current construction rules of the higher-level sciences tend to revolve around the prescription of certain mathematical forms for macrolevel laws; examples include the rules of population genetics, of transport theory in physical chemistry, and of neoclassical economics. I suspect that there is a certain element of art in choosing the right mathematical form, and that the art takes into account the nature of interactions between individual enions, and so involves further overlap with the kinds of construction rules specified above in connection with EPA. But the art tends not to be documented in textbooks, for which reason it is difficult to substantiate this claim here.

Philosophers of science occasionally remark that ecologists and workers in other higher-level sciences seem more interested in model-building techniques than in laws. The reason, I suggest, is that in ecology and many other sciences, the construction rules have a much wider scope than do individual laws. There is often little point in devoting large amounts of resources to learning laws that govern only a few systems, and that may change at any time. Better to learn the model-building techniques that constitute the first step in understanding or predicting the laws of a wide range of systems.

Sometimes the emphasis on the construction rules at the expense of the narrow-scope laws is taken to show that higher-level sciences are not really about laws. But if the picture presented in the previous paragraph is correct, this is a mistake or, at best, an oversimplification. The interest of the construction rules lies precisely in their role in illuminating a wide range of different narrow-scope laws, just as the interest of my argument in the previous four chapters lies in its power to explain the differing simple behavior of a wide range of systems. I will have more to say about this in the next section.

Let me now return to the question of the existence of wide-scope generalizations concerning complex systems. I have so far described two kinds of generalizations that might turn up in a higher-level theory. First, there are the narrow-scope laws, each governing a different kind of system in the theory's domain. Second, there are the rules of construction, which usually take the form of a very general method for building a model of any system in the domain. What of wide-scope laws? So far, there are none to be seen: the generalizations that have the form of laws do not have wide scope; the generalizations that have wide scope have the form of methodological rules rather than that of laws.

But one might reasonably suspect that, buried somewhere within the methodological prescriptions that make up the rules of construction, there are generalizations about complex systems to be found. The search for wide-scope generalizations of this sort is the topic of the next section.

5.24 Wide-Scope Laws

The methodological prescriptions that make up a domain's rules of construction owe their effectiveness to certain properties shared by all systems in the domain. A wide-scope generalization inherent in the rules of construction will articulate some such common property.

There are a variety of forms that wide-scope generalizations might take. In this section, I focus on the possibility that they may have the aspect of wide-scope laws. Some other possibilities are mentioned at the end of the section, including one, the possibility that many wide-scope generalizations concern relations of causal relevance, that is the subject of section 5.3.

Are there wide-scope laws about complex systems? Because all models of complex systems in a domain are built in roughly the same way, it seems reasonable to think that they will, at least in certain respects, behave in the same way. This expectation is borne out: systems in population genetics, for example, obey the Hardy-Weinberg law, and ecosystems tend to conform to the principle of competitive exclusion. For other examples of patterns of behavior with wide scope, the reader might simply turn back to the general trends given as examples of simple behavior in section 1.11.

Inevitably, though, the seeker after wide-scope laws in the higher-level sciences returns disappointed. Although such laws may be found, and are sometimes of considerable interest, they seem only ever to contain a part of the knowledge that there is to be had about a domain of systems, and in many cases, a small and rather arbitrary part.

The reason for this, I suggest, is the primacy of the rules of construction. Some elements of the rules of construction can be captured in the requirement that systems obey certain wide-scope laws, but many other elements cannot. To search only for wide-scope laws is therefore to search for only a part of the rules of construction. Even if the search were entirely successful, it could never turn up all the facts about complex systems necessary to understand why they obey the narrow-scope laws that they do, simply because some such facts do not have the form of laws. Attempts to understand complexity by way of very wide scope laws are, for this reason, not likely to succeed.

In arguing against the importance of wide-scope laws in the higher-level sciences I am agreeing with writers such as Beatty (1995). But Beatty and others who altogether discount the significance of laws in the higher-level sciences neglect the importance of narrow-scope laws. It is only because complex systems obey narrow-scope laws that the rules of construction, facts about causal relevance, and other pieces of non-lawlike information have any significance at all; thus, although the rules and so on are not themselves laws, they are always and everywhere *about* laws. In one sense, then, everything in the higher-level sciences depends on the narrow-scope laws.

To illustrate this last claim, consider Darwinian explanation in evolutionary biology. As summarized in section 1.22, a Darwinian explanation accounts for the predominance of a trait by citing the trait's superior fitness. For a trait to be selected, the ecosystem or ecosystems in which it evolved must give that trait a consistent advantage over its competitors for a certain duration of time. This requires that each ecosystem be governed by a narrow-scope law of a particular form for the duration in question. The law may change over the duration, but it must not change so much that the trait loses its advantage.[4] The existence and relative stability of the narrow-scope laws governing each ecosystem, then, is necessary for Darwinian explanation: it is only because of the stability of the laws that there are stable facts about fitness. In this sense, the whole of the explanatory power of natural selection is predicated on the simple, stable behavior of complex systems, where that simple, stable behavior is due to the nomological authority of the narrow-scope laws.

Try to frame a wide-scope generalization to capture the role of the narrow-scope laws in explaining natural selection, and you end up with a useless truism such as *survival of the fittest*. This slogan points at the importance, in natural selection, of a stable property of fitness, but it cannot begin to explain why such a property should exist.

To summarize the argument presented above:

1. Laws are paramount in understanding the higher-level sciences of complex systems, but the laws that matter are narrow-scope laws.
2. In most higher-level sciences, it is not fruitful to pursue knowledge of the narrow-scope laws directly, because there are too many and they change too frequently. Instead, what is pursued is knowledge as to how

to understand or predict, for any given system in a domain, whatever narrow-scope law governs that system. This knowledge has the form of rules of construction.

3. The rules of construction may imply various wide-scope laws, but wide-scope laws contain only a part, and often not a very useful part, of the information in the rules of construction.

I conclude by asking whether there is some form of wide-scope knowledge, other than wide-scope lawhood, suitable for representing the facts about a domain inherent in the rules of construction. The answer has to be, I think, that there are several such forms, some complementary, some—in a friendly and fruitful way—competing. One very interesting idea is that wide-scope knowledge (and much knowledge of not so wide a scope) can take the form of knowledge of invariant generalizations, for which see Woodward (2000). Another possibility is that wide-scope knowledge takes the form of knowledge of causal relevance relations. It is to this possibility that I next turn.

5.3 Causal Relevance

In the previous section, I discussed the possibility of a wide-scope approach to high-level domains, that is, an approach that seeks to generalize about many or all systems in a domain. Where such generalizations can be made, I proposed, they are true in virtue of the rules of construction for the domain. The generalizations that arise out of the rules of construction may, but need not, have the form of laws. In this section I will discuss one very important species of non-lawlike generalization, statements of causal relevance. Statements of causal relevance can, of course, have varying scope. The following discussion is primarily but not exclusively concerned with statements of reasonably wide scope.

Let me give some examples of important, though perhaps untrue, claims of causal relevance:

1. Whether a child lives with both parents or just a single parent is causally relevant to the child's success at school.
2. The degree to which a nation is politically free is causally relevant to the life expectancy of its inhabitants.
3. The independence of a nation's central bank is causally relevant to the nation's rate of inflation.

Sometimes degrees of causal relevance are compared:

4. The means by which goods are produced is far more important than matters of the intellect in shaping cultural change.
5. Differences in social behavior were more important than differences in linguistic ability in the dominance of modern humans over Neandertals.

Claims of causal relevance are important because they are often available even when circumstances preclude—at least temporarily—our discovering the narrow-scope macrolevel laws of a domain, as is so often the case. There are a number of different kinds of circumstances under which claims of causal relevance may be our best or only generalizations about a system or a domain of systems, four of which are as follows.

Narrow-scope laws are as yet unknown: In order to learn efficiently the as yet unknown macrolevel laws governing a complex system, it is important to have some knowledge of relations of causal relevance, so that the right statistics can be compiled and used to estimate the parameters in the right kind of law.

Narrow-scope laws are forever unknown: Sometimes it is not possible ever to learn the laws of a system, as when, for example, the system is a historical entity that no longer exists, such as a Cambrian ecosystem or the economy of seventeenth-century France. Even when there is too little evidence to reconstruct the law, there may be enough to recover some relations of causal relevance.

Narrow-scope laws are complicated: Though a system's macrolevel laws are known, they may be so complicated that it is difficult, for practical reasons, to use the knowledge for explanatory or predictive ends. In particular, it may be difficult to ascertain the values of all the macrovariables, if there are many, or it may be difficult to compute the course of macrolevel evolution, if the macrovariables are related in computationally intricate ways. In such cases, both prediction and explanation may be more effectively pursued by way of facts about causal relevance.

Narrow-scope laws are fluid: Some systems—in particular, perhaps some economic and social systems—have macrolevel laws that tend to change rapidly, not just with respect to the values of parameters, but with respect to mathematical form. If the change is fast enough, the macrolevel laws will be so ephemeral as to be of little interest, even if they can be caught on the wing. But certain features of the laws will likely endure even as the details change. Often the kind of description that best captures what endures is a claim of causal relevance. Even in an ideal science, then, with unlimited access to the values of macrovariables and vast computational resources, knowledge of the lasting properties of certain classes of complex systems may take the form of knowledge of relations of causal relevance.

Causal relevance has an epistemological and physical hardiness that more delicate arrangements, in particular narrow-scope laws, lack. But one may reasonably question the basis of this apparent hardiness, asking:

1. Why is it that relations of causal relevance can be known when other general facts, such as narrow-scope macrolevel laws, cannot?
2. How can relatively simple relations of causal relevance exist in systems with complicated laws?
3. How can relations of causal relevance persist when the laws in which they are founded are constantly changing?

I will not discuss question (1) here.[5] The remainder of this section concerns questions (2) and (3).

Questions (2) and (3) arise, I think, from a well-known worry about the significance of causal relevance claims about complex systems, which I will call the Keynesian worry after Keynes (1921); see Cartwright (1989, §4.3) for discussion. The Keynesian worry is as follows. What happens to a complex system or its parts is determined by many variables. What happens to an individual enion, in particular, is determined by perhaps thousands of microvariables. A variable plays a role in determining what happens in a complex system only in conjunction with all the other variables. Thus, in order to say what effects a given variable will have, one must know the values of all the other variables. From this fact, the Keynesian infers that, in order to say *roughly* what effects a given variable will have, one must know the *rough* values of all the other variables. I will call this the Keynesian premise.

The Keynesian premise entails that any claim about the rough effects of one variable can be made only relative to facts about the rough values of all the other variables. In particular, there are no true statements of the form

> A variable x is positively (negatively, strongly, weakly, and so on) causally relevant to an outcome e,

only true statements of the form

> A variable x is positively (negatively, strongly, weakly) causally relevant to e provided that $y \approx a$, $z \leq c$, . . .

where y, z, and so on are other variables of the system. Call the conditions appearing after the *provided that* the claim's provisos. The worry, then, is that causal relevance claims about complex systems will always have long lists of provisos, from which one may well conclude that there are no interesting, useful, and true claims of causal relevance, since any claim of causal relevance sweeping enough to be interesting or useful is also sweeping enough to be false.

In what follows, I accept the conventional wisdom that something like the following is true:

> A variable x is causally relevant to an outcome e just in case x affects the probability of e.[6]

In chapters two, three, and four I argued that the probabilities attached to the behavior of complex systems and their parts depend on many fewer aspects of the system than might be expected, and certainly, many fewer aspects of the system than need to be taken into account to make exact predictions about a system's behavior. Given the probabilistic understanding of causal relevance, this suggests that relevance relationships will hold with many fewer qualifications than the Keynesian premise assumes.

Let me focus on cases of macrovariable/enion relevance. A claim of macrovariable/enion relevance is a claim that a certain macrovariable is relevant to an outcome to which an enion probability is attached, that is, to the fate of a given enion, as, for example, in the claim that political freedom is relevant to a person's longevity.

It has been an aim of this study to show that enion probabilities can be defined in such a way that they depend on macrovariables alone, so that a system's microvariables cannot affect the dependence one way or another. Where this aim is achieved, relations of probabilistic, hence causal, relevance will be far simpler than the Keynesian imagines: the number of variables in their provisos cannot exceed the number of macrovariables.

In many cases, one may reasonably hope for something even better than this: a relevance relation with no provisos, as in the claim that the number of foxes is positively causally relevant to rabbit death. Such relations will hold if the influence of the macrovariable on the enion probability is independent of the influence of any other macrovariables.

Total independence is perhaps a rare thing, but partial independence is not, where the influence of a macrovariable is partial if the approximate magnitude of the influence—that is, of the probability change—or, more weakly, its direction, is independent of the values of other macrovariables. It is plausible, for example, that the probability of rabbit death over the course of a month is always raised by an increase in the fox population, regardless of the values of the other macrovariables.[7] Where there is partial independence, there will be a qualitative or even an approximate quantitative relevance relation with no provisos.

A sufficient condition for partial independence suggested by the rabbit/fox case is as follows. Suppose that a macrovariable m measures the number of enions of a certain type F in a system (think of the fox population). Suppose that the chance that some given enion r (think of a rabbit) meets an enion of type F increases with the frequency of Fs, and that the probability of r's experiencing some event e (think death) increases with the number of Fs it meets. Then an increase in m will result in an increase in r's probability of experiencing e, regardless of the values of the other macrovariables in the system. These other macrovariables may affect the size of the increase, but not the fact of the increase itself. Thus the relevance of m to e is partially independent of the system's other variables.[8] There are, of course, other sufficient conditions for partial independence.[9]

Besides claims of macrovariable/enion relevance, there are claims of microvariable/enion relevance and macrovariable/macrovariable relevance. A claim of microvariable/enion relevance is a claim that a certain microvariable is relevant to an outcome to which an enion probability is attached, as, for example, in the claim that a child's home environment is relevant to the child's success at school. A claim of macrovariable/macrovariable relevance is a claim that a certain macrovariable is relevant to the value of another macrovariable, as, for example, in the claim that the means of production is highly relevant to cultural change.

I have nothing to say about claims of macrovariable/macrovariable relevance, except to note that the existence of simple macrolevel laws provides the foundation on which such claims rest.

Concerning microvariable/enion relevance, I will make one remark. Except in special circumstances, such as those that create the problem of rabbit health, the results of this study show that enion probabilities do not depend on the values of microvariables. Thus, microvariables are not causally relevant to the fates of enions. For example, rabbit position is not causally relevant to rabbit death (except in the special case where the position is inches from a fox's jaws). This provides a powerful filter for use in determining causal relevance relations. I would guess that this filter grounds many of our tacit negative judgments of causal relevance, judgments which allow us to reduce the number of candidate causes, in many situations, from a hopelessly large to a manageable number.

5.4 The Social Sciences

To what social systems might enion probability analysis be successfully applied? One way to answer this question is to survey the actual use of EPA or techniques similar to EPA in the social sciences, and to evaluate such techniques' accuracy and generality. The survey would be a book-length undertaking, and would in any case not necessarily reveal the underlying reasons for EPA's success, my abiding concern in this study.

What I propose to do instead, in this short discussion, is to examine the question whether there are any promising choices for randomizing variables in complex social systems. The notion of a randomizing variable, recall, brings together two essential parts of the argument for the microconstancy and independence of enion probabilities. A randomizing variable must fulfill the following two conditions:

1. The values taken on by the variable for different enion interactions must be probabilistically patterned in the right sort of way, and
2. The probabilistically patterned values must have the causal efficacy to propel the enions, and so the system, on a random walk through the space of possibilities.

In what follows I survey some possible choices of randomizing variables. Before moving on to the survey, though, let me say something about each of these two requirements in turn.

The first requirement demands that instantiations of a randomizing variable be probabilistically patterned, and in particular, that they be stochastically independent. There is a deep and a shallow way to search for microdynamic variables in social systems having this property. The deep approach proceeds as I did in the discussions of kinetic theory and population ecology, by looking for variables that satisfy the sufficient conditions for microdynamic independence stated in section 4.5. The shallow approach looks for variables that are probabilistically patterned, without asking why they are so patterned.

The deep approach is, of course, explanatorily more satisfying. However, our knowledge of the underlying mechanics of social interaction is so poor, despite our familiarity with the interactions themselves, that it is impossible to take the deep approach at all securely. I will therefore take the shallow approach.

The second requirement on a randomizing variable is that it affect the goings-on in a complex system sufficiently strongly to wash out the influence of a system's initial microstate in the time allowed for an enion probabilistic experiment. I will not try to make any systematic argument that the potential randomizing variables canvassed below meet this requirement; it is important, though, that all the variables considered are major determinants of a particular enion's trajectory.

My subject, then, is systems of human beings in which social interactions have strong effects, but in which actors move in fairly independent ways. The goal is to find some microdynamic variable or variables that have a considerable effect on the paths that actors take through social systems, and which are probabilistically patterned. Is there any reason at all to think that the microdynamics of human social interaction can supply variables that conform to these demands? In what follows, I will suggest four ways in which social interaction or individual decisions might take on a probabilistic form.

First, as noted in section 4.9, humans as well as animals make decisions by probability matching. It is unclear how far this behavior extends, but it is not confined to the exploitation of food resources. For example, the Yale students whose probability-matching behavior is described by Gallistel (1990, 351) were participating in a guessing game where nothing material was at stake. Probability matching might introduce a probabilistic element into almost any kind of human decision making where there is a genuine choice to be made; perhaps the most obvious example is that of economic systems, in which actors must choose from a wide variety of opportunities to profit.

Second, the social environment may affect the mind in other, perhaps arational, probabilistic ways. Very little is known about this. But the fact that probabilistic strategies can be evolutionarily stable (see note 48 of chapter 4) provides, at the very least, good reason not to rule out this possibility. Perhaps—this is pure speculation—the differing reactions of different people to stressful or resource-poor environments, both before and after birth, are due to some such strategy.

Third, humans may react deterministically to many kinds of social scenarios, but the nature of the scenarios themselves may be probabilistically determined, hence the reactions may be distributed probabilistically. The clearest cases are those where the probabilistically distributed element of the scenario is physically rather than socially determined, but this is perhaps rare. An example might be the insurance company's branch of social studies, where incidents that result in claims are often caused by low-level processes that are amenable to a probabilistic representation. More speculatively, what makes a difference between a vulnerable person's attempting or not attempting suicide might, in some cases, be an accident or a series of accidents.

Fourth, a coarse-grained representation of a deterministic system may be probabilistic in the sense required by EPA. The treatment of kinetic theory in section 4.8 provides the paradigm. It is very difficult to say which aspects of that treatment will carry over into the social realm. An example of a system in which coarse-graining might provide a conceptual basis for a set of stochastically independent, microdynamic probabilities is the study of patterns of road traffic (of which there are already many probabilistic models).

Perhaps the most salient objections to the attempt to ground social regularities in probabilistically patterned individual variation arise from the observation that humans control or adapt to variations in the environment so as to diminish their effects.

One such objection is a version of the homeostasis problem discussed in section 4.92. Human reaction to environmental disturbances, it is noted, will frequently erase the effects of those disturbances. This makes finding a randomizing variable hard in the same way as in the case of population ecology.

To see this, recall the two conditions that a variable must satisfy in order to count as randomizing:

1. The values taken on by the variable for different enion interactions are probabilistically patterned, and
2. The probabilistically patterned values have the causal efficacy to propel the enions, and so the system, on a random walk through the space of possibilities.

Intelligent and flexible human action makes both conditions (1) and (2) difficult to satisfy. Condition (1) is made difficult to satisfy because, as explained in section 4.92, goal-directed behavior may result in the deflationary IC-evolution of all potential random variables, so undermining the conditions required for stochastic independence. Condition (2) is made difficult to satisfy because human compensation for the jostlings of a potential random variable will tend to neutralize the ability of the variable to give actors the sort of shove required to send them on a random walk.

The first difficulty was discussed at length in section 4.92, and I have nothing to add to the solutions to the problem described there. The second difficulty, which was dismissed very quickly in the treatment of population ecology, might, however, be thought to be more acute in the case of social systems, on the grounds that humans have, and choose to exert, the ability to regulate many more aspects of their lives than other animals.

This is a very interesting and very difficult problem for those who seek to understand social systems. One kind of solution to the problem might begin by surveying all the ways in which humans fail to fully compensate for life's rough treatment.

Another kind of solution begins with the suggestion that human goals themselves may be among the randomizing variables. If each person's short- or medium-term goals are fixed in part probabilistically, then, no matter how relentlessly and predictably those goals are pursued, the pursuer is engaged, in the technical sense, in a random walk from goal to goal.

Let me consider two objections to this proposal. First, it is not obvious that the randomization of goals is sufficient to take a system on the kind of random walk necessary to derive the microconstancy and independence of enion probabilities.

Second, it is, of course, abhorrent to suppose that important choices such as the setting of goals are not decided by extended rational deliberation. It may, nevertheless, be an abhorrent truth. And perhaps not so abhorrent after all: unless talent, proclivity, remuneration, and so on completely determine, say, the ideal occupation for each person—and except in Socrates' ideal city, they do not—an individual probabilistic choice among the most attractive options, taking into account the aforementioned factors and then applying a method such as the probability-matching rule, may be the most efficient means we have for labor allocation.

The adaptability of human behavior also raises a rather different concern about EPA in the social sciences. In population ecology, EPA has the advantage of providing the only good explanation of macrolevel simplicity that we have. Thus, wherever there is macrolevel simplicity in population dynamics, there is some prima facie reason to think that EPA explains the simplicity.

In social systems, by contrast, there is always the possibility that human adaptability alone explains simple macrolevel behavior (see, for example, Jervis 1997, 263–282). Sometimes the adaptability manifests itself in the negotiation of binding conventions, as in our agreement to drive on one side of the road; sometimes the coordination is by subtler means, as in a market where posted prices guide behavior without explicit agreement. While human ingenuity is perhaps never entirely sufficient for order at the macrolevel—ingenuity cannot impose, but must always merely enable and amplify some propensity to order inherent in a system—it is possible that the role of EPA in explaining such order is in many cases minor or non-existent. My own guess is that EPA will sometimes explain social regularities, sometimes not. But it is difficult to say more as long as our ignorance of the mechanics of social behavior is so great.

5.5 The Mathematics of Complex Systems

The mathematics of complex systems consists of whatever mathematical techniques help us to understand the behavior of complex systems. Enion probability analysis is one such technique; no doubt there are others, some as yet undiscovered. In this section I summarize the properties of EPA that make it especially apt for understanding the relation between the microdynamics and the macrodynamics of a complex system. I then generalize, proposing that these features are essential for any kind of understanding of complex microdynamics.

Because a complex system is made up of many independently articulated parts, each with a number of intrinsic properties, it is a matter of combinatorics that any particular type of complex system can be instantiated in a vast number of ways. A technique for understanding the relation between microdynamics and macrodynamics in a given type of system must encompass explicitly all this possible microlevel variability. It is widely believed, I think correctly, that the best way, perhaps even the only practical way, for a technique to take into account microlevel variability is through *compositionality*. That is, the technique must assign some property or properties to the individual elements of the system, in particular, to the enions, and it must then provide some set of rules for systematically inferring macrolevel properties from the arrangement of the elements.[10]

It is also widely believed that the systematic set of rules will be manageable only if interactions between the individual elements in the system are minimal. This leads to an apparent impasse: to understand systems of many parts, a compositional theory is required, but no tractable compositional theory is possible if the parts interact. The micro/macro relation in systems of many interacting parts must be understood, it would seem, in some non-compositional way, or not at all.[11]

As EPA shows, however, a compositional theory of the micro/macro relation is quite feasible. Despite the interactions between parts, the composition rules need not be at all complicated. Why has it been thought otherwise? Because it has been assumed that the properties assigned to the individual elements of a system are intrinsic properties of the elements. There is nothing in the intrinsic properties to represent the effects of interactions between parts, therefore the effects of interactions must be taken account of entirely in the composition rules. Since the possibilities for interaction are extremely complicated, the composition rules are also extremely complicated.

If this line of reasoning is correct, then a compositional theory of complex systems must either be hopelessly complicated, or it must assign *extrinsic* properties to individual elements—extrinsic for the reason that they somehow already encompass the net effect of the interactions that an element will experience. Assigning extrinsic properties to the elements of a system does not, of course, guarantee the existence of a simple set of composition rules. But it makes it possible, and EPA shows that this possibility can be realized.

The success of EPA, I suggest, is due to the fact that enion probability is just the kind of extrinsic property that makes simple rules of composition possible. Call a theory that shares this advantage—a compositional theory that ascribes extrinsic properties to the parts so as to achieve simple rules of composition—an *introjective* theory. I remarked in section 1.24 that EPA has its power because enion probabilities are extrinsic properties that can be aggregated as though they were intrinsic properties. I might better have said: enion probabilities are extrinsic properties that can be aggregated in a system with interactions as though they were intrinsic properties in a system with no interactions.

This explanation of the success of EPA has a parallel in analytical mechanics (although the systems that can be feasibly treated by analytical mechanics are not in general complex in my sense). Unlike Newtonian, or vectorial, mechanics, analytical mechanics builds some of the effects of the interactions between parts of a system into the coordinate system with which the system is represented. That is, the coordinate system of an analytical representation implicitly constrains the system in the same way that explicitly represented forces constrain the system in the vectorial representation. If the right interactions are built into the coordinates in the right way, fewer interactions need to be taken into account explicitly, and problems become much easier to solve.

The most powerful mathematical techniques for representing and understanding the behavior of complex systems—hence some of the most powerful mathematical techniques in the higher-level sciences—will, I conjecture, be introjective theories. At this point in time, EPA may be the only introjective technique of great generality that we have. It is up to philosophers, mathematicians, and other scientists working on complex systems to give us more.

5.6 Are There Simple Probabilities?

To the best of our knowledge, the fundamental laws of nature are the laws of quantum mechanics. On the standard interpretation of quantum mechanics, these laws contain probabilities. Since the probabilities are at the fundamental level, they must be simple probabilities. Therefore not all probabilities are complex; whatever philosophical account of them one wishes to give, simple probabilities exist.

Yet—or so I will argue in this section—there is a good reason to think that quantum probabilities are not simple. If this is true then quantum mechanics cannot, after all, be a fundamental-level theory.

The argument that quantum mechanical probabilities are complex is as follows. We are confronted, at the quantum level, by a familiar phenomenon— the probabilistic patterning of events—that is explained, everywhere else it is found, by microconstancy. (At least, it is explained by microconstancy if it is explained at all.) We have a choice. Either we can assume that, when they turn up at the quantum level, the probabilistic patterns have the same explanation they do everywhere else, or we can posit an entirely new explanation. All things being equal, it is more reasonable to assume that the old explanation applies to the new case.

If there were no reason at all to believe that the laws of quantum mechanics were fundamental laws, then all things would indeed be equal, and the argument that quantum probabilities are complex probabilities would, I think, be very strong. It is not, of course, deductive, but it is founded on a robust inductive principle: similar phenomena ought to be given similar explanations.

But there is a good reason to think that quantum mechanics *is* fundamental, namely, the fact that nothing more fundamental has been discovered. What we have, then, are two conflicting inductive arguments. The first is an inductive argument that quantum mechanics is fundamental, from which it follows (more or less deductively) that quantum mechanical probabilities are simple probabilities. The second is an inductive argument that quantum mechanical probabilities are complex probabilities, from which it follows (more or less deductively) that quantum mechanics is not fundamental.[12]

Clearly, this conflict cannot, at this time, be resolved conclusively. But it is possible to ask which of the two arguments is the stronger. The first argument has the weight of long tradition behind it, and there is no need for me to praise it here. What I would like to do is to suggest that the results presented in this study give considerable weight to the second argument. I will not presume that I can convince the reader that the second argument overpowers the first; I will be happy to make a case that the second argument ought to be taken very seriously, and that the fundamental status of quantum mechanics and the existence of simple probabilities are therefore not simply to be taken for granted.

The second argument, I asserted above, derives its force from the following rule: posit similar explanations for similar phenomena. While this rule exerts a strong influence on the scientific mind—witness, for example, the many attempts from the early modern period on to find a single account of the nature of the different forces—it might not unreasonably be claimed to be more like a rule of thumb than an indisputable law of inductive logic. Certainly, the rule does not always lead to the truth. The planets go round the sun, and race cars go round a track, but the explanations of these two varieties of circular motion are rather different.

The argument that quantum probabilities are complex has at its service, however, another important rule of scientific procedure: do not explain complicated phenomena with simple dispositions.[13] (A simple disposition is a disposition that is a fundamental metaphysical building block, like a simple probability.) This rule, in conjunction with the other, provides a much stronger inductive argument than either would furnish alone.

Let me illustrate this claim with the help of an example. Suppose that we were to discover (excuse the fancy) that quarks are intelligent. They talk to one another, write poetry, and hold symposia on the nature of probability. How to explain this behavior? One possibility is to attribute the behavior to a simple disposition to behave intelligently. To adopt this strategy is to posit the existence of a metaphysically basic property P, possession of which causes quarks to behave in intelligent ways.

It would follow that there are two quite metaphysically distinct explanations for intelligence in our world. Some entities—namely, quarks—are intelligent because they possess P, and it is a fundamental law of nature that entities with P act intelligently. Call this *simple intelligence*. Other entities—for example, dogs, humans, and dolphins—are intelligent because their behavior is mediated by enormously complicated, yet-to-be-understood structures of a certain kind. Call this *complex intelligence*.

I take it that to react to the surprising eventuality of quark intelligence by positing the existence of "simple intelligence" is an obvious mistake. It is a mistake because it flouts the two procedural rules stated above:

1. Do not explain complicated phenomena with simple dispositions, and
2. Posit similar explanations for similar phenomena.

Rule (1) alone calls into serious question the decision to explain intelligence by means of a simple disposition. But what makes the decision seem ludicrous is the fact that intelligent behavior occurs elsewhere, and in all cases has a non-simple explanation. It is rule (2) that is behind this impression of ludicrousness.

I suggest that the postulation of two varieties of probability, simple and complex, is, though more plausible, not enormously more plausible than the postulation of simple and complex intelligence. Probability is too complicated a phenomenon to be explained by a simple disposition. It is, of course, much less complicated than intelligence. But it is much more complicated than the kinds of behavior normally attributed to simple dispositions, such as inverse square law forces.

Here the advocate of simple probability will, with the full backing of the philosophical tradition, protest that probabilities, or rather probabilistic processes, are in fact very simple things. But this view is, I think, based on a mistake. (Some of the views that follow were earlier advanced in Strevens (2000, §5).) It is a consequence of too close an identification of probabilism with the failure of determinism. The perpetrators of this mistake—from Lucretius to Paul Humphreys[14]—think of a probabilistic process as being a deterministic process with something taken away. The laws of nature and the boundary conditions dictate every detail of the unfolding of a deterministic process;[15] by contrast, according to these authors, such details are left undetermined by a probabilistic process. How these details turn out, then, is a matter of *chance*.

Chance in this sense is a "mere negative word," "literally nothing." It is an ongoing incontinence in Nature, a lack of control, the fact of the laws' failure to govern a certain domain of activity. As such it is surely a rather simple thing.

But this "chance," a metaphysical blank, is not the same thing as probability. That certain details of a process are unconstrained by the laws says much less about the process than does the imposition of a particular probability distribution over the same details. In particular, the absence of nomological constraint has no implications for the existence or otherwise of probabilistic patterns in the details.

To take a simple example, suppose that the laws of nature dictated that a tossed coin lands either heads or tails, but said no more. With what frequency would a tossed coin land heads? Or with what frequency would every prime-numbered toss land heads? The laws do not answer these questions. If, by contrast, the laws also stipulated that the probability of heads is, say, one third, then the answers follow immediately. A probability distribution, even just a single probability, provides an enormous amount of complicated information about the outcome of a probabilistic process; the probabilistic patterns are in this sense a complex phenomenon.

It follows, according to rule (1), that we ought to be wary of attributing the existence of the probabilistic patterns to a simple disposition. Given the existence of an alternative explanation of the patterns, rule (2) strongly suggests that, in any case, we ought to look to that alternative first. I conclude that there are strong reasons to seek to explain the probabilistic patterns observed in quantum mechanical phenomena by appealing to some kind of microconstant process operating at the sub-quantum level.

It is interesting, in this respect, to note that the probabilistic element of kinetic theory, according to my account in section 4.86, manifests itself as a series of momentary probabilistic interventions in an otherwise deterministic process, just as does, on the standard interpretation, the probabilistic element of quantum mechanics.

The argument against simple quantum probabilities, I would say, is at least as strong as the argument in favor. I do not expect all readers to agree. But suppose that I am right. Then, given all we know of the world so far, probability ought to be regarded as essentially a high-level phenomenon, moreover, a high-level *deterministic* phenomenon. The propensity theorists are exactly wrong; all probabilities are complex.

Notes

Glossary

References

Index

Notes

1. The Simple Behavior of Complex Systems

1. Strictly speaking, I should say that population levels remain stable provided that external factors such as the severity of the winters and so on remain constant in their effects.

2. How should one classify a behavior such as a stability in suicide rates? For an approach to this question, see website section 1.1A.

3. Almost all, not all, because a dynamic law describes a system's approach to equilibrium as well as its behavior once it gets there. A system might reliably approach a fixed point, but it might do so along trajectories that are irredeemably complex. That said, as far as we know, real systems seldom behave this way.

4. There remains considerable doubt as to whether chaotic behavior in complex systems can be attributed to simple dynamic laws. (In simple systems, by contrast, such as forced pendulums and certain electric circuits, the laws are already known to be simple.) For this reason, I do not include any examples in my survey of simple behavior. I will merely note that the mathematical techniques of chaos theory have been applied to systems as various as financial markets, ecosystems, and the weather. Perhaps I am being too cautious: given the fact that so many complex systems exhibit a transparently simple dynamics, it requires no great leap to suppose that others have simplicity of the hidden sort.

5. The word *enion* is my own invention, a combination of two Greek elements. *En-* is a variant of *hen*, the neuter form of the Greek word for "one" (related to Latin *unus*); *-ion* is derived from the verb "to go." An enion, then, is component of a system with its own motion.

6. Although, even in embryological development, certain processes appear to be accomplished without much coordination.

7. On this subject and other issues discussed in this section, see Porter (1986), Daston (1988), and Hacking (1990).

8. An exception to my comment about biology is the neutral theory of molecular evolution, in which probability is responsible for a constant rate of genetic change over time. Having descended to the level with which statistical physics is concerned,

363

biologists find it easier to admit an explanatory role for probability similar to the role that it plays in statistical physics.

9. Note that this characterization of complex behavior assumes some restriction on what counts as a variable and on the mathematical techniques for representing laws, for the same reason as does the definition of simple behavior given in section 1.12.

10. After Maxwell and Boltzmann, inquiry into the foundations of statistical physics followed several diverging paths (Sklar 1993). Most of these treatments do not conform to EPA (section 1.25). Although none has been entirely successful, I should note that the applicability of EPA does not preclude their use. My point here is that EPA can be, and has been, applied to the behavior of gases. It may yet turn out to provide the best treatment for many purposes (section 4.8).

11. Here and elsewhere I talk about independence as a relation between probabilities. This is not at all standard: it is normal to talk about independence as a relation between events. My reasons for flouting convention are as follows. First, independence is, in reality, clearly a relation between probabilities rather than events. Consider: two events, such as two coin tosses, may be independent relative to the relevant physical probabilities, but not independent relative to the subjective probability distribution of a person who subscribes to the gambler's fallacy. This has nothing to do with the events in question and everything to do with the probability distributions. Second, talking in the conventional way would involve constant circumlocution as I would be forced again and again to mention events so as to talk about their probabilities: rather than "enion probabilities must be independent," I would have to say "the outcomes attached to enion probabilities must be independent." This is annoying and long-winded, as well as conveying the misleading impression that a requirement is being imposed on events rather than on a joint probability distribution.

12. In principle, it is possible to aggregate the effects of enion behavior without making the independence assumption, for reasons explained at the end of section 4.44. In practice, independence is always assumed; without the independence assumption, none of the current scientific applications of EPA could proceed.

Even if some future use of EPA to explain simple behavior were not to invoke the independence of enion probabilities, it would still have to invoke, as explained in section 4.4, the stochastic independence of certain lower-level probabilities that I call the microdynamic probabilities. It is no distortion, then, to say that stochastic independence is a property of central importance both in the vindication of EPA and in the explanation of the simple behavior of complex systems. (The independence of microdynamic probabilities, I show in section 4.44, entails the independence of enion probabilities in most complex systems, thus this conjectured future use of EPA will, in any case, get enion probability independence "for free.")

13. I stress that the probabilities assumed to be independent here are probabilities conditional on all relevant macrolevel information. It is not required, then, that information about individual enions has no impact on enion probabilities, only that what impact such information has is entirely due to its impact on macrovariables. Or to put it another way, it is required that the only information in the microvariables that affects enion probabilities is information that is also contained in the macrovariables. This is a part of, but only a very small part of, the complete microlevel information.

14. For the exceptions, see section 4.6.

15. But see notes 12 and 13.

16. This is a rather loose formulation, and would need to be cleaned up considerably to play anything more than the expository role accorded it here. For example, it would have to be weakened so as not to require—as it apparently does in its current form—that the enion probability of an event *a* be unaffected by conditioning on the information that *a* occurred.

17. As I noted in section 1.23, the supercondition does not always have to hold completely in order to apply EPA: a small amount of microlevel dependence can be eliminated using other techniques, as discussed in section 4.6.

18. With the possible exception of E. T. Jaynes's epistemic approach (Jaynes 1983).

19. What, exactly, counts as *straightforward*? This is not a particularly important question in the context of this study, and in any case the answer will depend to some degree on the way that probabilities appear in fundamental physics. Let me offer, however, the following working definition of a simple probability, which assumes a standard collapse interpretation of quantum mechanics. A probability is simple just in case the process to which it is attached fits the following description: the process consists of a period of evolution that is deterministic at the quantum level (that is, evolution determined entirely by Schrödinger's equation or its relativistic equivalents), followed by a single measurement. The simple probabilities are then the probabilities assigned by the collapse postulate to the outcomes of the measurement. The preliminary period of deterministic evolution can be arbitrarily complex, as in the case of the Schrödinger's cat experiment.

20. A quasi-deterministic law is a probabilistic law in which all probabilities are very close to zero or one. It is often supposed that, as they apply to everyday objects, the probabilistic laws of quantum mechanics are quasi-deterministic.

21. It is unclear where "impure" complex probabilities, that is, probabilities that depend in part but not wholly on simple probabilities, ought to be placed in such a scheme of classification.

22. Various attempts have been made, in connection with the frequentist approach to the metaphysics of probability (von Mises 1957) or the formal definition of the notion of randomness (Fine 1973; Earman 1986), to offer a rigorous definition

of what counts as a probabilistic pattern. I need no more than an informal characterization.

23. Or at least, no reference to probabilities of the sort that were supposed to be defined. Subjectivism defines "physical" probabilities in terms of subjective probabilities.

24. Propensity theorists allow that real probabilities may exist where the statistical dispositions of fundamental physics "percolate up" to higher levels. These "percolated probabilities" might be simple (as is, I would say, the probability of Schrödinger's cat's death), but they might also be quite complex. The propensity literature is very vague on this matter. I presume, however, that many propensity theorists would count what I call simplex probabilities as real. The conclusion I draw in the main text applies to the propensity account's treatment of these probabilities, too: the propensity account makes no attempt to understand the physical processes that produce the outcomes to which the probabilities are attached. In particular, there is no account of how statistical dispositions percolate up or why this is important to our understanding of the systems in which they do so.

25. A subjective probability of an outcome is, intuitively, an individual's level of expectation that the outcome will occur. For two more sophisticated discussions of the notion of subjective probability, see Ramsey (1931) and Osherson, Shafir, and Smith (1994).

2. The Physics of Complex Probability

1. Not all metaphysical theories, however, claim to provide a unique description of their subject matter. The logical positivist approach to metaphysics, exemplified in the case of probability by Reichenbach (1949), is the great tradition in philosophy that spurns uniqueness. But then, the positivists saw themselves as anti-metaphysical.

2. In what follows, I take probability distributions to be functions of events, not of propositions.

3. This claim was asserted in section 1.24; it will be vindicated over the course of the next three chapters, but see especially section 2.21.

4. It is not so clear what might count as the mechanism pertaining to the probability of, say, a person's contracting tuberculosis over the course of a year, since there seems to be no physical limit to the facts that might affect the outcome. The person may, for example, come in contact with people who are visiting from other parts of the world. Are these far-flung regions part of the mechanism, or do they just contribute to the initial conditions?

   I discuss a policy for distinguishing mechanisms from initial conditions in the case of central importance to this study, that of enion probabilities in complex systems, in section 4.2. Is there a policy that applies to all cases? Perhaps not. But there is no reason to think that the notion of complex probability is useful only if there is an unambiguous criterion for individuating mechanisms. It may be that different criteria suit different problems. I therefore endorse a pragmatic, context-relative approach to the problem of distinguishing the mechanism to which a probability is attached. The same approach is appropriate for the more general task of individuating the parts of a probabilistic network (section 2.4).

There is one important requirement, however, that any approach to the individuation of mechanisms must satisfy: in a process governed by underlying deterministic laws, the mechanism together with the initial conditions must determine whether or not the designated outcome occurs. In a process governed by underlying probabilistic laws, the mechanism and initial conditions must determine a definite nomic probability that the designated outcome occurs; for further discussion of this case, see section 2.6.

5. An IC-variable is a "variable" in the same sense as a random variable. To be precise, an IC-variable is not a variable in the mathematical sense, but is rather a function from certain physical quantities to a set of numbers. (Note that I take the function's domain to be properties of the real world, not elements of a model of the real world.) This introduces a notational problem familiar from probability theory: sometimes it is necessary to introduce a variable (in the mathematical sense) that takes on possible values of the IC-variable. In more informal contexts I use the symbol for the IC-variable itself, always a Greek letter, for this purpose. In more formal contexts I use a lowercase italic letter.

6. Technically, the required condition on $h_e(\zeta)$ is that it be a Borel function, or equivalently, that the set of all e-values of ζ be a Borel set. Observe that $h_e(\zeta)$ has the structure of what mathematicians call an indicator function for e. That cprob(e) is well defined follows from the mathematical fact that the indicator function for a Borel set is a random variable.

The value of cprob(e) in no way depends on the fact that I use the standard measure in the quantification of complex probability. The standard measure only appears as a middle term, as it were. I have used the standard measure to represent the IC-variable distribution as a density $q(\zeta)$, so I must use the same measure to unpack the distribution for the purpose of extracting the probability of an e-value. The standard measure simply provides a way of encoding probabilistic information in a density function. Any other encoding system would work as well, provided that the same system is used both to encode and decode the probabilities.

It follows that no gerrymandered or grue-like redescription of a probabilistic setup can alter the probability of a designated outcome.

7. Provided, that is, that the wheel is set up properly. Bass (1985) describes the attempts of a group of scientists to exploit systematic biases that change from roulette wheel to roulette wheel, caused by the wheels' being less than perfectly level. Knowledge of the biases was not enough, however; the scientists also needed

concealed electronic equipment to observe and analyze information about the initial conditions of each spin.

8. U should be a Borel set. That the set of e-values in U is measurable then follows from the earlier requirement that the evolution function is a Borel function.

9. Strictly speaking, of course, the probability will not be absolutely well-defined: there will be an element of fuzziness that comes from the fact that the equality between microconstant probability and strike ratio is only approximate.

10. I say *more or less* because even relatively high level information about the IC-densities of microvariables in EPA—I have in mind in particular, of course, the information that they are macroperiodic—is information about microvariables that exceeds the information in a macrostate. Thus the demand that enion probabilities be independent of the values of microvariables has not, strictly speaking, been met. But suppose that it is true that the IC-densities of complex systems are almost always macroperiodic. (This assumption will later be made explicit.) Then the fact of macroperiodicity may be considered a part of the background in all reasoning about complex systems, rather than something that needs to be built in as an explicit premise to particular arguments. When reasoning about macrostates, then, we will conditionalize not only on the information conveyed by the macrovariables, but also on all relevant background information, hence on the fact of initial condition macroperiodicity. This endows enion probabilities with all the microlevel independence needed to explain EPA.

11. To explain the very special short-term disorder characteristic of the Bernoulli and other probabilistic patterns, it is necessary to examine the properties of the joint densities that give the distributions over the sets of IC-values for entire sequences of trials. I do so in sections 3.4 and 3.5; the comments here should be regarded as merely preliminary.

12. Here I assume, of course, that IC-variables with macroperiodic distributions will tend to give rise to sets of IC-values that are themselves macroperiodically distributed.

13. Advocates of the subjectivist interpretation of probability have long appreciated this point.

14. The sets cannot be made micro-sized for very low values of ω and τ. For further comments, see the very end of this section (p. 69).

15. This is trivially true given the form of the generalization; just let $j = k$ in the formal definitions in section 2.C2.

16. Formally, the notion of a strike set is defined first (see definition 2.13), and an IC-variable is said to be eliminable in favor of a set just in case the set is a strike set and does not contain the variable. The notion of eliminability is provided for expository purposes.

17. For an actual frequentist, of course, such a distribution *is* a probability distribution. I do not wish to take sides here for or against frequentism.

18. It is not quite right to say that the basic IC-values of the traffic network have a non-probabilistic source; they arrive from other traffic networks, networks that are just as probabilistic, presumably, as the network under consideration. This illustrates the fact that the traffic network is really a fragment of a larger network. In practice, science tends to deal with network fragments rather than with complete networks.

19. The work is done by the periodic cosine function, which is tuned to oscillate as the evolution function oscillates, so that τ^* races through the times that yield a red outcome but crawls through the times that yield a black outcome. Division by π ensures that the periodic part of the function never decreases faster than τ^* is increasing, thus guaranteeing that the relation between τ and τ^* is one to one.

20. As I remarked a few paragraphs back, cprob(e) is the same no matter how the IC-variables are measured; see note 6.

21. Even more specifically, a macroperiodic joint density over sequences of perturbations is required. This is still not quite enough for the arguments sketched in the main text to go through, but it is close enough, given the relatively low level of precision I aspire to in this section.

22. If time is racked in the same way as distance, then mv^* is conserved and IC-values spend equal amounts of $time^*$ in stretched and squashed regions. But the argument for macroperiodicity does not go through, for with this new measure of time it is no longer the case that values are equally likely to stop at any moment. The same problem undercuts any other variant of this strategy. (Note also that the conservation of mv^* is less interesting than it sounds—a moment's thought shows that, with distance and time racked in the same way, v^* is none other than standard velocity and mv^* standard momentum.)

23. This is not quite true. As with any complex probabilistic experiment, if the IC-values for such an experiment are probabilistically patterned, the outcomes will, in most cases, be probabilistically patterned. But microconstancy does not explain this fact; what does the explaining is the pattern in the initial conditions.

24. The properties of microconstancy discussed in section 2.2 can also be understood by way of the notion of a critical IC-value; see example 3.5 of section 3.B4. The ideal formal treatment of the material presented in this book would, I think, begin with the definition of a critical IC-value, and would use the notion to explain and prove every major result.

25. Thus the number 19, say, would not be an effective IC-value at all, as it does not pick out any of the stated intervals; in particular, it does not pick out either [18, 19), which is picked out by 18, or [19, 21), which is picked out by 20. I will shortly introduce a slightly different scheme for individuating effective IC-values in the microconstant case, on which [19, 20) and [20, 21) correspond to distinct effective IC-values, and on which 19 will count, according to the present numbering scheme, as an effective IC-value, picking out [19, 20).

26. The function $\hat{q}(z)$ is a "discrete density" in the sense that, for any set of values V of ζ^e, the probability of a value of ζ^e falling into V is $\sum_{z \in V} \hat{q}(z) M(z)$.

27. Use the following relations involving the discrete density. The complex probability of an event e is $\sum_{\zeta^e \in V} h(\zeta^e) \hat{q}(\zeta^e) M(\zeta^e)$, and the strike ratio for e over a member U of a constant ratio partition for e is $\sum_{\zeta^e \in U} h(\zeta^e) M(\zeta^e)$ (using the set inclusion symbol rather loosely!).

28. One other use for the notion of a discrete density is to provide an alternative treatment of frequency-based ic-densities. My earlier treatment (section 2.33) required a "smearing" operation to transform statistical information into a continuous density. A more principled approach to the problem is to encode the information as a discrete density. No smearing is then required. The drawback is that a discrete density is always relative to a partition, so there is no such thing as *the* discrete density corresponding to a body of statistical information.

29. The facts about the distribution of effective information, note, do not determine whether an experiment's full ic-variable has a macroperiodic density, only whether the effective ic-variable has a macroperiodic density, in the sense defined in section 2.72.

30. Because the rolling ball experiment is microconstant, I use the microconstant scheme for assigning effective ic-values, according to which intervals corresponding to effective ic-values must fall entirely within an optimal constant ratio partition member.

31. This scheme will work only when the critical ic-value is the highest level part of the effective ic-value, a possibility mentioned below.

32. In some other examples, however, Poincaré comes closer to considering initial conditions and evolution functions.

33. For experiments with complicated evolution functions, it might be useful to go even further, and to divide the experiment's domain into curves (or the higher-dimensional equivalents), but this will introduce measure-theoretic difficulties that I do not want to discuss here.

3. The Independence of Complex Probabilities

1. The reason for the hedge is to exclude from consideration "cheating" selection rules that pick pairs of trials by looking at, say, their ic-values or their outcome.

2. It must not be a "cheating" selection rule; see note 1.

3. This is not strictly true, since I allowed in section 2.43 that basic ic-values may have a non-probabilistic causal history. I will simply stipulate that basic ic-values do not have a causal history of the sort that might interfere with independence. This leaves open the question as to what kind of history might or might not interfere with independence, but as the reader will soon see, it is precisely this sort of question that I want, eventually, to raise.

4. If the ancestries of two outcomes overlap, then they overlap with respect to a basic IC-value, thus two ancestries overlap if and only if they overlap with respect to a basic IC-value.

5. "Cheating" selections, as ever, do not count.

6. This is true, at least, in Newtonian gravitational theory.

7. The exact independence requirement is actually somewhat stronger than this, as will shortly become clear.

8. More exactly, the critical level information about the IC-values for one experiment must be both (a) independent of the critical information about the IC-values for the other, and (b) independent of the non-critical, effective information about both sets of IC-values. Clauses (a) and (b) here are weaker versions of clauses (a) and (b) in the definition of infracritical independence.

9. The reason is as follows. Let f be the simple collision transformation, which maps the two-dimensional space $\zeta_1 \times \zeta_2$ to itself. Then $h'(x) = h(f(x))$. So the point x in the collision outcome map has the same value (zero or one) as the point $f(x)$ in the non-colliding outcome map. The no-collision outcome map may therefore be transformed into the collision outcome map by taking each point $f(x)$ to x or, in other words, by taking x in the no-collision map to $f^{-1}(x)$ in the collision map. In this sense, the collision outcome map is obtained by applying the inverse of the simple collision transformation to the no-collision outcome map.

   Note that, in this formulation, $f^{-1}(x)$ is not a function; it is just the set $\{y \mid f(y) = x\}$. The inverse transformation copies the value of the outcome map at x—that is, $h(x)$—to all the points y that map to x. For example, if a transformation maps both y and z to x, its inverse, on the current understanding, maps point x on the outcome map (and thus the color of x, gray or white) to both y and z. These comments on the nature of the inverse become important when I investigate non-linear collision transformations in section 3.65.

10. Why, if this is a graph of a composite evolution function, does it not have the canonical structure illustrated in figure 3.3? What is seen in figure 3.5 is not the graph of the full composite evolution function, which is a four-dimensional structure representing a function in which ζ_1 and ζ_2 each appear twice as arguments (see section 3.2). It is a two-dimensional cross-section of this structure, representing all physically possible combinations of IC-values for the full composite evolution function. For a discussion of the same point using a simpler example, see note 27.

11. The situation is in fact a little more complicated than this. Rather than a side-on viewing, the given values yield an outcome map that is all white. This is because the transformation of the outcome map is the inverse of the transformation of the variables (see note 9). The inverse of the simple collision transformation for which a and b are both 1 maps the line $\zeta_1 = -\zeta_2$, which is all white, to the entire area of the map, so that the map is all white. (Here I make some additional assumptions

about the significance of negative speed.) Thus what is obtained is not a side-on viewing of the chess board, although it corresponds, in a certain sense, to the side-on case. A side-on viewing would yield an outcome map that had no physical interpretation.

12. But in certain circumstances, even absolute dependence may not destroy the stochastic independence of outcomes. For more details, see website section 3.6A.

13. A non-invertible linear transformation is, as the name suggests, a transformation that has no inverse. Non-invertible transformations are sometimes called singular transformations.

14. The determinant of the generic linear transformation above is $ad - bc$.

15. By the joint ic-density for the tosses, I mean the ic-density for spin speed relative to a selection that picks out pairs of tosses that result in collisions.

16. Here I assume that Y includes only points that have inverse images in X. Given that X includes only physically possible sets of ic-values, this is equivalent to the assumption that the outcome map includes only physically possible sets of transformed ic-values.

17. Here, as elsewhere, worm-shaped regions do not count as microregions; see p. 56.

18. Two complications in characterizing the g_is should be pointed out. First, it may be that some of the g_is are not defined over all of R. This is allowed, but I note that the action of such a g_i on R cannot be approximated by a linear function over R. Thus where such g_is exist, the outcome transformation will not count as microlinear in my sense.

    Second, if the only requirement on breaking up the inverse image of R is that the resulting R_is be contiguous, it is possible that the ic-variable transformation maps some of the R_is onto R in a way that is not one-to-one. In such a case, each set R_i must be further broken up into maximal sets that are each mapped one-to-one to R. Then the one-to-one g_is are guaranteed to exist. Readers interested in a formal handling of complications such as these should consult section 3.B6.

19. If a transformation f is close to another transformation that induces absolute dependence of the ic-variables, the inverse of f has the same property (although it is close to a different absolute-dependence inducing transformation). Thus, were condition (3) on outcome transformations to be explicitly retained, it would appear in the same form as a condition on ic-variable transformations.

20. The stated conditions ensure independence, but it should be noted that they do not guarantee that the probability of a given outcome will remain the same from trial to trial. If, for example, the velocities marked on an autophagous wheel satisfying the conditions proportionally favored those that give rise to a red outcome, the probability of obtaining a red outcome on the second and all subsequent trials in a chain would be higher than the probability of red on the very first trial. The stated conditions ensure, then, that the probability of an outcome of a chained trial is independent of any other outcome, but not that the probability is independent of

the position of the trial within the chain. In section 3.76, I give conditions sufficient for independence from position as well as from previous outcomes.

21. I have omitted one further requirement: the information at each level must be *well tempered*; see definition 3.15. Without this constraint, the requirement that the form of the function f_i be independent of the previous $i - 1$ outcomes has no real content.

22. I will later (section 3.77) consider chains of experiments in which the outcome of one trial determines the experiment on which the next trial is conducted. In such a chain, the IC-evolution function for the ith trial will depend on the outcomes of some or all of the previous $i - 1$ trials; as the reader might expect, I count this as a case in which IC-evolution is not outcome independent.

23. In this formulation, b is the difference between the minimum and maximum speeds marked in each section of the wheel, c is the minimum speed marked in each section, and a is the number of sections traversed per unit of speed multiplied by b. For the straight wheel, $b = 100$, $c = 0$, and $a = 100$ (since the wheel traverses exactly one section per unit of speed).

24. Strictly speaking, this is not a fold, because it is a discontinuous mapping. I am therefore using a rather loose notion of folding here and in what follows.

25. The usual example of a stretch-and-fold transformation is the so-called baker's transformation, which is very similar to $H(\zeta)$. Such transformations have often been mentioned in attempts to justify statistical mechanics by way of the notion of a deterministic Bernoulli system. Sklar (1993) explains the notion for philosophers. A related idea is briefly developed in Suppes (1987). The similarities and differences between the approach to statistical mechanics referred to in this note, on the one hand, and the approach I take to complex systems in this study, on the other, are discussed in sections 1.25 and 4.87.

26. But of course, in any system that has an inflationary IC-evolution function and whose variables are bounded, some sort of folding will occur eventually, at the point where the variables reach their physical limits.

27. A word of explanation: in section 3.2, a composite evolution function for two experiments, each with one IC-variable, was defined so as to have two IC-variables. Strictly speaking, a composite evolution function for two multi-spin experiments does have two IC-variables, but since the space of possible IC-values is confined to points where the values of the two IC-variables are equal—in effect, a one-dimensional cross section of the two-dimensional evolution function—it is convenient to represent the evolution function for the composite experiment as having just one IC-variable, representing the initial IC-value of the relevant chain. (I did the same thing with the colliding coins in section 3.6; see note 10.)

The composite evolution function $h(\zeta)$ for red outcomes on both the m and $m + 1$ spin experiments, then, is just the function

$$h(\zeta) = \begin{cases} 1, & \text{if } h_m(\zeta) = 1 \text{ and } h_{m+1}(\zeta) = 1; \\ 0, & \text{otherwise.} \end{cases}$$

where $h_m(\zeta)$ and $h_{m+1}(\zeta)$ are the evolution functions for red on m and $m+1$ spin experiments.

28. The zigzag wheel in fact satisfies the formal version of the sufficient conditions for independence stated in section 3.72, due to certain liberalities in my formal definition of low-level information. The wheel's simplicity, however, makes it ideal for discussing outcome-dependent ic-evolution, for which reason I will talk, in the main text, as though a broader treatment is needed to explain independence in the zigzag wheel.

29. It is not necessary, as it was in section 3.72, to require that the low-level information is well tempered.

30. I am assuming, then, that effective ic-values correspond to contiguous regions of full ic-values, as specified in the formal definition of an effective ic-value (section 2.71). On an alternative definition suggested in the discussion of the teleological scheme for assigning critical ic-values (section 2.73), effective ic-values would sometimes span larger, non-contiguous regions. If this definition were adopted, the inflationary requirement for independence stated in the next section would be easier to meet, but meeting the microlinearity requirement stated in the same section would be more difficult.

31. It is not the case, note, that the restricted ic-evolution functions for the zigzag wheel produce the same distribution regardless of the distribution on which they operate. It is important, for independence, that they are working on a uniform distribution. But given even marginally suitable input, they tend themselves to create the distribution they need, by inflation. That is, they inflate small segments of the ic-value distribution, which tend to be roughly uniform, so that those segments, stretched, become the entire (conditional) distribution for later trials. Or equivalently, they inflate low-level information, which tends to be uniformly distributed, to the level at which it determines the effective ic-value for the next trial.

32. That an ic-evolution function satisfies the weaker of the two microlinearity conditions does not entail that it is strongly, rather than weakly, inflationary. But even when ic-evolution is not, as I implicitly assume in the main text, strongly inflationary, weak inflation will anyway tend to smooth out imperfections in the relevant distributions, for somewhat complicated reasons that I will not go into here.

33. The relevant sense of macroperiodicity requires that the restricted ic-densities obtained by assigning arbitrary values to the ic-variables not in the strike set are all macroperiodic. See sections 2.B and 2.C2.

34. In the pseudorandom number generator, the arithmetic is all done with integers, whereas in a world where space is discrete, it is done with real numbers with limited decimal precision. There is an important difference between the two arithmetics: in fixed precision arithmetic, unlike integer arithmetic, information is lost, as it were, at both ends: folding eliminates very high level information while the fixedness of decimal precision eliminates very low level information. In integer arithmetic, by contrast, the linear congruential function and its non-linear generalizations involve loss only at the high level, due to folding. (An early pseudorandom number generator, the middle square method (Knuth 1998, 3–4), uses integer arithmetic in a way that involves loss at both ends.) I ignore this difference in the comments that follow.

35. As was the cosine distribution over relative impact angle; see section 4.85.

36. Both results require that the critical information be well tempered, and so neither result will go through on an ordinal scheme for assigning critical information (section 2.73).

37. This is true on the ordinal scheme for assigning critical ic-values, but not the teleological scheme (see section 2.73) . On the teleological scheme, the lower-level partition is a coarser partition in which members of the outcome partition belonging to the same optimal constant ratio partition member and mapping to the same outcome (that is, all mapping to gray or all mapping to white) are joined into a single set.

4. The Simple Behavior of Complex Systems Explained

1. Background variables that appear in macrolevel laws may also be counted as macrovariables. Thus, background variables that appear in both the microlevel and the macrolevel laws count as microvariables *and* macrovariables.

2. If the number of enions in a system can change as the system evolves (as in an ecosystem), different parts of the state space have different numbers of dimensions. The laws of microdynamics will determine when a system jumps from part to part of a composite space.

3. This may seem to be an impossible task. There are some microregions in state space that are obviously more dangerous for a rabbit than others. Such cases are discussed in section 4.43 below.

4. Of course the microstate plays *some* role in determining the single microstate probability: it determines the macrostate (the number of foxes and so on). But all microstates under consideration determine the same macrostate, the m in $cprob(e_m)$, so this residual role for the microstate cannot make one single microstate probability differ from another.

5. This does not mean that history is irrelevant, only that, in order to affect future evolution, an aspect of a system's history must be explicitly represented in a microvariable.

6. Except, obviously, "microvariables" that have the same value for every microstate in a macrostate.

7. Mathematically, the "washing out" of the initial microstate can be seen as a consequence of the ergodic theorem for constant Markov processes (see any book on random processes, for example, Grimmett and Stirzaker 1992). Conditions A and C entail that the microdynamics of the system effect a constant Markov process. Condition B entails a certain further condition required for the application of the ergodic theorem, namely, that either the microstates of state space form an irreducible set (no fences), or the system starts out in or is destined for a single irreducible set (no straddling). The other conditions required for the application of the ergodic theorem (non-null persistence and aperiodicity of the irreducible set) seem obviously to be fulfilled by all the (bounded) complex systems considered in this study.

   Yet in its sanguine temper, this note perhaps underestimates the probabilistic complexity of microdynamics. It is not straightforward to interpret microdynamics, directed as it is by multiple simultaneous probabilistic experiments as well as by a possibly quite significant deterministic element, as a single Markov process. Let me remark on two potential problems.

   First, how should the creation of a new enion be represented? In an ecosystem, for example, how to represent the birth of a new fox? It is best to assume, I think, that such events occur only at the end of enion probabilistic trials, so that the question does not arise. Because this is a reasonable assumption only if trials are relatively short, the assumption puts a constraint on the time allowed for washing out.

   Second, is it really true that the dynamics of any given complex system over the course of an enion probabilistic trial is aperiodic, for example, that there are no cyclic tendencies built into the probabilistic microdynamics of an ecosystem? Again, by assuming that the duration of an enion probabilistic trial is not too long, some major sources of worry can be removed, for example, our knowledge that certain natural populations do cycle in the long term. But even in the short term, it is not entirely clear, given the complexity of microdynamics, how a general argument for aperiodicity might be constructed.

   There is much more to be said about the conditions for washing out, then, but I do not intend to say it here. This is, in part, because the issues are fairly well understood and the subject of some rather sophisticated mathematics in the theory of random processes, and in part because a proper consideration of these issues would require a descent into specific facts about specific systems.

   What, then, is left of my claim that the ergodic theorem underwrites a general washing-out result? Properly understood, the claim stands. There is a general tendency for Markovian microdynamics to wash out the importance of a system's initial microstate, and the ergodic theorem, and other such results, uncover the deep mathematical reasons for this tendency. This does not constitute a proof that there must be washing out in every case, but it does provide the basis for a broad understanding of the phenomenon of washing out, and so for an understanding of the steady-state microconstancy of enion probabilities.

8. As noted in section 1.25, the approach in this section can be compared to the various attempts to found the probabilities of statistical physics in modern ergodic theory (Sklar 1993, chaps. 6 and 7). I appeal to some of the same kinds of properties of systems as do the ergodic theorists; however, my philosophical scaffolding is rather different. See section 4.87 for some further comments.

9. Although conditions A, B, and D can be satisfied by choosing the right descriptive apparatus, the necessary apparatus may be so cumbersome as to be effectively useless. An example was given in the discussion of the impassable wall above (condition B). Thus the necessity of satisfying A, B, and D does, for practical purposes, put some constraints on the systems to which EPA may be applied.

10. The answers: In the case of fox proximity, D fails, in the sense that the rabbit may die before there is time for the influence of the initial microstate to be washed out. In the case of poor health, B fails, for reasons to be discussed below.

11. To see this: Let $r_1, r_2, \ldots$ be the trajectories leading to rabbit death and $f_1, f_2, \ldots$ the trajectories leading to fox death. Then the probability of both the rabbit and the fox dying is $P(r_1 f_1) + P(r_1 f_2) + \cdots$. The product of the individual probabilities of rabbit and fox death is

$$(P(r_1) + P(r_2) + \cdots)(P(f_1) + P(f_2) + \cdots)$$
$$= P(r_1)P(f_1) + P(r_1)P(f_2) + \cdots$$

So independence holds if $P(r_1 f_1) = P(r_1)P(f_1)$ and so on.

12. There will likely be a large probability that the conservation law is violated, but also a large probability that it holds *approximately* true.

13. This technique, note, in no way solves the problem raised by rabbit health in section 4.43, since it, too, depends on condition B, the no-fences assumption.

14. The argument is not so simple when X and Y are not microconstant. For the non-microconstant case, see website section 4.5A.

15. The assumption will not hold if the macrolevel laws are sensitive to initial conditions, that is, sensitive to small differences in the values of the macrovariables figuring in the laws.

16. There are other conditions under which an average probability can be used successfully. If the population structure is varying cyclically, for example, an averaging approach will work for longer-term predictions (but not short-term predictions) provided that the average takes into account the cyclical variations.

In effect, in the rabbit case, a one-month probability of survival would be back-formed from the average probability of survival over an entire cycle. This sort of case is important because in many respects the structure of biological populations varies seasonally.

17. This is a considerable simplification of the actual system, but one that is often made.

18. Despite its name, the mean field approach is not, strictly speaking, what I have called an averaging strategy, because it keeps track of the relevant (in this case very simple) population structure.

19. Because each atom is the neighbor of more than one other atom, the subpopulations will overlap; the structured population approach allows this.

20. I have not mentioned in the main text a very important feature of the renormalization strategy: systems that are quite different at the microlevel may have derived models that are almost identical (or that *are* identical if one goes all the way to the limit). In this way the renormalization theory explains the fact that many different substances obey the same laws when they undergo phase transitions; also explained is the fact that the macrolevel laws derived by way of renormalization do not mention even the small number of subpopulations in the derived model, and so are even simpler than the description in the main text would lead the reader to expect. For a recent philosophical discussion of renormalization in statistical physics, see Batterman (2002).

21. Or sometimes, probabilities are not mentioned, but probabilistic patterns are assumed, in the guise of an assumption about the frequencies of different kinds of collisions, what Rudolf Clausius called the *Stosszahlansatz*.

22. Because the identity of M's partner in the $(i + 1)^{\text{th}}$ collision is in part determined by the ith collision, there is not one partner but a number of potential partners. The positions and velocities of all the potential partners are ic-variables of the experiment I am describing. This issue is discussed further, and then rendered irrelevant, in section 4.84.

23. Of course, the positions and velocities of the molecules involved in a collision *determine* the relative angle of impact for the collision. Thus positions, velocities, and the relative angle of impact cannot be an independent set of ic-variables for the experiment. It is important, however, that the relative angle of impact be an ic-variable, so what I am really suggesting is not that the positions and velocities are ic-variables for the experiment, but that certain parameters are ic-variables that, together with the relative angle of impact, fully determine the positions and velocities. These parameters might include, for example, the magnitude of the relative velocity of the molecules and the absolute velocity and position of their center of mass. In the main text, I continue to speak as though it is the positions and velocities that are ic-variables; this makes for a more accessible exposition.

24. This will not be true in an extremely dense gas, in which a rebounding molecule is almost certain to hit one of its closest neighbors. Indeed, none of the claims that follow applies to such a gas, which might have to be approached either by finding some set of non-microconstant microdynamic probabilities, or by treating the gas in the same way as a solid (see section 4.88).

25. How small the members of the partition need to be to count as micro-sized is, of course, dependent on the shape of the relevant ic-density. The nature of the density is to be discussed in section 4.85, thus, a full justification of the claim that the collision partition is micro-sized must wait until that section.

26. A technical way of stating this property is as follows. Define the outcome map for a member U of the collision partition as the outcome map for just those ic-values that fall into U. I claim that the relation between the outcome maps for any two members of the collision partition is captured by a linear, hence strike ratio–preserving, transformation. The claim is true, note, only because the variable with respect to which it is formulated, relative impact angle, has been carefully chosen.

27. One might dispute the claim, arguing that it cannot be true that the strike ratio for any impact angle is the same in any member of the collision partition, for the following reason: some members of the collision partition will correspond to situations in which the molecule that is being hit is partially occluded. If one side of the molecule is occluded, only impact angles corresponding to hits on the other side are possible outcomes of the collision; impact angles corresponding to hits on the occluded side must have a strike ratio of zero in such a partition member.

Given that the diameters of molecules are small relative to the distances between them, not many members of the collision partition will correspond to hits on occluded molecules. For this reason alone, they might be ignored. A more elegant solution to the problem is obtained by appealing to the coarseness of grain of Boltzmann-level microstates; on this solution, occlusion turns out not, after all, to stand in the way of the collision partition's being a constant ratio partition. For the details, see note 42.

28. As I observed in note 23, it is not, strictly speaking, position and velocity that are the ic-variables and so are assigned fixed values, but related quantities that are independent of (in the mathematical sense, that is, not determined by) the relative angle of impact.

29. Except the very first collision, which may have a different macroperiodic distribution over its impact angle.

30. In the context of the debate about the source of the apparently temporally asymmetric character of statistical mechanical dynamics, this assumption is a philosophically very interesting one. While no one denies that some assumption of this sort about the initial conditions of the systems treated by statistical mechanics is in fact true, there is considerable controversy as to why it is true. For further discussion see Reichenbach (1956); Sklar (1993); Price (1996).

31. An exception: no matter how coarsely impact angles are grained, if the molecule involved in the $(i + 1)^{th}$ collision is sufficiently close to the site of the ith collision, some impact angles for the $(i + 1)^{th}$ collision will be unreachable. The reader will note that in the discussion of microlinearity, too, I assume that successive collisions occur at a reasonable distance. But this assumption will not always hold, for which reason impact angles will not always be entirely independent of one another. Fortunately, these brief and relatively rare departures from independence will not compromise, in any significant way, the washing out of microlevel information necessary for the microconstancy and independence of the Maxwell-Boltzmann probabilities, except for a very dense gas (see note 24).

32. This assumption does not impose a genuine limitation on the scope of the demonstration, for the reason given in note 35.

33. The mean free path for nitrogen at room temperature is about 200 times its effective diameter.

34. For mathematical purposes I am measuring angles in radians, then, though for expository reasons, the accompanying commentary uses degrees.

35. Let me now explain how to generalize this result to the case where the rest frame of reference for M's partner in the $(i + 1)^{th}$ collision, that is, N, is not also the rest frame for the center of mass of M and its partner in the ith collision. Take as the frame of reference the rest frame for N. Because the center of mass of the molecules involved in the ith collision is moving in this frame, it is no longer true that the angles θ and α in figure 4.8 stand in the simple linear relationship $\alpha = 2\theta$. Rather, they will stand in some more complicated, non-linear relationship $\alpha = f(\theta)$.

   Now, because the multiplying factor x/d in the ic-evolution function is very large—because ic-evolution is quite inflationary—only a small range of values of the impact angle for the ith collision, that is, of θ, will result in M's colliding with N. For the purpose of assessing microlinearity, we are interested in the behavior of $f(\theta)$ only over this very small range (less, perhaps much less, than two degrees). But over such a small range, $f(\theta)$ is very close to being linear. (The result is not surprising, so I spare the reader the calculations. It is also important, I note, that $f(\theta)$ is at most slightly deflationary.) Thus, for the purpose of assessing the microlinearity of ic-evolution, $f(\theta)$ may be regarded as linear, from which it follows that ic-evolution in the general case will be of the same form as that shown in figure 4.9; it will have the form of the arc sine function.

36. Readers of the notes know why; that is, they understand how the derivation of the cosine density can be generalized to cases in which N is moving relative to the center of mass of M and its partner in the ith collision. As shown in note 35, ic-evolution has the arc sine form in the general case. Thus, the conditional distribution over impact angle in the general case has the cosine density.

   This conclusion tells us something about the evolution function for the corresponding microdynamic experiment, that is, the experiment spanning the time between the ith and $(i + 1)^{\text{th}}$ collisions. I claimed in section 4.84 that, within any member of the collision partition for the microdynamic experiment, that is, within any contiguous set of IC-values leading to a collision with the same molecule, the strike ratio for any designated outcome is the same. The designated outcome is, of course, a Boltzmann-level impact angle for the next collision, say, the event of the impact angle falling in the interval I. Now, the strike ratio for this event, over a given collision partition member U, is equal to what the probability of the event would be, given a uniform distribution over U (and assuming that the distribution over U sums to one). What I have just shown, in the main text and in the first part of this note, is that for any collision partition member, this probability is the same, namely, $\frac{1}{2} \int_I \cos \varphi \, d\varphi$. Thus, for any collision partition member, the strike ratio for the outcome is the same, as promised.

An even stronger result can be stated. The discussion in the main text assumes that all relevant microvariables except the impact angle for the ith collision have fixed values. Even given this assumption, the strike ratio for any designated outcome is the same. It follows that all other IC-variables of a microdynamic experiment can be eliminated in favor of the impact angle, as claimed in section 4.84.

37. Strictly speaking, the question is whether each segment of the cosine density corresponding to a single coarse-grained impact angle is macroperiodic, since the IC-density of interest in the $(i + 1)^{\text{th}}$ trial is the density conditional on the outcome of the ith trial, that is, conditional on the impact angle with which the $(i + 1)^{\text{th}}$ trial begins. But this question can be addressed just as well by thinking about the macroperiodicity of the density as a whole, as I do in what follows.

38. This is the number for a gas in three dimensions.

39. Closer attention to the mathematical details vindicates this line of thought.

40. The reader might wonder if I am cheating here. I have supposedly eliminated all microdynamic IC-variables except the impact angle, but when a microdynamic IC-variable is eliminated, the effective IC-value for a trial fixes (rather unintuitively) a definite value for that variable. (This is explained in sections 2.74 and 2.B.) For the purpose of determining microdynamic probabilities, then, N's position is fixed exactly, not merely at the Boltzmann level.

I am indeed cheating. A more honest approach would "uneliminate" at least one other IC-variable, say, the position IC-variable of M. The positions of all of M's potential collision partners remain eliminated. Thus, the strike set for a microdynamic experiment corresponding to M's colliding with another molecule contains the impact angle of the collision and the position of M at the time of the collision. This introduces the necessary slack into the effective IC-variable. I ought now to go back and show that the IC-evolution of position is inflationary and microlinear; I will not, but the arguments presented earlier in this section show well enough how such a demonstration would proceed.

41. More exactly, it is the weighted average. Because, as I will later explicitly assume, the probability distribution over position is macroperiodic with respect to the partition of position space induced by the Boltzmann coarse-graining, the distribution over position within any particular Boltzmann-level microstate is uniform. Thus the weighted average is just the average.

42. The same considerations will dissolve the occlusion problem (described in note 27). A specification of the positions and velocities of molecules at the Boltzmann level will not determine which part of an occluded molecule is inaccessible, thus all parts will be accessible.

43. Maxwell and Boltzmann's work is, though, hardly the last word on the subject. Research continues on the problem of deriving probability distributions over molecular position and velocity from molecular chaos in a rigorous way, especially in non-equilibrium situations; see, for example, Kac (1959).

44. For many systems treated by the ergodicists, this hope is not quite realized. What is shown instead is that the system's evolution is equivalent to a *semi-Markov* chain (Ornstein and Weiss 1991, 38). In a semi-Markov chain, the probability distribution over the $(i + 1)^{\text{th}}$ state is independent of the ith state, in contrast to a Markov chain, but the time taken for the transition to the $(i + 1)^{\text{th}}$ state depends on the ith state.

45. By partitioning time into durations equal to or greater than the length of a cycle, the ergodic approach might explain some aspect of the behavior of a cycling system, but not the fact of its cycling.

46. As the example of rabbit health in section 4.43 shows, this condition does not require that every microvariable be fully randomized. A small number of non-randomized variables may be taken care of at the aggregation stage, as explained in section 4.6.

47. This probabilistic rule, though the primary rule underlying probability matching, is not the only rule. Having made their original decision, animals may switch food sources, this switching tending, as one would expect, in the direction of proportionally less exploited sources.

48. I will say something here about the advantages of the two varieties of probabilistic foraging described above. In both cases, probabilistic decision rules can be shown to provide a very simple and robust solution to a coordination problem, either that of how a number of organisms should efficiently search a large area (the ants), or that of how a number of organisms should efficiently exploit a number of food resources (probability matching). Fretwell (1972) has shown that, under certain conditions, the probability matching rule is an evolutionarily stable strategy for solving the resources problem. An evolutionarily stable strategy is one which does not create a selection pressure against itself. An example of an unstable strategy is the alternative rule, *always choose the most abundant food supply*. If all animals followed this rule, the less abundant supplies would be left unattended, creating

a selection pressure in favor of organisms that preferred to choose smaller food sources. This provides the beginning of an evolutionary explanation of probability matching.

49. For further examples of probabilistically patterned foraging, see Glymour (2001). Glymour suggests a neurobiological source for the randomness, as I do in section 4.94.

50. I am not sure, however, how far researchers have gone in the statistical testing of decision patterns. The data reported in the psychology literature mostly concerns frequencies.

51. Softky and Koch's version of the temporal coding view highlights what is most important for my discussion here, but omits mention of one of the most important themes of other versions of the view, that a temporal code is essential to coordinating the activity of a large number of neurons so as to solve, among other problems, what is known as the binding problem (König, Engel, and Singer 1996).

Also not considered explicitly here is the debate, closely connected to the debate between rate coding and temporal coding theorists, as to whether neurons are best considered as *integrate-and-fire* devices or as *coincidence detectors*. The reader will find the two sides of this debate laid out in Shadlen and Newsome (1994) and König, Engel, and Singer (1996).

52. For a study of the neural basis of probability matching in monkeys, see Platt and Glimcher (1999). Platt and Glimcher's results, though suggestive, neither support nor undercut the suggestion sketched in this paragraph.

53. The probability of release varies considerably not only for different neurons but for different synapses of the same neuron. Even for the same synapse at different times, the probability depends on the recent firing history and other facts. Thus the statement in the main text, that release is a Bernoulli process, is an oversimplification.

54. I am side-stepping the question of how, exactly, neurotransmitter release probabilities might contribute to the Poisson pattern of spike trains. For further discussion, see website section 4.9A.

5. Implications for the Philosophy of the Higher-Level Sciences

1. The picture is *physicalist* because it presumes that the lowest level is always the level of fundamental physics; however, this assumption plays no role in what follows.

2. A number of philosophers have argued, as I do in what follows, though for different reasons, that multiple realizability is no obstacle to reduction; see, for example, Sober (1999). For some very interesting recent views on reduction that are centered to some extent around the problem of multiple realizability, see Lange (2000) and Batterman (2002). Batterman's view is quite close, in certain respects, to the view offered in this section.

3. Some philosophers are reluctant to call narrow-scope generalizations *laws,* even when the generalizations inherit the full nomological force of the lower-level laws on which they systematically depend. Terminological wrangles in philosophy often disguise substantive issues, but the case of the possibility of narrow-scope laws seems not to exemplify this maxim.

4. This is a quantitative game: infrequent and non-catastrophic departures from superiority may not prevent a trait's being selected, in the long run. But whatever, exactly, is required in order that a trait drive its competitors to extinction, it is some property of the narrow-scope ecological laws.

5. For an approach to inferring probabilistic relevance relations that is based on the notion of microconstancy, see Strevens (1998).

6. Although the probabilistic theory of causal relevance is widely accepted, the details are debated. There is an ongoing dispute over how to go about choosing the neutral case with respect to which it is determined whether a probability has been raised or lowered. For an overview, see Hitchcock (1993).

7. Over longer time periods, things become more complicated: if the rabbit population is subject to boom and crash cycles, the judicious introduction of a few extra foxes into a system may, in the long term, *increase* a rabbit's life expectancy.

8. The stated condition is also, note, a sufficient condition for the use of the structured populations approach to aggregation; see section 4.62.

9. What if the enion probabilities in a system depend in part on microlevel information, in the way that the probability of rabbit death depends on the rabbit's health (section 4.43)? In such cases, the provisos of a relevance claim may have to include health-like microvariables, but a relationship without provisos will hold when the effect of the macrovariable whose relevance the claim asserts is partially independent of the other macrovariables *and* of the health-like microvariables.

10. For discussions of the role of compositional theories in understanding complex systems, see Bechtel and Richardson (1993), Simon (1996), and Wimsatt (1986).

11. A middle way that is becoming ever more popular is simulation. Rather than investigating the significance of the many interactions between enions in a complex system through mathematics, simulation investigates the effects empirically, by manipulating the interactions and observing the consequences. Because the vehicle of understanding is not mathematics, simulation is outside the scope of the remarks in this section.

12. It is possible that the probabilities of quantum mechanics are complex but that the quantum level is nevertheless the fundamental level. This is the picture presented by Bohm's interpretation of quantum mechanics (Albert 1992). Bohm's theory, however, makes no use of the notion of microconstancy, and as such does not explain the probabilistic patterns any more than does the notion of non-microconstant probability; rather, on Bohm's theory, it must be assumed that at the beginning of the universe, all particles are already distributed in a

probabilistically patterned way. This distribution is then passed on from physical interaction to physical interaction, as probabilistic patterns are passed on from non-microconstant probabilistic experiment to non-microconstant probabilistic experiment.

13. Of course, scientists have sometimes not followed this rule. But in all such episodes—vitalism in biology, epigenesis in developmental biology, certain kinds of dualism in psychology—the protagonists have turned out to have been wrong in their conviction that a complex phenomenon was not to be accounted for in terms of a complex physical structure.

14. Humphreys (1989, 141) says that chance is "literally nothing."

15. For a more careful discussion of the nature of determinism, see Earman (1986).

Glossary

This glossary is restricted to terms that are either introduced or given specialized definitions—as in the case of *complex system*—in the course of this study, and which appear in more than one place. Terms defined only for the sake of proofs are not glossed here. Definitions of terms not in the glossary may be found using the index.

Each gloss ends with a reference to the section in which the term is introduced. What follow are not canonical definitions but aids for the reader; canonical definitions, where necessary, are given in the main text.

Terms used in the glosses that are themselves glossed elsewhere are set in **boldface**. Obvious cross-references are omitted.

bad region A part of a **macroperiodic ic-density** where the density is not approximately uniform, as it is supposed to be; or a member of a **constant ratio partition** where the **strike ratio** is not approximately equal to that of the other members, as it is supposed to be. (§2.C1)

behavior, complex Behavior that can only be described by laws containing a large number of variables. (§1.15)

behavior, simple Behavior that can be described by laws containing only a small number of variables. (§1.12)

behavioral level In ecology, the level of description at which laws concerning the behavior of individual animals are formulated. (§4.31)

causal independence Two trials on **probabilistic setups** are causally independent just in case there is no **ic-value** that belongs to both trials' ancestries. The ancestry of a trial is the set of its ic-values, the ic-values that gave rise to those ic-values, and so on, down the **probabilistic network** to the relevant values of the **basic ic-variables**. (§3.33)

chain, deterministic A series of trials on a **probabilistic experiment** in which the outcome of any trial determines the **ic-values** of its successor. (§3.71)

complex system A system consisting of many, somewhat independent, interacting parts, or **enions**. (§1.13)

composite event   An event that is made up of two or more discrete events, such as the event of obtaining heads on a coin toss and a six on a die roll. When the discrete events are outcomes of trials on **probabilistic experiments**, the corresponding composite event is conceived of as the outcome of a **composite experiment**. (§3.2)

composite experiment   A **probabilistic experiment** that takes place on two or more distinct **mechanisms**, such as an experiment where two different coins are tossed; or, a probabilistic experiment that takes place on the same mechanism at two or more different times, such as an experiment where a single coin is tossed twice. An outcome of a composite experiment is a **composite event**. (§3.2)

constant ratio index (CRI)   Of a **constant ratio partition**, the size of the largest member of the partition. Of an **evolution function** or **probabilistic experiment**, with respect to some **designated outcome** e, the CRI of the **optimal constant ratio partition** for e. (§2.23)

constant ratio partition   With respect to a **designated outcome**, a partition of the space of possible IC-values for a **probabilistic experiment** into contiguous regions, such that each region has the same **strike ratio** for the outcome. (§2.23; definition 2.4, §2.C1; definition 2.10, §2.C2)

constant ratio partition, optimal   Of a **probabilistic experiment**, with respect to a **designated outcome**, the **constant ratio partition** with the smallest **constant ratio index**, or any of the smallest partitions, if there are more than one. (§2.23)

cprob(·)   A **complex probability**.

CRI   *See* **constant ratio index**

critical information, critical level   *See* **IC-value, critical**

designated outcome, designated set of outcomes   A designated outcome of a **probabilistic setup** is an outcome whose occurrence or non-occurrence interests us. The designated set of outcomes for a setup is the set of all designated outcomes. A setup and a designated set of outcomes together make up a **probabilistic experiment**. (§2.11; §2.14)

effective information   *See* **IC-value, effective**

elimination, of an IC-variable   *See* **IC-variable, eliminable**

endogenous variable   A variable representing a quantity that is determined by mechanisms lying entirely within the system under consideration. (§4.2)

enion   A part of a **complex system** that usually moves somewhat independently of, but sometimes interacts strongly with, other parts. (§1.13)

enion probabilistic experiment   A **probabilistic experiment** of the sort to which an **enion probability** is attached. (Not explicitly defined.)

enion probability   The probability that an **enion** of a **complex system** will be in a given state at a given time. Useful enion probabilities for the most part depend only on the values of **macrovariables**. For the differences between enion probabilities and **microdynamic probabilities**, see section 4.33. (§1.2)

enion probability analysis (EPA)   The method of predicting or explaining the **simple behavior** of a **complex system** by aggregating **enion probabilities** (assuming stochastic independence) to yield **simple macrolevel** laws. (§1.2)

enion probability, single microstate   The probability that an **enion** of a **complex system** will be in a given state at a given time, conditional on the system's starting out in a particular **microstate**. (§4.34)

EPA   *See* **enion probability analysis**

evolution function   The function that, given the initial conditions of a trial on a **probabilistic experiment**, determines whether or not a **designated outcome** occurs. (§2.13)

evolution function, complete   An **evolution function** for a **complex system** that takes as its IC-**variables** variables that give a complete description of the system at the **fundamental level**. (§4.41)

evolution function, probabilistic, explicitly and implicitly   An explicitly probabilistic evolution function represents the effects of **nomic probabilities** by setting each point $h_e(\zeta)$ in the **evolution function** equal to the nomic probability of the outcome, given the IC-**value** ζ. An implicitly probabilistic evolution function introduces additional IC-**variables** whose distributions represent the nomic probabilities. (§2.6)

exogenous variable   A variable representing a quantity that is determined by mechanisms lying partly or wholly outside the system under consideration. (§4.2)

experiment, microconstant   *See* **microconstancy**

explicitly probabilistic evolution function   *See* **evolution function, probabilistic, explicitly and implicitly**

feeder experiment   A **probabilistic experiment** that generates the IC-**values** of another experiment. (§2.32)

fundamental level   The level of description at which fundamental laws of nature are formulated, that is, the lowest possible level of description. (§4.31)

harmonic evolution functions   Two **evolution functions** are harmonic just in case an **optimal constant ratio partition** for one can be nested in the **outcome partition** for the other, that is, just in case any member of an optimal constant ratio partition for one is entirely contained in some member of the other's **outcome partition**. (§3.74)

high- and low-level information   High-level information gives approximate information about a quantity, for example, distance to the nearest meter. Low-level information is what must be added to high-level information to give more detail about the quantity, for example, distance in centimeters from the nearest whole number of meters. (§2.71; definition 3.4, §3.B2; definition 3.13, §3.B4)

ic-density   A probability density function over the values of an **ic-variable**. (§2.12)

ic-density, frequency-based   A density function that represents statistical information about **ic-values**. (§2.33)

ic-evolution   In a **deterministic chain**, the process by which the **full ic-value** of the ith trial is transformed into the full ic-value of the $(i + 1)^{th}$ trial. (Not explicitly defined.)

ic-evolution, inflationary   In a **deterministic chain**, inflationary ic-evolution tends to "stretch out" the set of **ic-values** for the ith trial, so that ic-values for the ith trial that are very close together map onto ic-values for the $(i + 1)^{th}$ trial that are further apart. Inflation is not a precise technical term in this study; for two ways to formalize the notion, see §3.73.

ic-evolution, outcome-dependent   In a **deterministic chain**, ic-evolution in which the transformation that determines the initial conditions of the $(i + 1)^{th}$ trial varies with the outcome of the ith trial. (§3.75)

ic-evolution, outcome-independent   In a **deterministic chain**, ic-evolution in which the transformation that determines the initial conditions of the $(i + 1)^{th}$ trial is the same no matter what the outcome of the ith trial. (§3.73)

ic-evolution function   In a **deterministic chain**, the function that, given the **full ic-value** of the ith trial, determines the full ic-value of the $(i + 1)^{th}$ trial. The process described by the ic-evolution function is **ic-evolution**. (§3.73)

ic-evolution function, restricted   Relative to an **effective ic-value** z, the function that determines the **full ic-value** of the $(i + 1)^{th}$ trial when the effective ic-value of the ith trial is z. (§3.75)

ic-value   A particular value of an **ic-variable**; hence, an initial condition of a particular probabilistic trial. (§2.12)

ic-value, ancestor   The ancestor ic-value of an **effective ic-value** in a **deterministic chain** is (a) that part of an earlier **full ic-value** that determines the effective ic-value, in **outcome-independent ic-evolution**, or (b) that part of an earlier full ic-value that plays a role in determining the effective ic-value, and that does not play a role in determining any earlier effective ic-values, in **outcome-dependent ic-evolution**. (§3.72)

ic-value, critical   A value encoding the critical information about an ic-value. The critical information about an ic-value is the information that determines whether the **full ic-value** falls into a gray part (causing the **designated outcome**) or a white part (not causing the designated outcome) of a given member of a specified **constant ratio partition**. (§2.73)

ic-value, effective   A value encoding the effective information about an ic-value. The effective information about an ic-value is the information that fixes the ic-value with a precision just sufficient to determine whether or not the **designated outcome** occurs. (§2.71)

ic-value, full   An **ic-value**, when distinguished from an **effective** or a **critical ic-value**. (Not explicitly defined.)

ic-value, raw   An **ic-value** whose **ic-density** lies inside a loop in a **probabilistic network**, but that is not produced by any **probabilistic experiment** inside (or outside) the loop. (§2.43)

ic-variable   A numerical representation of a physical quantity that is one of the initial conditions of a trial on a **probabilistic setup**. An ic-variable is, mathematically, a species of random variable. (§2.12)

ic-variable, basic   An **ic-variable** in a **probabilistic network** whose **ic-density** is not determined by a **probabilistic experiment**. (§2.43)

ic-variable, effective   The discrete variable that takes on the **effective ic-values** of a given **ic-variable**. (§2.72)

ic-variable, eliminable   An **ic-variable** which, although it makes a difference to the outcome of any particular trial, does not by itself affect the proportion of trials in which the **designated outcome** is produced. (§2.25)

ic-variable, initial   For a class of deterministic chains, an **ic-variable** ranging over **ic-values** for the very first trial of those chains. (§3.72)

ic-variable, standard   The sort of **ic-variable** we typically use to describe the world; a variable that induces a measure directly proportional to the si units; a variable with respect to which the laws of nature prescribe continuous, smooth change. (§2.52)

implicitly probabilistic evolution function   *See* **evolution function, probabilistic, explicitly and implicitly**

inflationary ic-evolution   *See* **ic-evolution, inflationary**

infracritical independence   Two **ic-variables** have infracritical independence if the **infracritical information** concerning each variable is stochastically independent of all other information about either variable. (§3.43; definition 3.12, §3.B3)

infracritical information The infracritical information about an ic-**variable** is all information about the variable at the **critical level** and below. (§3.43; definition 3.10, §3.B3)

macrodynamics The dynamics of the values of a system's **macrovariables**, that is, the dynamics of a system as described at the **macrolevel**. (§1.15; §4.1)

macrolevel The level of description at which **enion** statistics are represented. It is at the macrolevel that the **simple behavior** of **complex systems** emerges. (§1.15; §4.1)

macroperiodicity A macroperiodic probability density is approximately uniform over almost all **micro-sized regions**. (§2.23; definition 2.5, §2.C1; definition 2.14, §2.C2)

macro-sized region *See* **micro-sized and macro-sized regions**

macrostate A state of a **complex system** specified by a complete description of the system at the **macrolevel**, that is, a specification of the values of all of the system's **macrovariables**. (§4.1)

macrovariable A variable representing a **macrolevel** quantity, that is, a variable representing either an **enion** statistic or some aspect of the background environment. (§1.15)

mechanism The causal part of a **probabilistic setup**. The mechanism generates outcomes from initial conditions. Strictly speaking, a specification of a mechanism has two parts, a specification of a physical system and a specification of an operation on that system. (§2.13)

microconstancy An **evolution function** is microconstant just in case its domain can be partitioned into many small, contiguous regions each having approximately the same **strike ratio**. Claims of microconstancy are normally intended as claims of **standard microconstancy**. The **complex probability** of an outcome is microconstant just in case the evolution function for that outcome is microconstant. A **probabilistic experiment** is microconstant just in case the evolution functions for the members of the **designated set of outcomes** are all microconstant. (§2.23)

microconstancy, oscillating and steady-state An **evolution function** has oscillating microconstancy if it oscillates between zero and one at an approximately constant rate over every set in a micro-sized **constant ratio partition**. An **explicitly probabilistic evolution function** has steady-state microconstancy if it is approximately constant. An **implicitly probabilistic evolution function** has steady-state microconstancy if all other ic-**variables** can be **eliminated** in favor of the ic-variables representing the **nomic probabilities**. (§2.61)

microconstancy, standard A **probabilistic experiment** has standard microconstancy if it is **microconstant** when represented using **standard ic-variables**. (§2.52)

microconstant experiment   *See* microconstancy

microconstant probability   *See* microconstancy

microdynamic experiment   A **probabilistic experiment** to which a **microdynamic probability** is attached. The outcomes of microdynamic experiments determine the course of microlevel time evolution; the ic-**variables** of microdynamic experiments are at a lower level of description than the **microlevel**. (Not explicitly defined.)

microdynamic probability   A probability that appears in the laws of **microdynamics**. (§4.32)

microdynamics   The dynamics of a system's **microvariables**, that is, the dynamics of a system as described at the **microlevel**. (§1.15; §4.1)

microlevel   A level of description at which the states of individual **enions** are represented. (§1.15; §4.1)

microlinear transformation   A transformation that is approximately linear or linear plus constant, and invertible, over almost all **micro-sized regions** of its domain. (§3.65; definition 3.23, §3.B6)

micro-sized and macro-sized regions   These terms are perhaps best defined relative to each other: a micro-sized region is a region much smaller than any macro-sized region, and a macro-sized region is a region much larger than any micro-sized region. It is assumed throughout that such regions are contiguous, that is, connected. It is also assumed that a microregion does not have too unusual a shape. In particular, "worm-shaped" regions are excluded. (§2.23)

microstate   A state of a **complex system** specified by a complete description of the system at the **microlevel**, that is, a specification of the values of all of the system's **microvariables**. (§4.1)

microvariable   A variable representing a **microlevel** quantity. (§1.15)

multi-mechanism experiment   A **probabilistic experiment** consisting of several successive steps in a **deterministic chain**. (§3.74)

nomic probability   A probability occurring in a probabilistic law governing the operation of the **mechanism** of a **probabilistic setup**. (§2.6)

oscillating microconstancy   *See* microconstancy, oscillating and steady-state

outcome-dependent and -independent ic-evolution   *See* ic-evolution, outcome-dependent, and ic-evolution, outcome-independent

outcome map   A graph of an **evolution function**, showing which ic-**values** cause a **designated outcome**, considered as a mathematical object upon which transformations can be performed. (§3.62)

outcome partition   A partition of the set of the ic-values for a **probabilistic experiment** into maximal, contiguous sets, all members of which give rise to the same outcome. Normally, the partition is relative to a single **designated outcome**, in which case the members of a given set in the outcome partition either all give rise to the designated outcome or all do not. When dealing with a **microconstant experiment**, it is required in addition that members of the outcome partition each fit entirely into some member of a given **optimal constant ratio partition**. This results, in some cases, in a slightly finer partition. (§2.71)

population level   In ecology, the level of description at which laws concerning the behavior, in particular, the population dynamics, of ecosystems are formulated. (§4.31)

probabilistic experiment   A **probabilistic setup** together with a **designated set of outcomes**. (§2.14; definition 2.1, §2.C1; definition 2.6, §2.C2)

probabilistic network   A network of **probabilistic experiments** that supply one another with ic-values. (§2.41)

probabilistic patterns   The characteristic patterns of outcomes produced by **probabilistic experiments**, exhibiting short-term disorder and long-term order. (§1.32)

probabilistic setup   A type of process that generates outcomes probabilistically. A specification of a setup type consists of a specification both of a **mechanism** and of a probability distribution over **ic-variables**. (§2.11)

probabilistic trial   An instance of the probabilistic process that constitutes a **probabilistic setup**, in which a particular set of initial conditions produces a particular outcome. (§2.11)

probability, complex   A probability that is not a **simple probability** (§1.31). In the technical sense of this study, a probability for a **designated outcome** obtained by taking the measure, given by the **ic-density**, of the **ic-values** that produce the outcome. If the underlying laws are probabilistic, then each set of ic-values is weighted by the probability that it will produce the outcome (§2.14; §2.25; §2.61; definition 2.2, §2.C1; definition 2.7, §2.C2)

probability, microconstant   *See* **microconstancy**

probability, simple   A probability that appears in, or depends in a simple way on probabilities appearing in, the fundamental laws of nature. (§1.31)

probability, simplex   A **complex probability** that owes its probabilistic aspect entirely to underlying **simple probabilities**. Formally, a complex probability for which there exists a **strike set** of ic-variables with **simple probability** distributions. (§1.31; §2.63)

probability, true   A probability attached to a **probabilistic experiment** is true if either (a) it is **microconstant**, or (b) every route from a **basic ic-variable** of the experiment's **probabilistic network** to the experiment itself passes through a microconstant experiment. (§2.44)

quasi-deterministic law   A probabilistic law in which all probabilities are very close to either zero or one. (§1.31, note 20)

randomizing variable   A variable having the properties necessary to randomize the **microlevel** time evolution of a **complex system** sufficiently to make the system's **enion probabilities microconstant** and stochastically independent, as required for the argument in section 4.4. (§4.91)

selection rule   A rule that picks out certain pairs of trials on a pair of **probabilistic experiments,** or more generally, that picks out certain n-tuples of trials on a set of n probabilistic experiments. (§3.1)

shielding   An **ic-variable** of a **microdynamic experiment** is shielded if the **probabilistic experiments** affecting its value are **causally independent** of the rest of the system. (§4.54)

simple   *See* **behavior, simple**

srat($\cdot$)   A **strike ratio.**

steady-state microconstancy   *See* **microconstancy, oscillating and steady-state**

strike ratio   Of a region of **ic-values**, the proportion of **ic-values** in the region that produce a **designated outcome**. The strike ratio of an **evolution function** is the strike ratio of any of the members of its **constant ratio partitions**. (§2.23; definition 2.3, §2.C1; definition 2.9, §2.C2)

strike set   For a **probabilistic experiment**, a set of **ic-variables** in favor of which all of the experiment's other **ic-variables** can be **eliminated**. (§2.25; definition 2.13, §2.C2)

supercondition, probabilistic   The condition, required in order for EPA to succeed, that each **enion probability** be independent of **microlevel** information, or more exactly, that it be independent of information that goes beyond **macrolevel** information about (a) the initial conditions of any **enion probabilistic experiment**, and (b) the outcome of any enion probabilistic experiment other than itself. (§1.24)

trial   *See* **probabilistic trial**

weak independence   Two **ic-variables** with **macroperiodic ic-densities** are weakly independent relative to a **selection rule** just in case their joint density with respect to the selection rule is also macroperiodic. (§3.42)

References

Albert, D. Z. (1992). *Quantum Mechanics and Experience*. Harvard University Press, Cambridge, Mass.

———— (2000). *Time and Chance*. Harvard University Press, Cambridge, Mass.

Auyang, S. Y. (1998). *Foundations of Complex-System Theories in Economics, Evolutionary Biology, and Statistical Physics*. Cambridge University Press, Cambridge.

Bair, W. (1999). Spike timing in the mammalian visual system. *Current Opinion in Neurobiology* 9:447–453.

Bass, T. (1985). *The Eudaemonic Pie*. Houghton Mifflin, Boston.

Batterman, R. W. (2002). *The Devil in the Details: Asymptotic Reasoning in Explanation, Reduction, and Emergence*. Oxford University Press, Oxford.

Beatty, J. (1995). The evolutionary contingency thesis. In G. Wolters, J. G. Lennox, and P. McLaughlin (eds.), *Concepts, Theories, and Rationality in the Biological Sciences*. University of Pittsburgh Press, Pittsburgh.

Bechtel, W., and R. C. Richardson. (1993). *Discovering Complexity: Decomposition and Localization as Strategies in Scientific Research*. Princeton University Press, Princeton, N.J.

Breton, R. (1991). *Geolinguistics: Language Dynamics and Ethnolinguistic Geography*. University of Ottawa Press, Ottawa.

Cartwright, N. (1989). *Nature's Capacities and Their Measurement*. Oxford University Press, Oxford.

Chaitin, G. (1990). *Information, Randomness, and Incompleteness*. Second edition. World Scientific, Singapore.

Cowan, W. M., T. C. Südhof, and C. F. Davies (eds.). (2001). *Synapses*. Johns Hopkins University Press, Baltimore, Md.

Daston, L. (1988). *Classical Probability in the Enlightenment*. Princeton University Press, Princeton, N.J.

deCharms, R. C., and A. Zador. (2000). Neural representation and the cortical code. *Annual Review of Neuroscience* 23:613–647.

Driver, P. M., and D. A. Humphries. (1988). *Protean Behavior: The Biology of Unpredictability.* Oxford University Press, Oxford.

Dupré, J. (1993). *The Disorder of Things.* Harvard University Press, Cambridge, Mass.

Durkheim, E. (1951). *Suicide.* Translated by J. Spaulding and G. Simpson. Free Press, Glencoe, Ill.

Earman, J. (1986). *A Primer on Determinism.* D. Reidel, Dordrecht.

Endler, J. A. (1986). *Natural Selection in the Wild.* Princeton University Press, Princeton, N.J.

Engel, E. (1992). *A Road to Randomness in Physical Systems,* vol. 71 of *Lecture Notes in Statistics.* Springer-Verlag, Heidelberg.

Fetzer, J. (1971). Dispositional probabilities. *Boston Studies in the Philosophy of Science* 8:473–482.

Fine, T. (1973). *Theories of Probability.* Academic Press, New York.

Finetti, B. de. (1964). Foresight: its logical laws, its subjective sources. In H. Kyburg and H. Smokler (eds.), *Studies in Subjective Probability.* Wiley, New York.

Fodor, J. (1975). Special sciences. *Synthese* 28:97–115.

Fretwell, S. (1972). *Populations in Seasonal Environments.* Princeton University Press, Princeton, N.J.

Gallistel, C. (1990). *The Organization of Behavior.* MIT Press, Cambridge, Mass.

Giere, R. (1973). Objective single case probabilities and the foundation of statistics. In Suppes et al. (1973).

Glymour, B. (2001). Selection, indeterminism, and evolutionary theory. *Philosophy of Science* 68:518–535.

Grant, P. R. (1986). *Ecology and Evolution of Darwin's Finches.* Princeton University Press, Princeton, N.J.

Grimmett, G., and D. Stirzaker. (1992). *Probability and Random Processes.* 2d ed. Oxford University Press, Oxford.

Hacking, I. (1975). *The Emergence of Probability.* Cambridge University Press, Cambridge.

———— (1990). *The Taming of Chance.* Cambridge University Press, Cambridge.

Hempel, C. (1965). *Aspects of Scientific Explanation.* Free Press, New York.

Hitchcock, C. (1993). A generalized probabilistic theory of causal relevance. *Synthese* 97:335–364.

Hölldobler, B., and E. Wilson. (1990). *The Ants.* Harvard University Press, Cambridge, Mass.

Hopf, E. (1934). On causality, statistics, and probability. *Journal of Mathematics and Physics* 13:51–102.

Humphreys, P. (1989). *The Chances of Explanation*. Princeton University Press, Princeton, N.J.

Jaynes, E. T. (1983). *Papers on Probability, Statistics, and Statistical Physics*. Edited by R. Rosenkrantz. D. Reidel, Dordrecht.

Jervis, R. (1997). *System Effects: Complexity in Political and Social Life*. Princeton University Press, Princeton, N.J.

Jevons, W. S. (1882). The solar commercial cycle. *Nature* 26:226–228.

Kac, M. (1959). *Probability and Related Topics in Physical Science*. Wiley Interscience, New York.

Keller, J. (1986). The probability of heads. *American Mathematical Monthly* 93:191–197.

Keynes, J. M. (1921). *A Treatise on Probability*. Macmillan, London.

Kneale, W. (1949). *Probability and Induction*. Oxford University Press, Oxford.

Knuth, D. E. (1998). *The Art of Computer Programming: Seminumerical Algorithms*. 3d ed. Addison-Wesley, Boston.

König, P., A. K. Engel, and W. Singer. (1996). Integrator or coincidence detector? The role of the cortical neuron revisited. *Trends in the Neurosciences* 19:130–137.

Lange, M. (2000). *Natural Laws in Scientific Practice*. Oxford University Press, Oxford.

Lomolino, M. V. (1985). Body size of mammals on islands: The island rule reexamined. *American Naturalist* 125:310–316.

Mellor, D. H. (1971). *The Matter of Chance*. Cambridge University Press, Cambridge.

Mises, R. von. (1957). *Probability, Statistics, and Truth*. Dover Publications, New York.

Musk, L. (1988). *Weather Systems*. Cambridge University Press, Cambridge.

Nagel, E. (1979). *The Structure of Science*. Hackett, Indianapolis, Ind.

Ornstein, D. S., and B. Weiss. (1991). Statistical properties of chaotic systems. *Bulletin of the American Mathematical Society* 24:11–116.

Osherson, D., E. Shafir, and E. Smith. (1994). Extracting the coherent core of human probability judgment: A research program for cognitive psychology. *Cognition* 50:299–313.

Plato, J. von. (1983). The method of arbitrary functions. *British Journal for the Philosophy of Science* 34:37–47.

Platt, M. L., and P. W. Glimcher. (1999). Neural correlates of decision variables in parietal cortex. *Nature* 400:233–238.

Poincaré, H. (1896). *Calcul des Probabilités*. 1st ed. Gauthier-Villars, Paris.

Porter, T. M. (1986). *The Rise of Statistical Thinking, 1820–1900*. Princeton University Press, Princeton, N.J.

Price, H. (1996). *Time's Arrow and Archimedes' Point*. Oxford University Press, Oxford.

Prigogine, I. (1980). *From Being to Becoming*. W. H. Freeman, New York.

Putman, R. J., and S. D. Wratten. (1984). *Principles of Ecology*. University of California Press, Berkeley, Calif.

Ramsey, F. (1931). Truth and probability. Reprinted in *Philosophical Papers*. Edited by D. H. Mellor. Cambridge University Press, Cambridge.

Reichenbach, H. (1949). *The Theory of Probability*. University of California Press, Berkeley, Calif.

———— (1956). *The Direction of Time*. University of California Press, Berkeley, Calif.

Rieke, F., D. Warland, R. de Ruyter van Steveninck, and W. Bialek. (1997). *Spikes: Exploring the Neural Code*. MIT Press, Cambridge, Mass.

Savage, L. J. (1973). Probability in science: A personalistic account. In Suppes et al. (1973).

Shadlen, M. N., and W. T. Newsome. (1994). Noise, neural codes and cortical organization. *Current Opinion in Neurobiology* 4:569–579.

———— (1998). The variable discharge of cortical neurons: Implications for connectivity, computation, and information coding. *Journal of Neuroscience* 18:3870–96.

Simon, H. A. (1996). *The Sciences of the Artificial*. 3d ed. MIT Press, Cambridge, Mass.

Sklar, L. (1993). *Physics and Chance*. Cambridge University Press, Cambridge.

Skyrms, B. (1980). *Causal Necessity*. Yale University Press, New Haven, Conn.

Sober, E. (1984). *The Nature of Selection*. MIT Press, Cambridge, Mass.

———— (1999). The multiple realizability argument against reductionism. *Philosophy of Science* 66:542–564.

Softky, W. R., and C. Koch. (1993). The highly irregular firing of cortical cells is inconsistent with temporal integration of random EPSPs. *Journal of Neuroscience* 13:334–350.

Stevens, C. F., and A. M. Zador. (1998). Input synchrony and the irregular firing of cortical neurons. *Nature Neuroscience* 1:210–217.

Stewart, I. (1989). *Does God Play Dice? The Mathematics of Chaos*. Blackwell, Oxford.

Strevens, M. (1998). Inferring probabilities from symmetries. *Noûs* 32:231–246.

———— (2000). Do large probabilities explain better? *Philosophy of Science* 67:366–390.

Suppes, P. (1973). New foundations of objective probability: Axioms for propensities. In Suppes et al. (1973).

———— (1987). Propensity representations of probability. *Erkenntnis* 26:335–358.

Suppes, P., L. Henkin, G. C. Moisil, and A. Joja (eds.). (1973). *Logic, Methodology, and Philosophy of Science IV: Proceedings of the Fourth International Congress for Logic, Methodology, and Philosophy of Science, Bucharest, 1971*. North Holland, Amsterdam.

Tuckwell, H. C. (1989). *Stochastic Processes in the Neurosciences*. Society for Industrial and Applied Mathematics, Philadelphia.

Waters, C. K. (1998). Causal regularities in the biological world of contingent distributions. *Biology and Philosophy* 13:5–36.

Weber, M. (1999). The aim and structure of ecological theory. *Philosophy of Science* 66:71–93.

Wilson, K. (1979). Problems in physics with many scales of length. *Scientific American* 241:158–179.

Wimsatt, W. C. (1986). Forms of aggregativity. In A. Donagan, A. N. Perovich, and M. V. Wedin (eds.), *Human Nature and Natural Knowledge*. D. Reidel, Dordrecht.

Winnie, J. A. (1998). Deterministic chaos and the nature of chance. In J. Earman and J. D. Norton (eds.), *The Cosmos of Science*. University of Pittsburgh Press, Pittsburgh.

Woodward, J. (2000). Explanation and invariance in the special sciences. *British Journal for the Philosophy of Science* 51:197–254.

Yeomans, J. M. (1992). *Statistical Mechanics of Phase Transitions*. Oxford University Press, Oxford.

Zador, A. (1998). Impact of synaptic unreliability on the information transmitted by spiking neurons. *Journal of Neurophysiology* 79:1230–38.

Index

Italicized page numbers indicate definitions.